OXFORD
UNIVERSITY PRESS

ASPIRE
SUCCEED
PROGRESS

AF775000

Cambridge
Lower Secondary

Complete Mathematics 9

Second Edition

Deborah Barton

TEACHER HANDBOOK

OXFORD
UNIVERSITY PRESS

OXFORD
UNIVERSITY PRESS

Great Clarendon Street, Oxford, OX2 6DP, United Kingdom

Oxford University Press is a department of the University of Oxford.

It furthers the University's objective of excellence in research, scholarship, and education by publishing worldwide. Oxford is a registered trade mark of Oxford University Press in the UK and in certain other countries.

First published in 2022

British Library Cataloguing in Publication Data

Data available

978-1-38-201896-8

10 9 8

Paper used in the production of this book is a natural, recyclable product made from wood grown in sustainable forests.

The manufacturing process conforms to the environmental regulations of the country of origin.

Printed and bound by CPI Group (UK) Ltd, Croydon, CR0 4YY

Acknowledgements

The publisher and authors would like to thank the following for permission to use photographs and other copyright material:

Cover illustrations: vectoriart/iStockphoto

Artwork by Integra Software Services, and Oxford University Press.

® IGCSE is the registered trademark of Cambridge International Examinations.

Cambridge Assessment International Education bears no responsibility for the example answers to test-style questions which are contained in this publication. The questions, example answers, marks awarded and comments that appear in this book were written by the author. In examination, the way marks would be awarded to answers like these may be different.

Every effort has been made to contact copyright holders of material reproduced in this book. Any omissions will be rectified in subsequent printings if notice is given to the publisher.

This Teacher's Handbook refers to the Cambridge Lower Secondary Mathematics (0862) Syllabus published by Cambridge Assessment International Education.

This work has been developed independently from and is not endorsed by or otherwise connected with Cambridge Assessment International Education.

The manufacturer's authorised representative in the EU for product safety is Oxford University Press España S.A. of El Parque Empresarial San Fernando de Henares, Avenida de Castilla, 2 – 28830 Madrid (www.oup.es/en or product.safety@oup.com). OUP España S.A. also acts as importer into Spain of products made by the manufacturer.

Contents

About this book 4

Active learning resources

Expanding brackets dominoes activity 5

Bearings drawing activity 8

Place value detective mystery 10

Fractions jigsaw 17

Simultaneous equations Sudoku 22

Transformations card-sorting activity 26

Probability crossword 35

Different bases investigation 38

Resources for use with Student Book 9

Crossword puzzle 41

Tessellation cut-outs 43

Revision material

Multi-choice Exercise A 45

Multi-choice Exercise B 48

Multi-choice Exercise C 54

Answers to multi-choice Exercises A, B and C 60

Topic tests

Topic test – Chapters 1–5 62

Topic test – Chapters 6–10 70

Topic test – Chapters 11–15 80

Answers to topic tests 89

Progression test-style papers

Paper 1 (non-calculator) 93

Mark scheme for Paper 1 103

Paper 2 (calculator) 106

Mark scheme for Paper 2 116

Worked solutions

Worked solutions for some of the more difficult questions from Student Book 9 119

Answers

Answers for Student Book 9 126

Answers for Homework Book 9 171

kerboodle

Additional support can be found on Kerboodle. As well as extra assessments to accompany the Teacher Handbook, there are resources for each chapter in the Student Book, including interactives and worksheets for both formative and summative assessment.

About this book

This book is designed to provide teachers with useful resources to support teaching of the Cambridge Lower Secondary Stage 9 Mathematics curriculum framework in preparation for the Cambridge Checkpoint assessments and further study at IGCSE.

The active learning resources included ensure that there is plenty of scope for discussion and group work, which are valuable tools for reinforcing learning. Often students complete such activities thinking that they have done very little work as they have written little or nothing down. However, the work completed is often of greater value than a written task, as there is more scope for sharing knowledge and eliminating common misconceptions than when students work individually. There is less pressure than in a written exercise as these resources can have different starting points, meaning students can work in an order that suits them and their level of understanding.

Often there are no limits to learning, particularly when using the creative alternative approaches. You will find many students work at higher levels than expected, because of their natural competitiveness and desire to improve. Also, in group work and discussion, students are required to articulate sometimes quite complex mathematical ideas that wouldn't be covered when working individually. Not everyone learns best by writing things down. These activities address alternative learning styles, particularly if you find your classes often take the 'teacher gives examples, students do an exercise' approach.

All good resources should provide you with the opportunity to simplify or extend the resource as necessary. Explanations are provided here for how to differentiate the resources to cater for differing ability levels. Sometimes an activity takes less time than anticipated and this can be difficult to manage in lessons, so there are sections suggesting alternative ways to use the resources.

There are also resources for exercises in the Student Book that promote active learning; this ensures better understanding, and consequently better retention of the subject knowledge.

Revision material is included in the book in the form of multi-choice questions. These can be given to the students as worksheets at the end of a chapter or series of chapters, or they could be used in a more fun approach, for example as a quiz.

Assessment materials are provided in the form of topic tests and end-of-year progression tests. The assessment materials are balanced between all the content areas in the framework: number, algebra, geometry and measure, statistics and probability. They are also underpinned by thinking and working mathematically questions, providing a structure for the application of mathematical skills. Mental strategies are tested throughout. Mark schemes are included for all topic tests, and these are provided in great detail for the end-of-year progression tests (all mark schemes have been written and developed by the author).

There are worked solutions to some of the more difficult questions in the Student Book and answers to all of the exercises in the Student Book and the Homework Book and of course, don't forget Kerboodle, where you will find yet more resources.

Finally, I hope that all the material in this book is extremely useful for you, both in monitoring your students' progress and helping them increase their understanding and enjoyment of maths.

Deborah Barton

Expanding brackets dominoes activity

For use with Chapter 2 of Student Book 9

Teacher's notes

This dominoes activity is an interesting and enjoyable alternative to a written exercise, allowing students to discuss their ideas. If they are not sure how to do a particular question, students can have an alternative starting point. This dominoes activity is intended to be used to revise and consolidate work from Chapter 2 of the Student Book on finding the product of two linear expressions and simplifying the answer.

Introductory activity

There should be little need for an introduction as this is intended as a revision exercise. However, you may want to go through the following examples.

Example 1

$(x + 10)(x - 7) = x^2 + 10x - 7x - 70$

$= x^2 + 3x - 70$

Example 2

$(x + 5)(x - 5) = x^2 + 5x - 5x - 25$

$= x^2 - 25$

Example 3

$(x - 4)^2 = (x - 4)(x - 4)$

$= x^2 - 4x - 4x + 16$

$= x^2 - 8x + 16$

Example 4

$(x + 9)^2 - 10x = (x + 9)(x + 9) - 10x$

$= x^2 + 9x + 9x + 81 - 10x$

$= x^2 + 8x + 81$

Dominoes activity

Give each student (or group of students) a copy of the domino cards, cut out. Instruct students to join each question to the correct answer on the next domino to form a chain of joined dominoes.

Differentiation

To make the task easier you could provide some of the harder dominoes already paired. To make the task harder you could remove a few dominoes from somewhere in the chain, giving the students blank cards. Students should then complete the blank cards to fill in the gaps.

Alternative approaches

You could instruct students to use the blank dominoes to make an entirely new domino activity of their own or extend this chain by removing and replacing the domino that says 'finish' on it. You could produce larger copies of the dominoes to pin up on the walls in the classroom or around the school as a treasure hunt activity.

Blank dominoes

These can be copied for students to use:

Dominoes

$x^2 - 2x - 8$	Expand and simplify $(x - 3)(x + 5)$	Expand and simplify $(x + 3)(x - 5)$	$x^2 - 8x + 15$
Expand and simplify $(x - 4)(x - 2)$	$x^2 + 5x + 6$	Expand and simplify $(x - 3)(x + 3)$	Expand and simplify $(x - 3)^2$
$x^2 - 6x + 9$	Finish	Expand and simplify $(x + 4)(x + 2)$	$x^2 - 6x + 8$
Expand and simplify $(x + 3)^2$	$x^2 - 2x - 15$	Expand and simplify $(x + 3)(x + 2)$	$x^2 + 6x + 9$
$x^2 + 2x - 15$	$x^2 + 6x + 8$	Expand and simplify $(x + 3)^2 - 6x$	$x^2 - 9$
Expand and simplify $(x - 3)(x - 5)$	$x^2 + 9$	Start	Expand and simplify $(x - 4)(x + 2)$

Answers

Start	Expand and simplify $(x-4)(x+2)$

x^2-6x+8	Expand and simplify $(x+4)(x+2)$	x^2+6x+8	$x^2+2x-15$	Expand and simplify $(x-3)(x+5)$	x^2-2x-8

Expand and simplify $(x-4)(x-2)$	x^2+5x+6

Expand and simplify $(x+3)(x+2)$	x^2+6x+9	Expand and simplify $(x+3)^2$	$x^2-2x-15$	Expand and simplify $(x+3)(x-5)$	$x^2-8x+15$

Expand and simplify $(x-3)(x-5)$	x^2+9

Finish	x^2-6x+9	Expand and simplify $(x-3)^2$	Expand and simplify $(x-3)(x+3)$	x^2-9	Expand and simplify $(x+3)^2-6x$

Bearings drawing activity

For use with Chapter 3 of Student Book 9

Teacher's notes

This bearings drawing activity is intended to be a fun way to revise drawing bearings. In particular, students will revise that the angle should always be three figures and measured clockwise from the North line.

Introductory activity

This activity can be used as a plenary or revision exercise supporting the work completed in Chapter 3. You may find that little introduction is necessary. Alternatively, you may want to go through the following examples.

Example 1

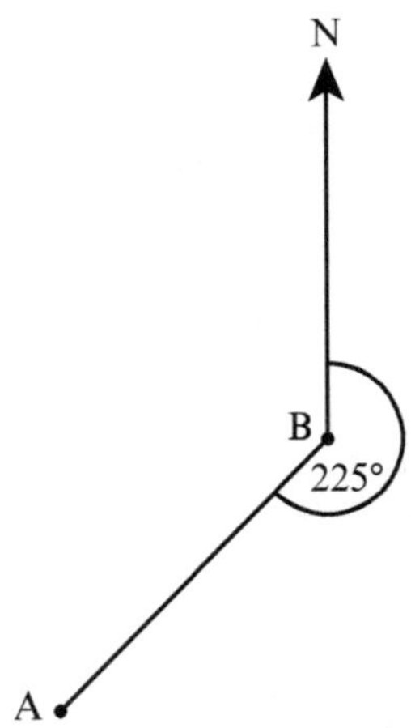

The bearing of A from B is 225°.

Example 2

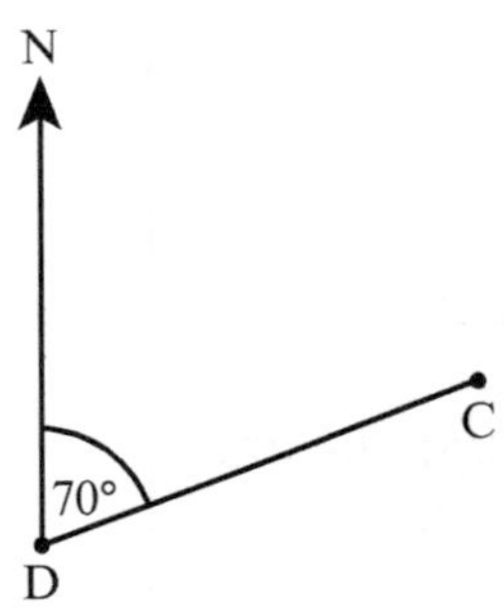

The bearing of C from D is 070°.

Drawing activity

Give each student a copy of the worksheet. Tell them to start at number 1 and to follow the instructions in each table to draw part of the picture. Each new line starts where the previous line ended and the bearing from the North line is shown in each case, along with how many units long each line should be (measured from one dot to another on the isometric grid). For each numbered North line there is a new table to start a new drawing.

Differentiation

To make the task easier you can give students this diagram to help:

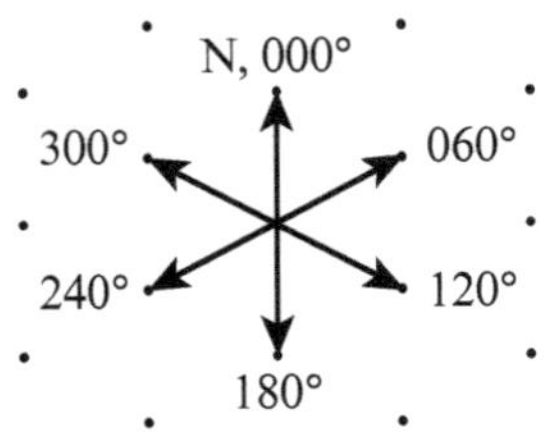

You could tell students that the image should be a cat, or leave them to discover this for themselves. You could also tell students that the image is symmetrical. To make the task harder, tell the students to add extra features to the drawing (e.g. a body or part of the body for the cat) along with the instructions for drawing any additional lines.

Alternative approaches

Instruct students to create their own drawing on an isometric grid with instructions for drawing all of the lines. Students could also create their own drawing, and instructions, on plain paper so they can use any angle and any length line.

Answer

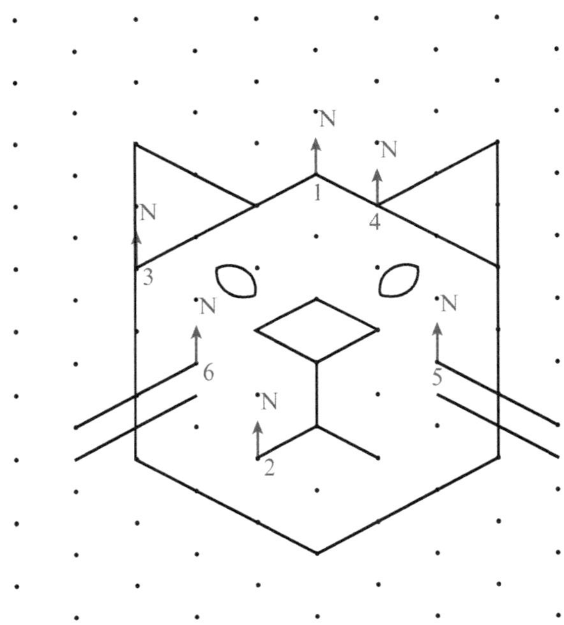

Worksheet

Do a different drawing for each number. Start at the North line and join the dots according to the bearings and lengths given. Start each new line where the previous line ends. Lengths are the number of gaps between dots in the given direction.

Number 1	
Bearing	**Length**
120°	3
180°	3
240°	3
300°	3
000°	3
060°	3

Number 2	
Bearing	**Length**
060°	1
000°	1
300°	1
060°	1
120°	1
240°	1
180°	1
120°	1

Number 3	
Bearing	**Length**
000°	2
120°	2

Number 4	
Bearing	**Length**
060°	2
180°	2

Number 5	
Bearing	**Length**
120°	2

Number 6	
Bearing	**Length**
240°	2

Place value detective mystery

For use with Chapter 4 of Student Book 9

Teacher's notes

This place value detective mystery is intended to be a fun way to revise and consolidate all of the work covered in Chapter 4 of the Student Book on place value, standard form and bounds. Students still need to solve questions as they would in an exercise. In this activity it is the context that provides an extra layer of interest.

Introductory activity

There should be little need for an introduction depending on whether or not this activity is done close to completing Chapter 4 or not. For more able students you may be able to just give them the activity and let them start. Weaker students may benefit from these examples.

Example 1

Simplify $10^2 \times 10^6 \times 10^{-1}$

Add the indices: $10^{2+6+-1} = 10^7$

Example 2

Work out 600×10^{-4}

This is the same as $600 \div 10^4 = 0.06$

Example 3

Work out $0.25 \div 10^{-3}$

This is the same as $0.25 \times 10^3 = 250$

Example 4

Write 4200 and 0.000378 in standard form.

$4200 = 4.2 \times 10^3$

$0.000378 = 3.78 \times 10^{-4}$

Example 5

52.4 is rounded correct to 1 d.p. Write an inequality to show the bounds for the number.

$52.35 \leq n < 52.45$

52.35 is the lower bound and 52.45 is the upper bound.

Detective mystery

Give each student (or group of students) a copy of the story and the clues. Tell them to play the part of the detective to solve the clues and determine who the thief is.

Differentiation

To make the task easier you could give some of the letters for some of the clues as examples.

To make the task harder you could set a time limit or add extra suspects and extra clues to solve.

Alternative approaches

Instruct students to race to see who can be first to solve the mystery. Tell them to think about whether or not every question needs to be answered to solve all of the clues. Students could then create their own detective mystery clues on this topic or any other topic. They can be as creative as they like.

Detective mystery - the story

The police are called to a Hollywood bank. An antique jewellery box, containing a diamond necklace belonging to a famous actor, has been stolen.

Detective Smith searches the bank and talks to witnesses. The detective finds out some useful clues and has sent the clues to you to decode.

The bank has a list of all the customers who visited the bank on the morning when the necklace was stolen. One of these customers is the thief.

Each clue you solve will eliminate some of the suspects. Can you help Detective Smith solve the clues and reveal the thief?

The customers who are suspects

Forename	Surname	Gender	Right / left handed	Wears glasses	Hair	Driver?	Guilty?
Mike	Dodds	M	Right	Yes	Brown	Yes	
Debbie	Loveridge	F	Left	Yes	Blond	No	
Geetha	Sinha	F	Right	No	Brown	No	
Sanath	Malinga	M	Left	No	Blond	Yes	
Daryl	Cooper	M	Left	Yes	Brown	Yes	
Mona	Chaterjee	F	Right	No	Blond	Yes	
Keren	Clayton	F	Right	Yes	Brown	No	
Katy	Howe	F	Right	Yes	Brown	Yes	
Lee	Badawi	M	Left	No	Black	No	
Kulwinder	Kendrick	F	Left	No	Blond	Yes	
Tyson	Kravitz	M	Right	Yes	Blond	Yes	
Nirmal	Paul	M	Right	Yes	Brown	Yes	
Agnetha	Forsberg	F	Right	No	Blond	No	
Alex	Gothard	M	Left	No	Brown	No	
Najib	Choo	M	Right	No	Blond	Yes	
Bhavani	Satish	F	Right	Yes	Black	No	
Angela	Nietzsche	F	Right	Yes	Brown	Yes	
Datuk	Hussein	F	Left	Yes	Brown	No	
Hayao	Suzuki	M	Right	No	Blond	No	
Edward	Jervis	M	Right	Yes	Black	Yes	
Rob	Fernandes	M	Right	No	Brown	No	
Christoph	Klum	M	Right	No	Blond	No	
Martin	Reagan	M	Right	No	Blond	Yes	
Grazia	Montessori	F	Right	No	Brown	Yes	
Ruth	Astley	F	Left	No	Blond	No	
Yoko	Osaka	F	Right	Yes	Black	Yes	
Tunku	Ibrahim	M	Right	Yes	Blond	No	
Ansua	Mitra	F	Right	No	Blond	No	
Chris	Toulson	M	Left	Yes	Brown	Yes	
Hans	Vettel	M	Right	No	Brown	Yes	
Somnath	Das	M	Right	No	Black	No	
Ingrid	Kamprad	F	Left	Yes	Black	Yes	

Detective mystery - the clues

Clue 1

One of the customers writes these five mathematical statements. The number of correct statements will determine the gender of the thief using the key below.

Multiplying by 10^{-4} is the same as dividing by 10^4.

Multiplying by 10^{-3} is the same as dividing by $\frac{1}{10^3}$.

Dividing by 10^5 is the same as multiplying by 10^{-5}.

Dividing by 10^{-8} is the same as multiplying by 10^8.

Dividing by 10^{-11} is the same as multiplying by $\frac{1}{10^{11}}$.

Key

Number of correct statements	Gender of the thief
0, 1 or 2	Male
3, 4 or 5	Female

Clue 2

The answer to each question is a number. Work out the number. Then translate the number into a letter using the alphabet code A = 1, B = 2, C = 3, and so on.

1 $200\,000 \div 10^4 = \ldots\ldots$

2 $37 \times 10^{-5} \times 10^{-3} = 37 \times 10^{-\square}$, $\square = \ldots\ldots$

3 $9800 \times 10^{-5} = 9800 \times \frac{1}{10^n}$, the value of n is ......

4 $0.0002 \times 10^{11} \times 10^{-6} = \ldots\ldots$

5 $720\,000 \times 10^{-x} = 0.0072$, the value of x is ......

6 $10^{-3} \times 10^{16} \div 10^4 = 10^m$, $m = \ldots\ldots$

7 $0.0411 \times 10^{\square} = 4100$, $\square = \ldots\ldots$

8 $17\,000 \div 10^{\square} = 0.017$, $\square = \ldots\ldots$

9 $1.2 \div 10^9 = 1.2 \times \frac{1}{10^n}$, the value of n is ......

10 $0.19 \times 10^6 \times 10^{-4} = \ldots\ldots$

11 $81\,300 \div 10^{-18} = 81\,300 \times 10^{\square}$, $\square = \ldots\ldots$

12 $0.000074 \times 10^m \div 10^n = 74\,000$, $m - n = \ldots\ldots$

13 $9.49 \times 10^{-5} \div 10^{-12} = 9.49 \times 10^{\square}$, $\square = \ldots\ldots$

14 $0.0027 \times 10^m \div 10^{-n} = 270\,000$, $m + n = \ldots\ldots$

15 $0.02 \div 10^{-3} = \ldots\ldots$

16 $4.6 \div 10^6 = 460 \times \frac{1}{10^p}$, the value of p is ......

17 $1\,000\,000 \times 10^{-6} = \ldots\ldots$

18 $0.0014 \div 10^{-4} = \ldots\ldots$

19 $0.315 \div 10^{-x} = 3150$, the value of x is ......

20 $230\,400 \div 10^{\square} = 2.304$, $\square = \ldots\ldots$

21 $0.00023 \div 10^{-2} = 230 \times \frac{1}{10^p}$, the value of p is ......

Clue 3

Convert each of these answers into correct standard form.

A	B	C	D	E
4 500 000	0.045	73 200	0.000732	450

F	G	H	I	J
0.00732	45	0.00045	45×10^{-8}	450 000

K	L	M	N	O
0.0045	7320×10^{2}	0.732×10^{2}	0.0732	0.732

P	Q	R	S	T
$0.732 \times 10^{8} \times 10$	732 000	0.45	732	4500

U	V	W	X	Y or Z
7320×10^{4}	45×10^{-6}	450×10^{6}	0.00000732	45 000

Now decode the clue.

4.5×10^{3}	4.5×10^{-4}	4.5×10^{2}	4.5×10^{3}	4.5×10^{-4}

4.5×10^{-7}	4.5×10^{2}	7.32×10^{-3}	4.5×10^{8}	4.5×10^{2}

4.5×10^{6}	4.5×10^{-1}	7.32×10^{2}	4.5×10^{1}	7.32×10^{5}

4.5×10^{6}	7.32×10^{2}	7.32×10^{2}	4.5×10^{2}	7.32×10^{2}

Clue 4

Use the clues to find the hair colour of the thief.

Number	Rounded to the nearest:	Lower bound	Upper bound	Choose this letter if only the lower bound is correct	Choose this letter if only the upper bound is correct	Choose this letter if both bounds are correct	Choose this letter if neither bound is correct
60	10	55	65	F	R	B	H
300	100	299.5	300.5	E	A	L	R
58.1	1 d.p.	57.95	58.15	D	O	A	I
25 000	1000	24 500	25 050	W	N	C	R
45.0	1 d.p.	44.95	45.05	D	K	N	R

Clue 5

A number, n, rounded correct to the nearest 100 is 4000.

Statements about the number n	True or false	If true, shade all squares containing this number:
n could be 4050		1
$3500 \le n < 4500$		2
$4000 \le n < 4000.5$		3
n could be 3999		4
$3950 \le n \le 4050$		5
n could be 3990		6
n could be a decimal		7
$350 \le n < 450$		8
$3950 \le n < 4050$		9
If you round the number n correct to the nearest 1000 you get the same answer as if you round the number n correct to the nearest 100		10

Now decode the clue.

4 T	1 R	9 H	6 E	8 S	3 E	4 T	7 H	2 T	4 I
5 H	9 E	3 I	7 F	3 C	6 C	1 N	8 T	9 A	5 S
2 E	5 H	6 N	1 F	8 I	10 N	4 O	8 S	1 W	9 T
5 T	2 H	10 D	1 V	4 R	2 E	5 I	5 T	2 H	3 G
3 S	2 L	8 S	5 R	6 I	1 T	7 V	8 H	3 B	7 E

Answers

Answers to clue 1 – female

Multiplying by 10^{-4} is the same as dividing by 10^4. Correct.

Multiplying by 10^{-3} is the same as dividing by $\frac{1}{10^3}$. Incorrect.

Dividing by 10^5 is the same as multiplying by 10^{-5}. Correct.

Dividing by 10^{-8} is the same as multiplying by 10^8. Correct.

Dividing by 10^{-11} is the same as multiplying by $\frac{1}{10^{11}}$. Incorrect.

Number of correct statements	Gender of the thief
3	Female

Answers to clue 2 – right-handed

1	$200\,000 \div 10^4 = \ldots\ldots$	Answer 20 = T
2	$37 \times 10^{-5} \times 10^{-3} = 37 \times 10^{-\square}$, $\square = \ldots\ldots$	Answer 8 = H
3	$9800 \times 10^{-5} = 9800 \times \frac{1}{10^n}$, the value of n is	Answer 5 = E
4	$0.0002 \times 10^{11} \times 10^{-6} = \ldots\ldots$	Answer 20 = T
5	$720\,000 \times 10^{-x} = 0.0072$, the value of x is	Answer 8 = H
6	$10^{-3} \times 10^{16} \div 10^4 = 10^m$, $m = \ldots\ldots$	Answer 9 = I
7	$0.0411 \times 10^{\square} = 4100$, $\square = \ldots\ldots$	Answer 5 = E
8	$17\,000 \div 10^{\square} = 0.017$, $\square = \ldots\ldots$	Answer 6 = F
9	$1.2 \div 10^9 = 1.2 \times \frac{1}{10^n}$, the value of n is	Answer 9 = I
10	$0.19 \times 10^6 \times 10^{-4} = \ldots\ldots$	Answer 19 = S
11	$81\,300 \div 10^{-18} = 81\,300 \times 10^{\square}$, $\square = \ldots\ldots$	Answer 18 = R
12	$0.000074 \times 10^m \div 10^n = 74\,000$, $m - n = \ldots\ldots$	Answer 9 = I
13	$9.49 \times 10^{-5} \div 10^{-12} = 9.49 \times 10^{\square}$, $\square = \ldots\ldots$	Answer 7 = G
14	$0.0027 \times 10^m \div 10^{-n} = 270\,000$, $m + n = \ldots\ldots$	Answer 8 = H
15	$0.02 \div 10^{-3} = \ldots\ldots$	Answer 20 = T
16	$4.6 \div 10^6 = 460 \times \frac{1}{10^p}$, the value of p is	Answer 8 = H
17	$1\,000\,000 \times 10^{-6}$	Answer 1 = A
18	$0.0014 \div 10^{-4} = \ldots\ldots$	Answer 14 = N
19	$0.315 \div 10^{-x} = 3150$, the value of x is	Answer 4 = D
20	$230\,400 \div 10^{\square} = 2.304$, $\square = \ldots\ldots$	Answer 5 = E
21	$0.00023 \div 10^{-2} = 230 \times \frac{1}{10^p}$, the value of p is	Answer 4 = D

Answers to clue 3 – wears glasses

4.5×10^3	4.5×10^{-4}	4.5×10^2	4.5×10^3	4.5×10^{-4}
T	H	E	T	H

4.5×10^{-7}	4.5×10^2	7.32×10^{-3}	4.5×10^8	4.5×10^2
I	E	F	W	E

4.5×10^6	4.5×10^{-1}	7.32×10^2	4.5×10^1	7.32×10^5
A	R	S	G	L

4.5×10^6	7.32×10^2	7.32×10^2	4.5×10^2	7.32×10^2
A	S	S	E	S

Answers to clue 4 – hair colour brown

Number	Rounded to the nearest:	Lower bound	Upper bound	Choose this letter if only the lower bound is correct	Choose this letter if only the upper bound is correct	Choose this letter if both bounds are correct	Choose this letter if neither bound is correct
60	10	55	65	F	R	B	H
300	100	299.5	300.5	E	A	L	R
58.1	1 d.p.	57.95	58.15	D	O	A	I
25 000	1000	24 500	25 050	W	N	C	R
45.0	1 d.p.	44.95	45.05	D	K	N	R

Answers to clue 5 – cannot drive

Statements about the number *n*	True or false
n could be 4050	False
$3500 \le n < 4500$	False
$4000 \le n < 4000.5$	False
n could be 3999	True
$3950 \le n \le 4050$	False
n could be 3990	True
n could be a decimal	True
$350 \le n < 450$	False
$3950 \le n < 4050$	True
If you round the number *n* correct to the nearest 1000 you get the same answer as if you round the number *n* correct to the nearest 100	True

4 T	1 R	9 H	6 E	8 S	3 E	4 T	7 H	2 T	4 I
5 H	9 E	3 I	7 F	3 C	6 C	1 N	8 T	9 A	5 S
2 E	5 H	6 N	1 F	8 I	10 N	4 O	8 S	1 W	9 T
5 T	2 H	10 D	1 V	4 R	2 E	5 I	5 T	2 H	3 G
3 S	2 L	8 S	5 R	6 I	1 T	7 V	8 H	3 B	7 E

The solution – the guilty customer is Keren Clayton.

Fractions jigsaw

For use with Chapter 7 of Student Book 9

Teacher's notes

This jigsaw activity is an interesting group work alternative to a written exercise. The advantages of a jigsaw activity over a written exercise are that it creates opportunities for students to discuss their ideas, and they can find an alternative starting point if they are not sure how to do a particular question. Students are also able to start with the answer and work backwards, which is sometimes an interesting way to tackle a question.

This activity requires students to be able to add, subtract, multiply and divide fractions, including mixed numbers. They are also required to be able to simplify and square fractions.

Introductory activity

This activity can be used as a plenary or revision exercise supporting the work completed in Chapter 7. You may find that little introduction is necessary. Alternatively, you may want to go through the following examples.

Example 1 – Simplifying fractions, dividing by common factors

Simplify $\frac{156}{260}$

Divide numerator and denominator by 4:

$$\frac{156}{260} = \frac{39}{65}$$

Be careful, as there is still a factor of 39 and 65, namely 13:

$$\frac{156}{260} = \frac{39}{65} = \frac{3}{5}$$

Students often miss large prime numbers as common factors so a few are included in the jigsaw.

Example 2 – Adding and subtracting fractions, including mixed numbers, and finding a common denominator

$$5\frac{3}{4} + 1\frac{4}{5} - 3\frac{7}{10}$$

Convert all fractions to fractions with common denominators; leave them as mixed numbers for now:

$$5\frac{15}{20} + 1\frac{16}{20} - 3\frac{14}{20} = 6\frac{31}{20} - 3\frac{14}{20} = 3\frac{17}{20}$$

You may wish to provide further examples, particularly if students are expected to borrow or convert to improper fractions to aid subtraction.

Example 3 – Multiplying fractions, including mixed numbers

$$4\frac{3}{8} \times 3\frac{3}{7}$$

Convert to improper fractions first:

$$4\frac{3}{8} \times 3\frac{3}{7} = \frac{35}{8} \times \frac{24}{7}$$

Cancel before multiplying:

$$\frac{\overset{5}{\cancel{35}}}{\underset{1}{\cancel{8}}} \times \frac{\overset{3}{\cancel{24}}}{\underset{1}{\cancel{7}}} = \frac{15}{1} = 15$$

Example 4 – Dividing fractions, including mixed numbers

$$6\frac{2}{3} \div 2\frac{10}{15}$$

Convert to improper fractions first:

$$6\frac{2}{3} \div 2\frac{10}{15} = \frac{20}{3} \div \frac{40}{15}$$

Then multiply by the reciprocal of the second fraction:

$$\frac{20}{3} \times \frac{15}{40}$$

Cancel before multiplying, and give the answer as a mixed number:

$$\frac{\overset{1}{\cancel{20}}}{\underset{1}{\cancel{3}}} \times \frac{\overset{5}{\cancel{15}}}{\underset{2}{\cancel{40}}} = \frac{5}{2} = 2\frac{1}{2}$$

Note that some of the questions in the jigsaw are harder than these examples. For example, there are calculations involving squaring, finding square roots and doing inverse operations. You may choose to cover some harder examples, as these cover only the basics.

Jigsaw activity

Organize the students to work in small groups. Copy the two pages of jigsaw pieces onto card to make a complete set for each group. Do not tell the students what shape to expect from the jigsaw as this adds to the challenge. This jigsaw has no blank edges so it is harder for your students to complete.

Differentiation

To make it more difficult there are 'questions' and 'solutions' around the outside edges of the jigsaw. These extend into slightly more challenging calculations and include common incorrect answers to other questions in the jigsaw. Students do not need to do these, of course. To make it easier you could indicate outside edges by marking them.

Alternative approaches

You could simply instruct students to complete the jigsaw, or you could make use of the fact that the outside edges are not blank. You could make copies of the blank jigsaw pieces, provided, and tell students to use these pieces to extend the jigsaw by finding solutions to the 'outside' questions, and writing questions with the given solutions. Alternatively, students could be creative and make an entirely new jigsaw of their own.

Answers

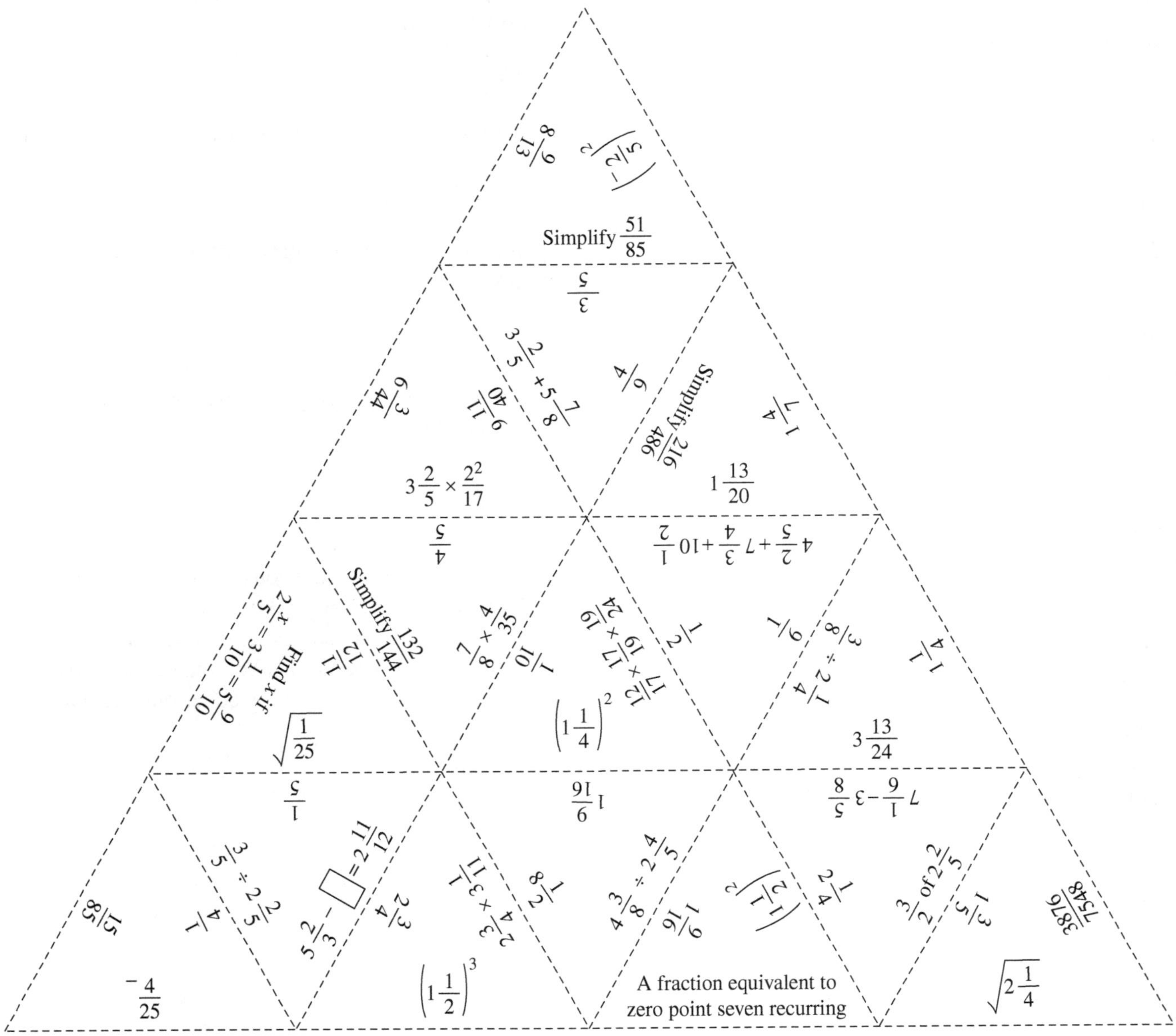

Jigsaw pieces

Page 1 of 2

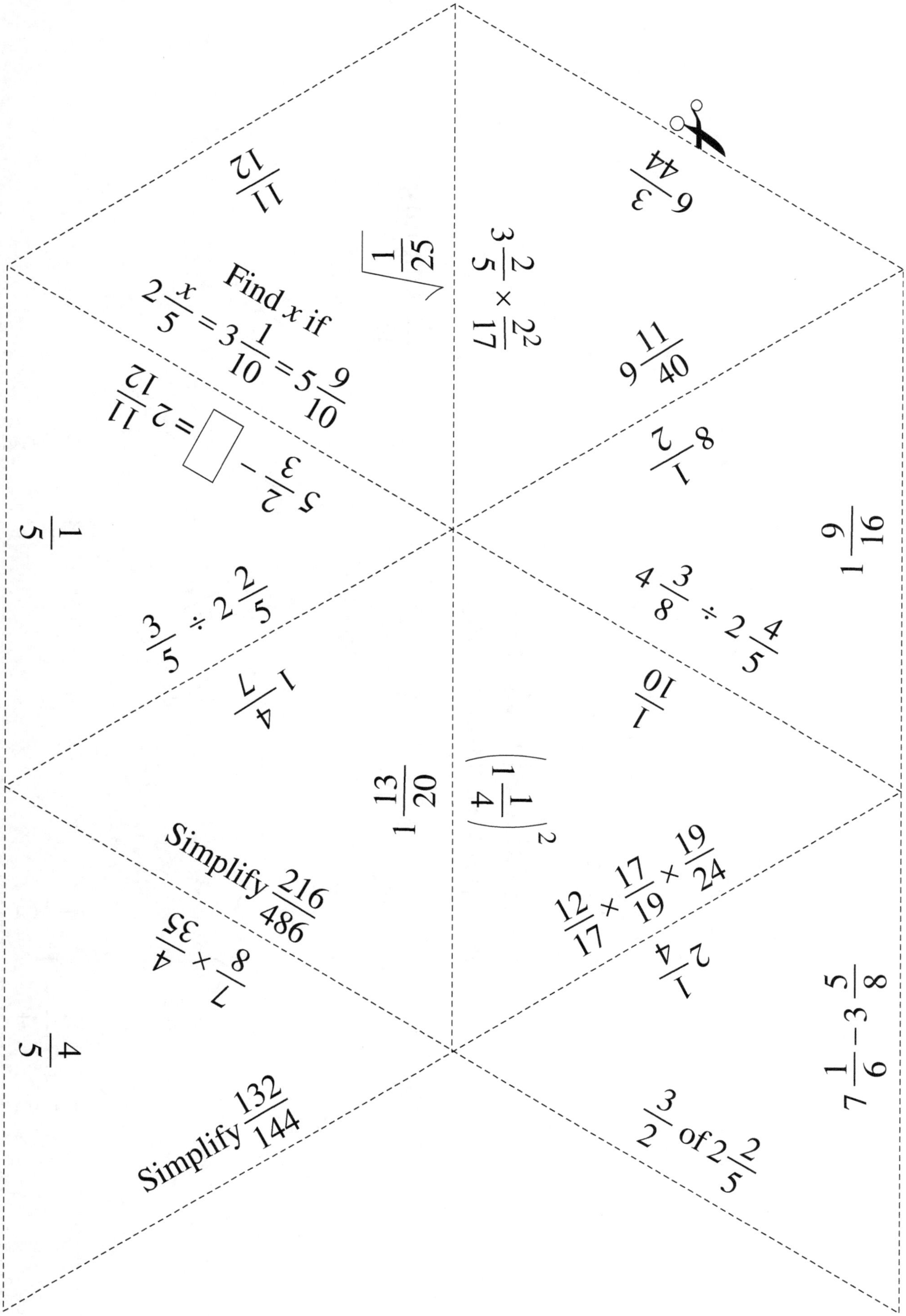

Jigsaw pieces

Page 2 of 2

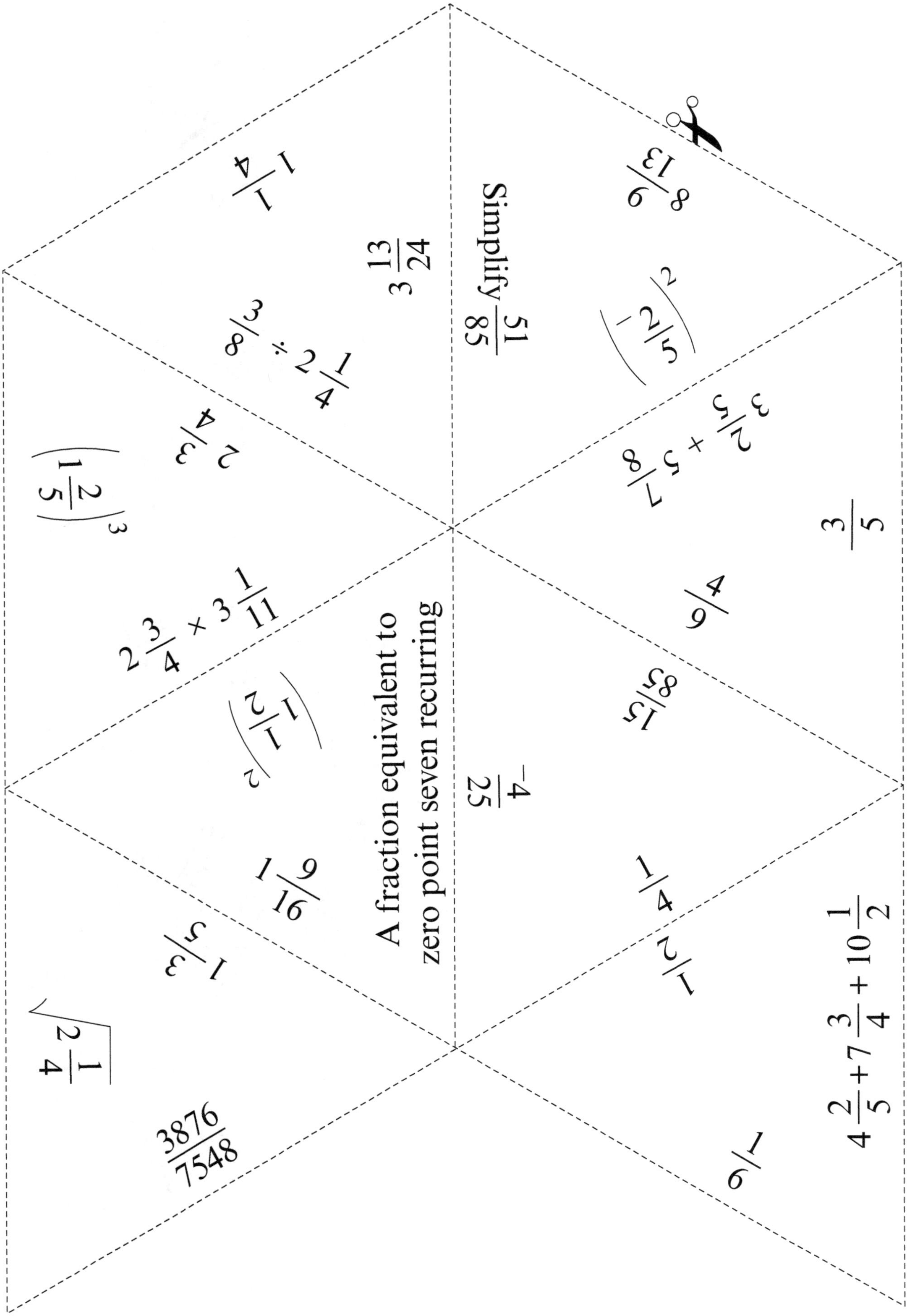

Jigsaw pieces – blank set

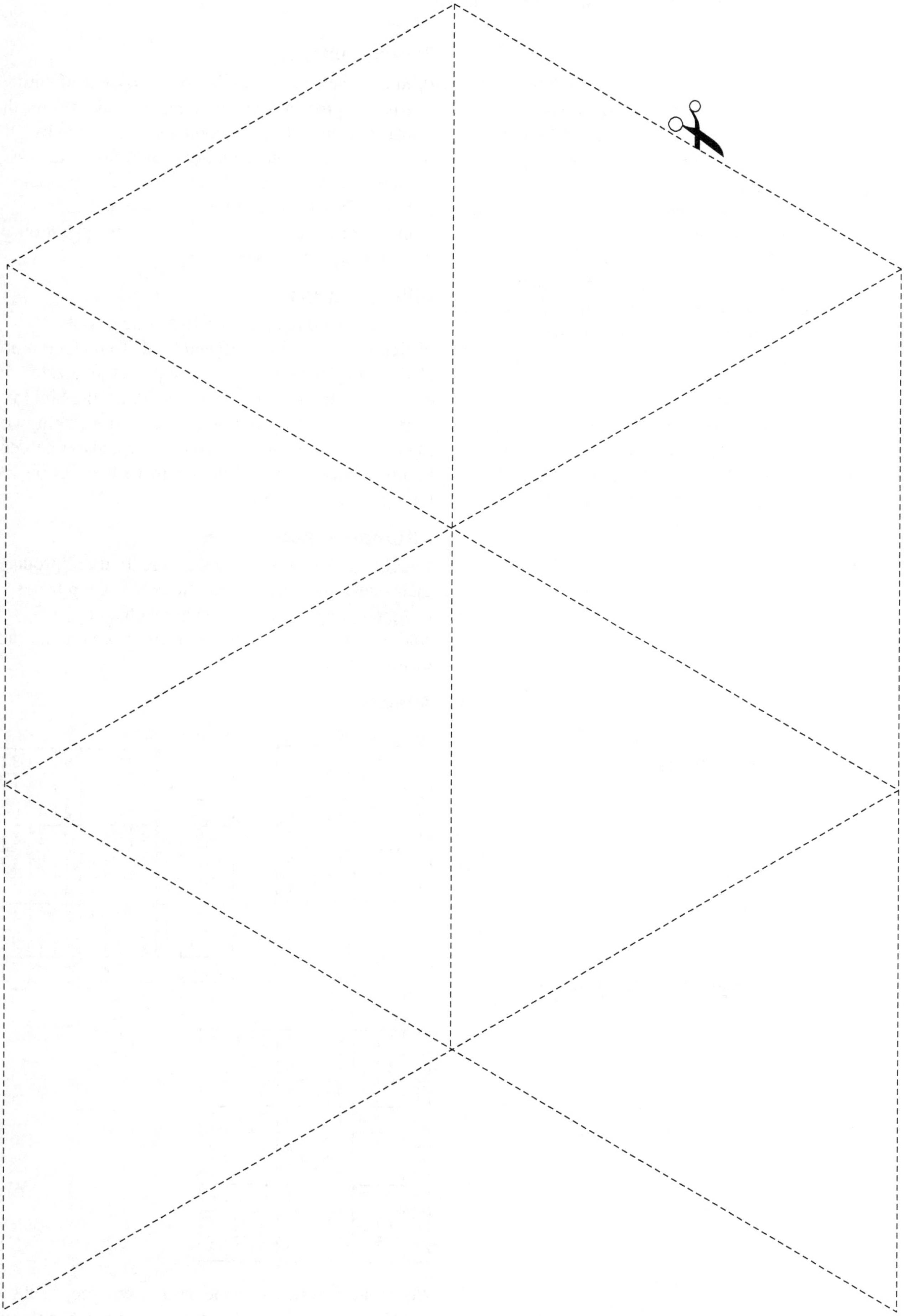

Simultaneous equations Sudoku

For use with Chapter 8 of Student Book 9

Teacher's notes

This activity is an extension activity for students who already know how to solve simultaneous equations with two unknowns (covered in Chapter 8). This puzzle is more challenging, as it covers equations with more than two unknowns. It requires students to find and solve a pair of equations with two unknowns, then to substitute these values into other equations where there may be more than two unknowns, to simplify and solve them. Note that solving simultaneous equations with more than two unknowns is not in the Cambridge Secondary 1 curriculum framework.

Introductory activity

You may choose to go through the following example which shows how to deal with simultaneous equations with more than two unknowns. Or, for more able students, you might choose to see how they cope without first seeing an example.

Example

Here is a set of simultaneous equations to solve:

[1] $2w + x + 3y + z = 23$
[2] $2w + 5z = 30$
[3] $w + x + y + z = 14$
[4] $2w + 3z = 22$

Look for a pair of equations with the same two unknowns: equations [2] and [4].
Solve these in the usual way:

[2] $2w + 5z = 30$
[4] $2w + 3z = 22$
[2] – [4] $2z = 8$, so $z = 4$

Substitute into [4] $2w + 12 = 22$

$2w = 10$, so $w = 5$

Then substitute $z = 4$ and $w = 5$ into equations [1] and [3]:

[1] $10 + x + 3y + 4 = 23$
[3] $5 + x + y + 4 = 14$

These simplify to give:

[1] $x + 3y = 9$
[3] $x + y = 5$

Solve these in the usual way:

[1] – [3] $2y = 4$, so $y = 2$

Substitute into [3] $x + 2 = 5$, so $x = 3$

Sudoku activity

Print out one copy of the Sudoku puzzle grid with instructions for each student, and a blank grid for their answers. Tell students to complete the puzzle by solving simultaneous equations, then filling in one number (from 1 to 9, or 1 to 6 for the easier puzzle) per cell. They need to follow the rule that each number from 1 to 9 (or 1 to 6) can only appear once in each row, column or region.

Differentiation

There are three puzzles at different levels of difficulty: easier, medium and hard. To make it more challenging, tell students to complete the puzzle against the clock. You can give students the hint to start with a pair of simultaneous equations with two matching unknowns to solve first, or make it harder by not giving this hint. You can make it easier by telling students the value of one letter.

Alternative approaches

Encourage students to use their creativity to produce their own simultaneous equations Sudoku puzzles using the blank grids (this is quite challenging – students could get a Sudoku puzzle from the internet to start with).

Answers

Easier puzzle

2	1	5	6	3	4
4	6	3	2	1	5
5	3	1	4	6	2
6	4	2	1	5	3
1	5	4	3	2	6
6	2	6	5	4	1

Medium puzzle

5	4	6	1	2	7	9	3	8
2	8	9	6	4	3	5	7	1
3	1	7	5	8	9	2	4	6
1	6	3	4	9	2	7	8	5
4	5	2	8	7	6	1	9	3
9	7	8	3	5	1	4	6	2
6	9	5	7	1	8	3	2	4
7	3	4	2	6	5	8	1	9
8	2	1	9	3	4	6	5	7

Hard puzzle

3	5	1	8	9	6	7	4	2
7	4	2	3	1	5	6	9	8
8	6	9	2	4	7	3	5	1
2	7	4	1	5	3	8	6	9
9	8	6	4	7	2	1	3	5
1	3	5	9	6	8	2	7	4
4	2	7	5	3	1	9	8	6
5	1	3	6	8	9	4	2	7
6	9	8	7	2	4	5	1	3

Note that if students do the hard puzzle they will be expected to solve simultaneous equations like these:

[1] $v + x = 8$ and [2] $v + x + r = 16$, therefore $r = 8$

Simultaneous equations Sudoku

Page 1 of 3

Easier puzzle

		a	$3c$	d		14
b						
a					c	7
e					d	9
1					e	
	c	e	a			13
15		11	11		11	

Medium puzzle

m		n				p		j	
			$2h$	$2f$	h				13
	i						g		
		h		$3h$		k			19
g		f				i		h	10
		$4f$		m		g			17
	$3h$						f		
			f	n	m				
$2g$		i				n		k	22
		20	8	24		27		18	

Rules of ordinary Sudoku

Sudoku is played in a 9×9 grid divided into nine 3×3 sub-grids called regions. Sudoku begins with some of the grid cells already filled with numbers. The objective is to fill the other empty cells with numbers between 1 and 9 (one number in each cell) according to the following rules:

- A number from 1 to 9 can appear only once in a row
- A number from 1 to 9 can appear only once in a column
- A number from 1 to 9 can appear only once in a 3×3 sub-grid

To summarise, no column, row or 3×3 region can contain the same number twice.

For the easier puzzle here, the same rules apply, but only the numbers 1 to 6 are used.

Rules of simultaneous equations Sudoku

This Sudoku involves an extra challenge. You will notice that there are no numbers in the cells. Instead, there are only letters. The number at the end of a row gives the total of the letters in that row. For example, in the easier puzzle:

- the third row tells us $a + c = 7$
- the sixth row tells us $c + e + a = 13$

Similarly, the number at the bottom of a column gives the total of the letters in that column.

For example, in the medium difficulty puzzle:

- the third column tells us $n + h + f + 4f + i = 20$ (hint: simplify this)
- the fourth column tells us $2h + f = 8$

After the simultaneous equations have been solved to find the values of the letters, the puzzle is solved by using the rules of ordinary Sudoku.

Simultaneous equations Sudoku

Page 2 of 3

Hard puzzle

To solve this puzzle, as well as solving simultaneous equations, you will have to use logical reasoning.

Hard puzzle

v			8		u			t	
			v		x				8
	u			s			x		15
$2z$		$4z$				r		w	
	r						v		11
z		5				t		s	
	t			v			r		
			u		$3v$				
u			y		s			v	20
20	16							16	

Rules of ordinary Sudoku

Sudoku is played in a 9×9 grid divided into nine 3×3 sub-grids called regions. Sudoku begins with some of the grid cells already filled with numbers. The objective is to fill the other empty cells with numbers between 1 and 9 (one number in each cell) according to the following rules:

- A number from 1 to 9 can appear only once in a row
- A number from 1 to 9 can appear only once in a column
- A number from 1 to 9 can appear only once in a 3×3 sub-grid

To summarise, no column, row or 3×3 region can contain the same number twice.

Rules of simultaneous equations Sudoku

This Sudoku has an extra challenge. You will notice that it has few numbers in the cells. Instead, there are mostly letters. The number at the end of a row gives the total of the letters in that row. For example:

- the second row tells us $v + x = 8$
- the fifth row tells us $r + v = 11$

Similarly, the number at the bottom of a column gives the total of the letters in that column. For example:

- the first column tells us $v + 2z + z + u = 20$ (hint: simplify this)
- the second column tells us $u + r + t = 16$

After the simultaneous equations have been solved to find the values of the letters, the puzzle is solved by using the rules of ordinary Sudoku.

Simultaneous equations Sudoku – blank

Page 3 of 3

Use these grids for your answers, or to make your own puzzles.

Easier puzzle

$a =$ .... $b =$ ....

$c =$ .... $d =$ ....

$e =$ ....

Medium puzzle

$f =$ .... $g =$ .... $h =$ ....

$i =$ .... $j =$ .... $k =$ ....

$m =$ .... $n =$ .... $p =$ ....

Hard puzzle

$r =$ .... $s =$ .... $t =$ ....

$u =$ .... $v =$ .... $w =$ ....

$x =$ .... $y =$ .... $z =$ ....

Transformations card-sorting activity

For use with Chapter 13 of Student Book 9

Teacher's notes

This card-sorting activity can be used as a plenary or revision exercise. The aim is to match the diagrams of transformations with the cards that describe the transformations. Students are expected to know what is needed to give a precise description of a reflection, rotation, translation or enlargement. Students also need to understand combinations of transformations.

Introductory activity

You may wish to go through the following (or, if you want to make the activity more challenging, give no introduction at all):

- To describe a reflection – give the equation of the mirror line.
- To describe a rotation – give the angle and direction of the rotation and the coordinates of the centre of rotation.
- To describe a translation – give the vector.
- To describe an enlargement – give the coordinates of the centre of enlargement and the scale factor of the enlargement.

Card-sorting activity

Organize students to work in pairs or small groups. You will need one set of transformation cards, copied and cut out, for each pair or group of students. Instruct students to group the cards into sets, each set containing a diagram and description of a transformation. Inform them that on the diagrams the object is labelled with an 'O' and the image is labelled with an 'I'.

Note that sets will have different numbers of cards, depending on the type of transformation and whether the cards describing combined transformations are being used. Students will not know this, and you can decide whether or not to explain it to them.

Differentiation

To make the task easier you can use fewer cards, e.g. only two of each type of transformation instead of four. To make the task harder, include the extension cards, which give combinations of transformations including where one of the transformations is an enlargement (which is beyond the syllabus). Note that half of the diagrams can be described with cards for both a single transformation and a combination of transformations.

Alternative approaches

You could give students a blank set of cards and encourage them to use their creativity to devise their own completely different set(s) of equivalent cards. Alternatively, you could ask them to add new cards to each set, particularly those sets that do not have a card for a combination of transformations. You could also ask students to think of alternative ways to describe the transformations, and which transformations it is easier to do this for.

Answers

Single and combined transformations

Single transformation	Combined transformations	Diagram
ROTATION clockwise through 90° (or through 270° anticlockwise) centre (0, 3)	A rotation clockwise through 90°, centre ($^{-}1$, 1), followed by a translation through the vector $\begin{pmatrix} ^{-}1 \\ 3 \end{pmatrix}$	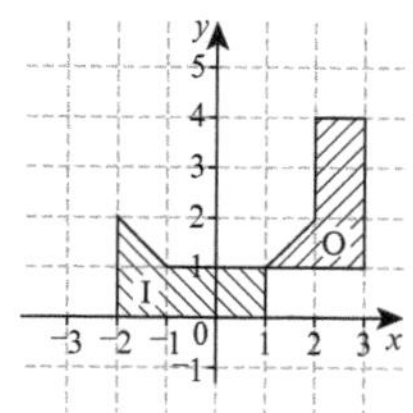
ENLARGEMENT centre (1, 1) scale factor 2	An enlargement centre (2, 0), scale factor 2, followed by a translation through the vector $\begin{pmatrix} 1 \\ ^{-}1 \end{pmatrix}$	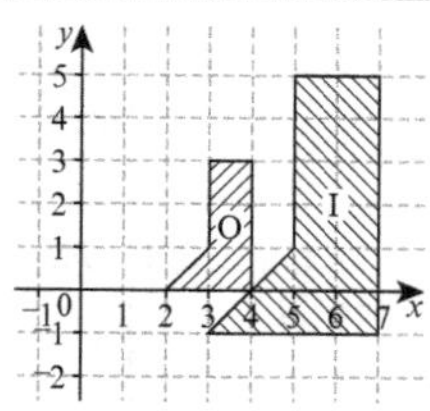
ROTATION through 180° centre (1, 3)	A rotation clockwise through 90°, centre (1, 3), followed by a rotation anticlockwise through 270°, centre (1, 3)	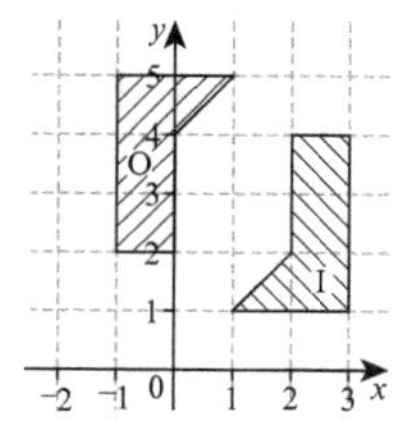
ROTATION clockwise through 90° (or through 270° anticlockwise) centre ($^{-}1$, 3)	A rotation through 180°, centre (1, 2), followed by a reflection in the line $y = x$, followed by a reflection in the y-axis	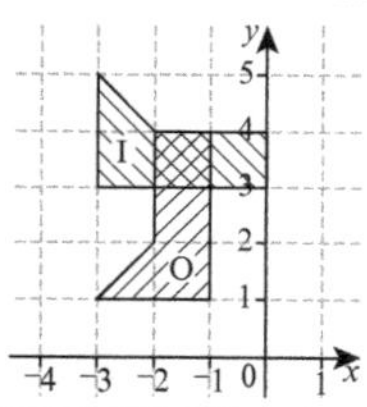
REFLECTION in the line $y = 2$	A reflection in the line $y = 5$, followed by a translation through the vector $\begin{pmatrix} 0 \\ ^{-}6 \end{pmatrix}$	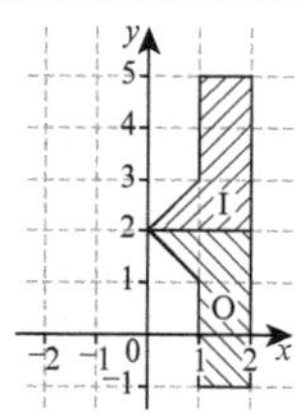
ROTATION anticlockwise through 270° (or through 90° clockwise) centre ($^{-}1$,$^{-}2$)	A translation through the vector $\begin{pmatrix} 3 \\ 1 \end{pmatrix}$, followed by a rotation clockwise through 90°, centre (0, 0)	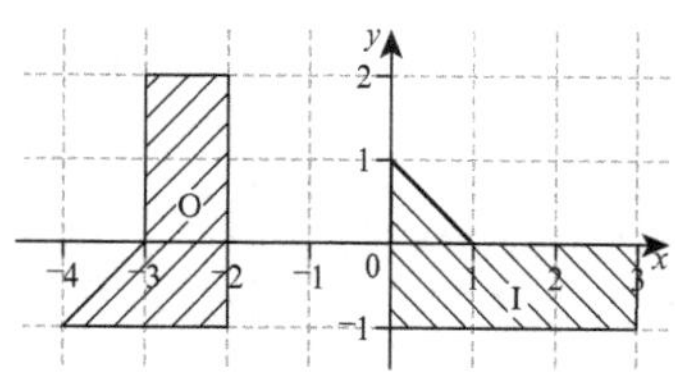
REFLECTION in the line $x = 2$	A translation through the vector $\begin{pmatrix} 2 \\ 0 \end{pmatrix}$, followed by a reflection in the line $x = 3$	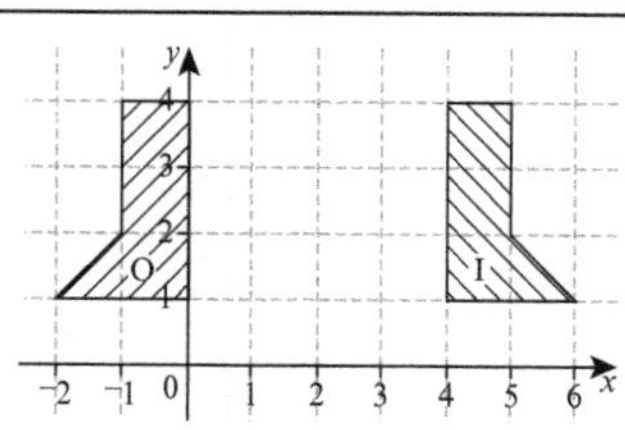
ENLARGEMENT centre (3, $^{-}2$) scale factor 3	A translation through the vector $\begin{pmatrix} ^{-}2 \\ 0 \end{pmatrix}$, followed by an enlargement, centre (0,$^{-}2$), scale factor 3	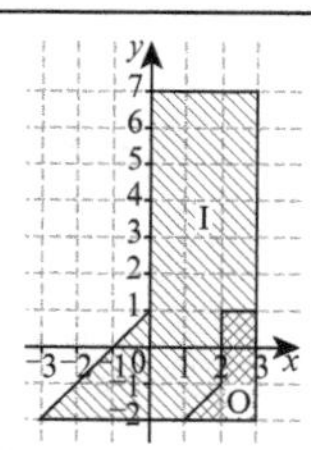

Single transformations only

Single transformation	Diagram
TRANSLATION through the vector $\begin{pmatrix} 1 \\ 3 \end{pmatrix}$	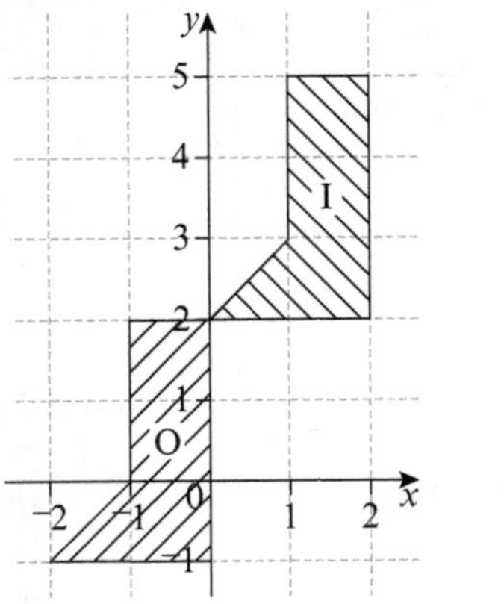
REFLECTION in the line $y = {}^-x$	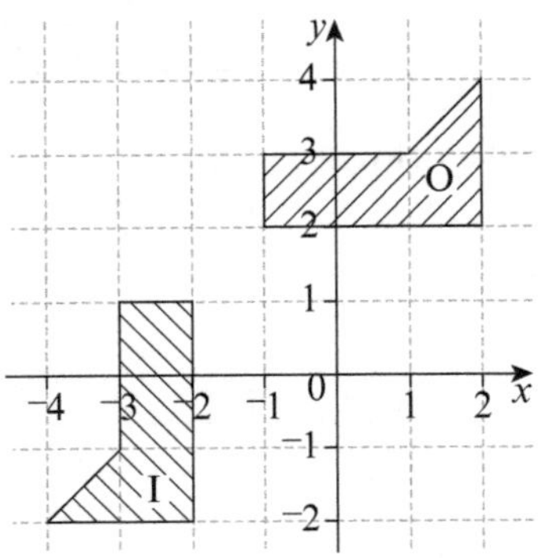
REFLECTION in the line $y = x$	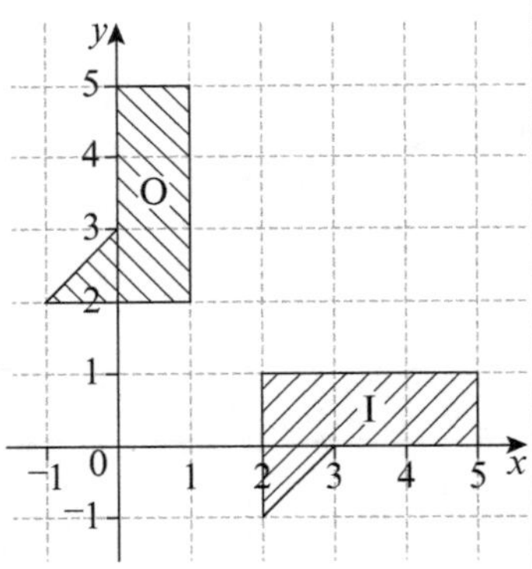
ENLARGEMENT centre (⁻2, 3) scale factor 5	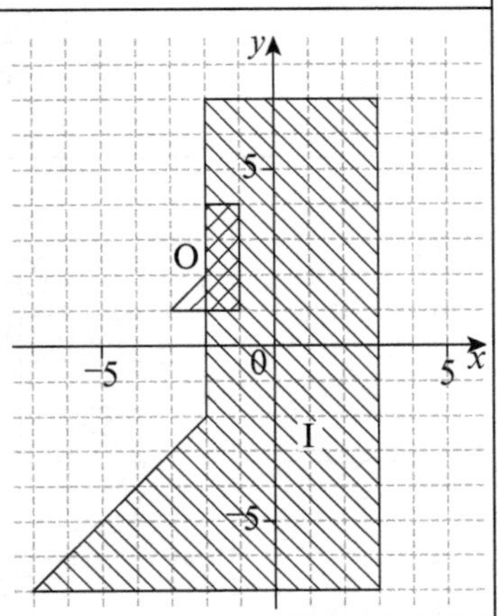

Single transformation	Diagram
TRANSLATION through the vector $\begin{pmatrix} 3 \\ 1 \end{pmatrix}$	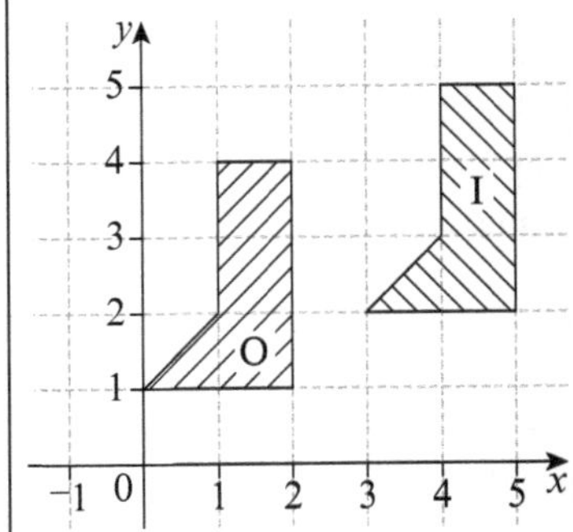
TRANSLATION through the vector $\begin{pmatrix} 2 \\ 3 \end{pmatrix}$	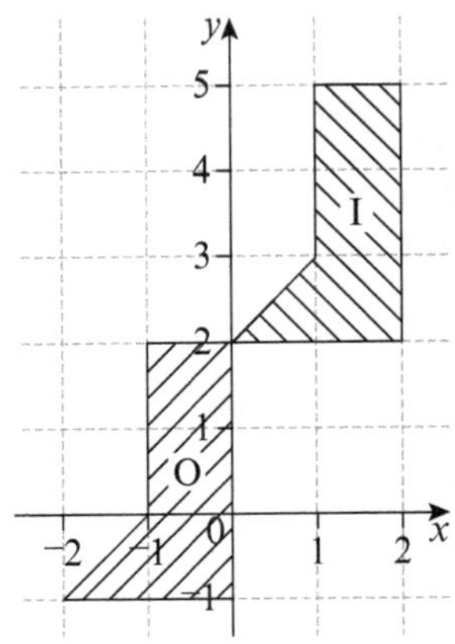
TRANSLATION through the vector $\begin{pmatrix} {}^-3 \\ 2 \end{pmatrix}$	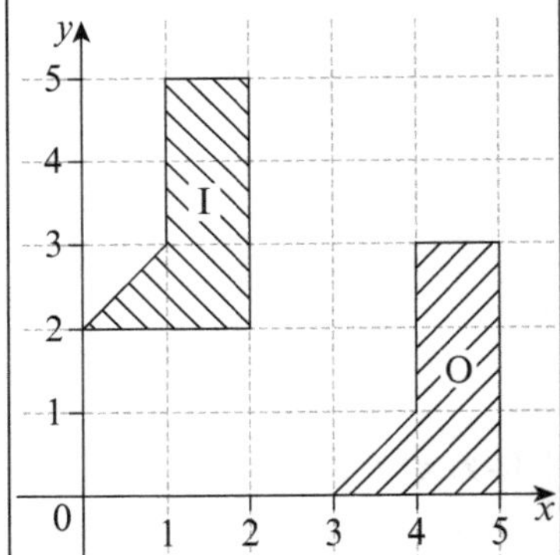
ENLARGEMENT centre (4, 9) scale factor 4	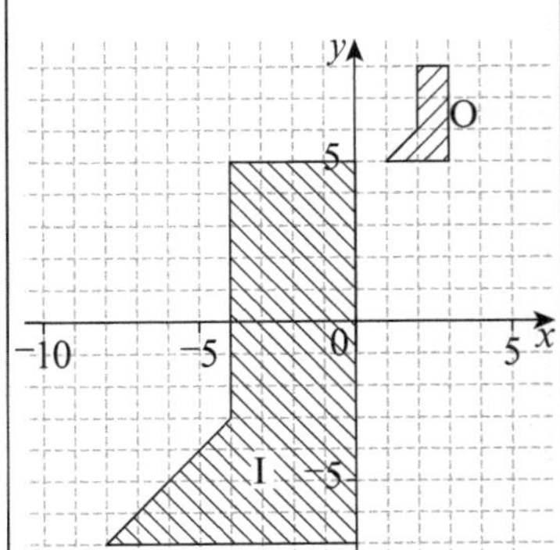

Transformation cards

Page 1 of 6

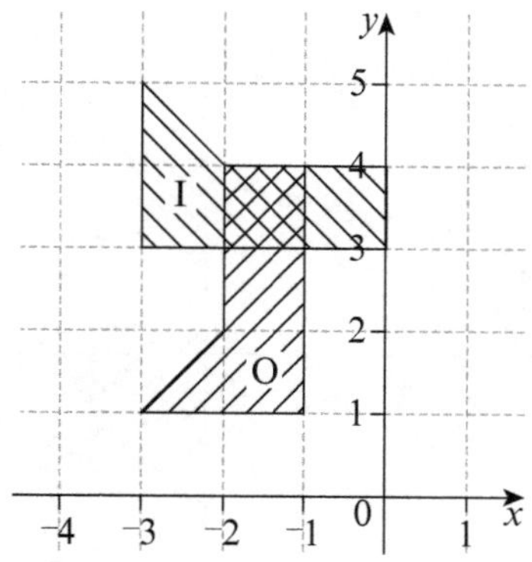

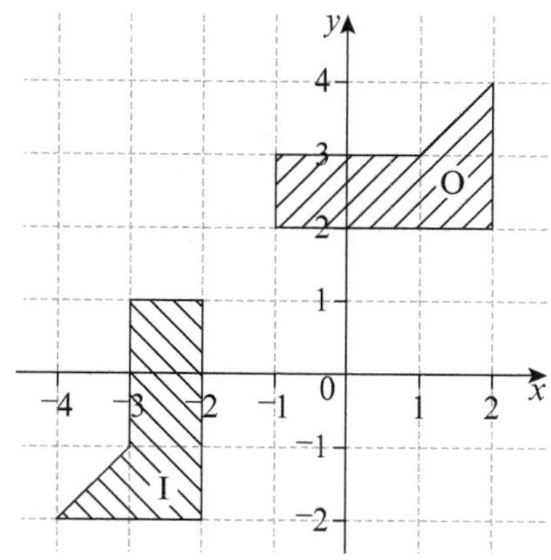

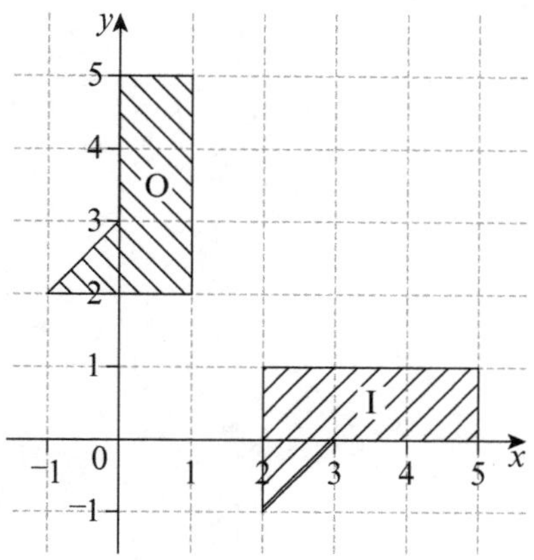

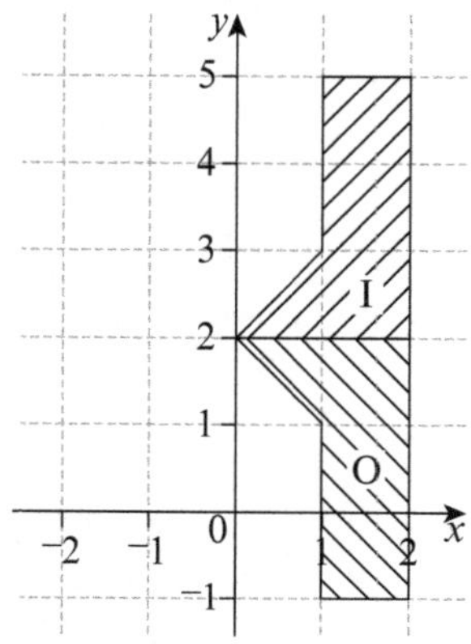

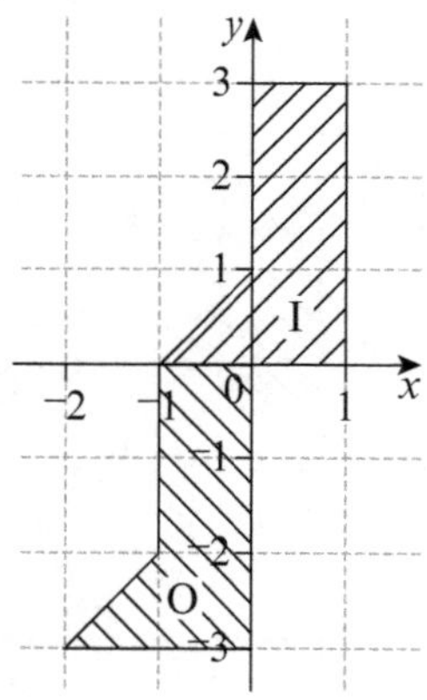

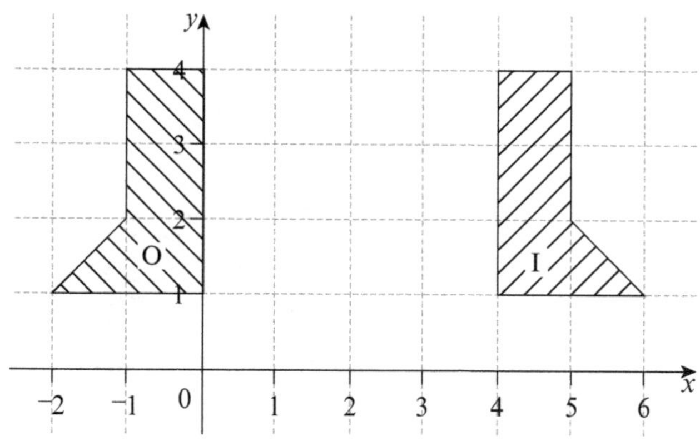

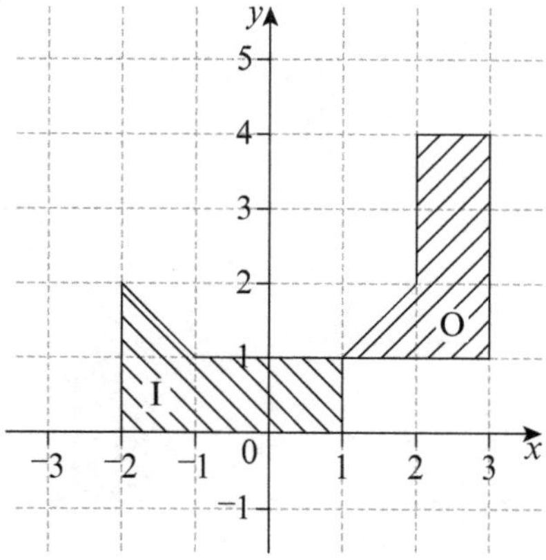

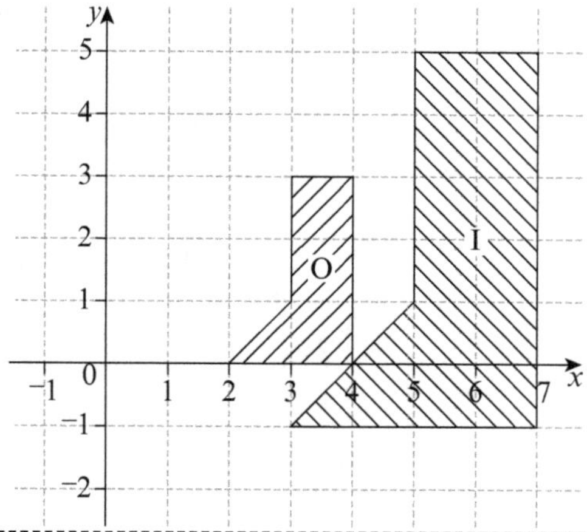

Transformation cards

Page 2 of 6

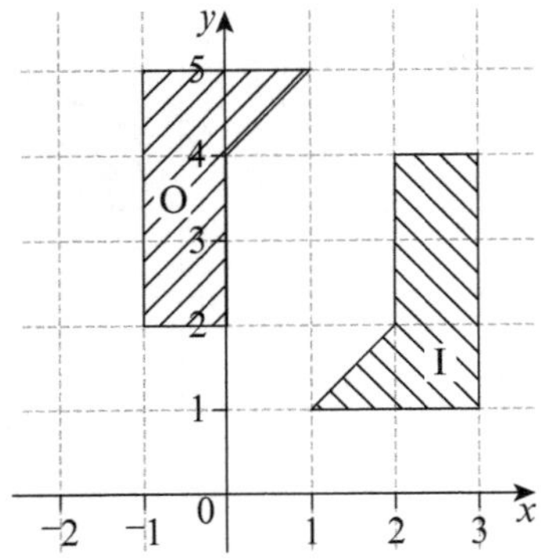

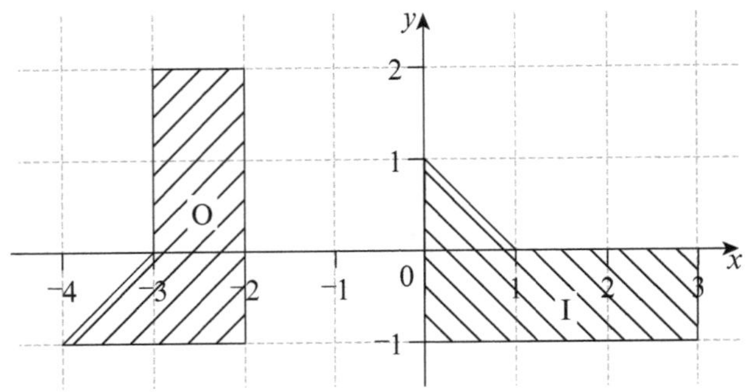

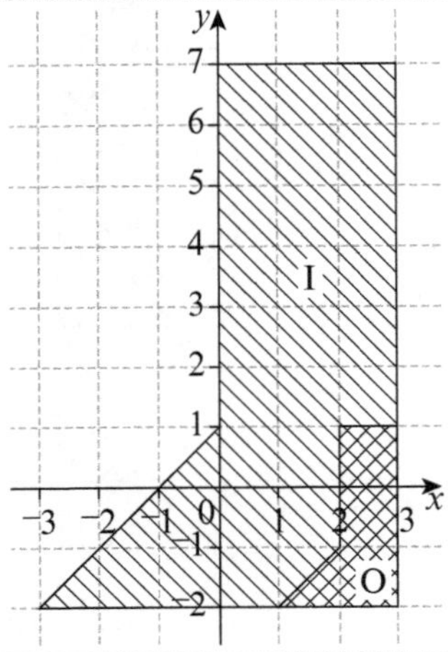

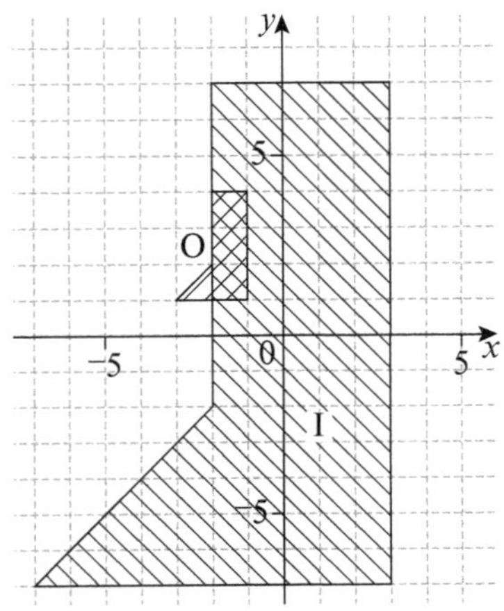

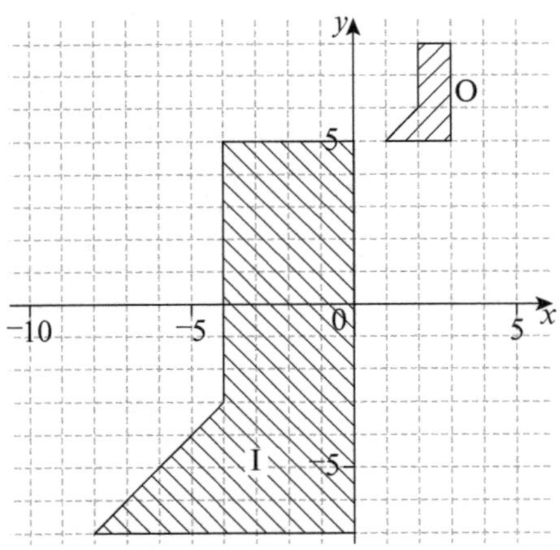

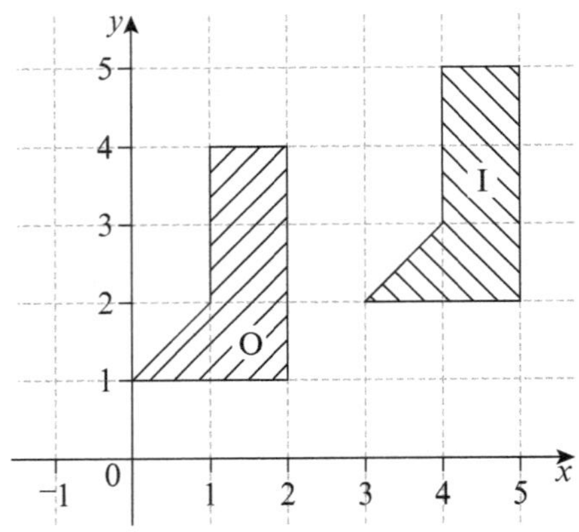

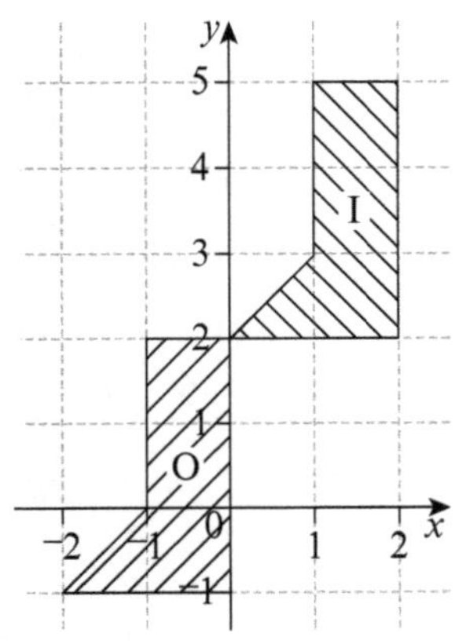

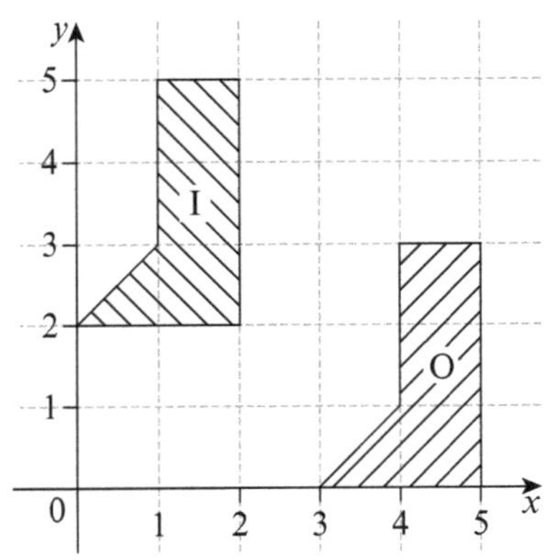

Transformation cards

Page 3 of 6

REFLECTION	REFLECTION
REFLECTION	REFLECTION
ROTATION	ROTATION
ROTATION	ROTATION
ENLARGEMENT	ENLARGEMENT
ENLARGEMENT	ENLARGEMENT
TRANSLATION	TRANSLATION
TRANSLATION	TRANSLATION

Transformation cards

Page 4 of 6

in the line $y = x$	in the line $y = {}^{-}x$
in the line $y = 2$	in the line $x = 2$
through the vector $\begin{pmatrix} 3 \\ 1 \end{pmatrix}$	through the vector $\begin{pmatrix} 1 \\ 3 \end{pmatrix}$
through the vector $\begin{pmatrix} {}^{-}3 \\ 2 \end{pmatrix}$	through the vector $\begin{pmatrix} 2 \\ 3 \end{pmatrix}$
scale factor 2	scale factor 3
scale factor 5	scale factor 4
centre (1,1)	centre (3,⁻2)

Transformation cards

centre (1,3)	centre (⁻1,⁻2)
centre (⁻1,3)	centre (⁻2,3)
centre (0,3)	centre (4,9)
through 90°	through 90°
through 270°	through 180°
anticlockwise	clockwise
clockwise	

Transformation cards

Page 6 of 6

Extension – combined transformations

A rotation clockwise through 90°, centre ($^{-}1$, 1), followed by a translation through the vector $\begin{pmatrix} ^{-}1 \\ 3 \end{pmatrix}$

An enlargement centre (2, 0), scale factor 2, followed by a translation through the vector $\begin{pmatrix} 1 \\ -1 \end{pmatrix}$

A rotation clockwise through 90°, centre (1, 3), followed by a rotation anticlockwise through 270°, centre (1, 3)

A rotation through 180°, centre (1, 2), followed by a reflection in the line $y = x$, followed by a reflection in the y-axis

A reflection in the line $y = 5$, followed by a translation through the vector $\begin{pmatrix} 0 \\ ^{-}6 \end{pmatrix}$

A translation through the vector $\begin{pmatrix} 3 \\ 1 \end{pmatrix}$, followed by a rotation clockwise through 90°, centre (0,0)

A translation through the vector $\begin{pmatrix} 2 \\ 0 \end{pmatrix}$, followed by a reflection in the line $x = 3$

A translation through the vector $\begin{pmatrix} ^{-}2 \\ 0 \end{pmatrix}$, followed by an enlargement, centre (0,$^{-}$2), scale factor 3

Probability crossword

For use with Chapter 15 of Student Book 9

Teacher's notes

This activity is for students who already know most facts from the probability section of the syllabus. It is a revision activity, covering different areas of the Cambridge Secondary 1 curriculum framework from Chapter 15 in Book 9. There are a number of keywords, as listed below, which students may not have seen before. One of the aims of this activity is to help students become more familiar and confident with this terminology.

Keywords

- mutually exclusive
- impossible
- theoretical probability
- data
- experiment
- biased
- equally likely
- successive events
- sample space diagram
- tree diagram
- outcome
- experimental probability
- logical
- random
- relative frequency
- probability
- certain
- event
- chance

Introductory activity

There should be little need for any introduction, as this is intended as a revision exercise.

Crossword activity

Print out one copy of the crossword grid and clues for each student or pair of students. Instruct them to complete the crossword by filling in one letter per box.

Differentiation

To make the task easier you can write some or all of the keywords on the board.

To make the task a little harder, or to add an element of competition, instruct students to complete the crossword against the clock.

Alternative approaches

Ask students if they can devise alternative clues for some of the words.

Encourage students to be creative, and to produce their own crossword using some or all of these keywords.

Answers

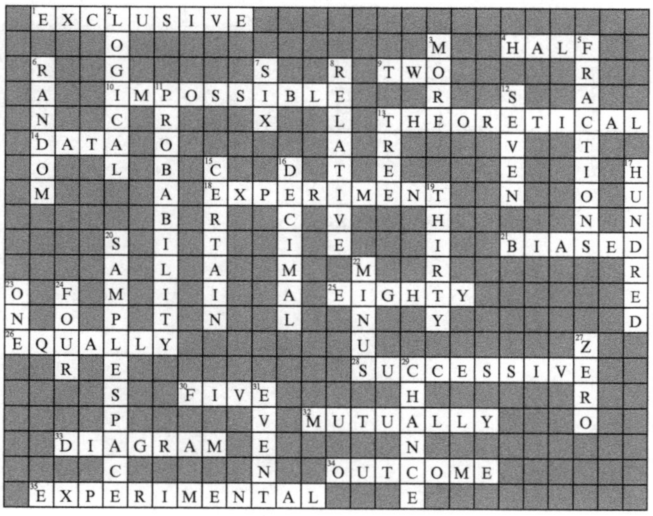

Probability crossword

Page 1 of 2

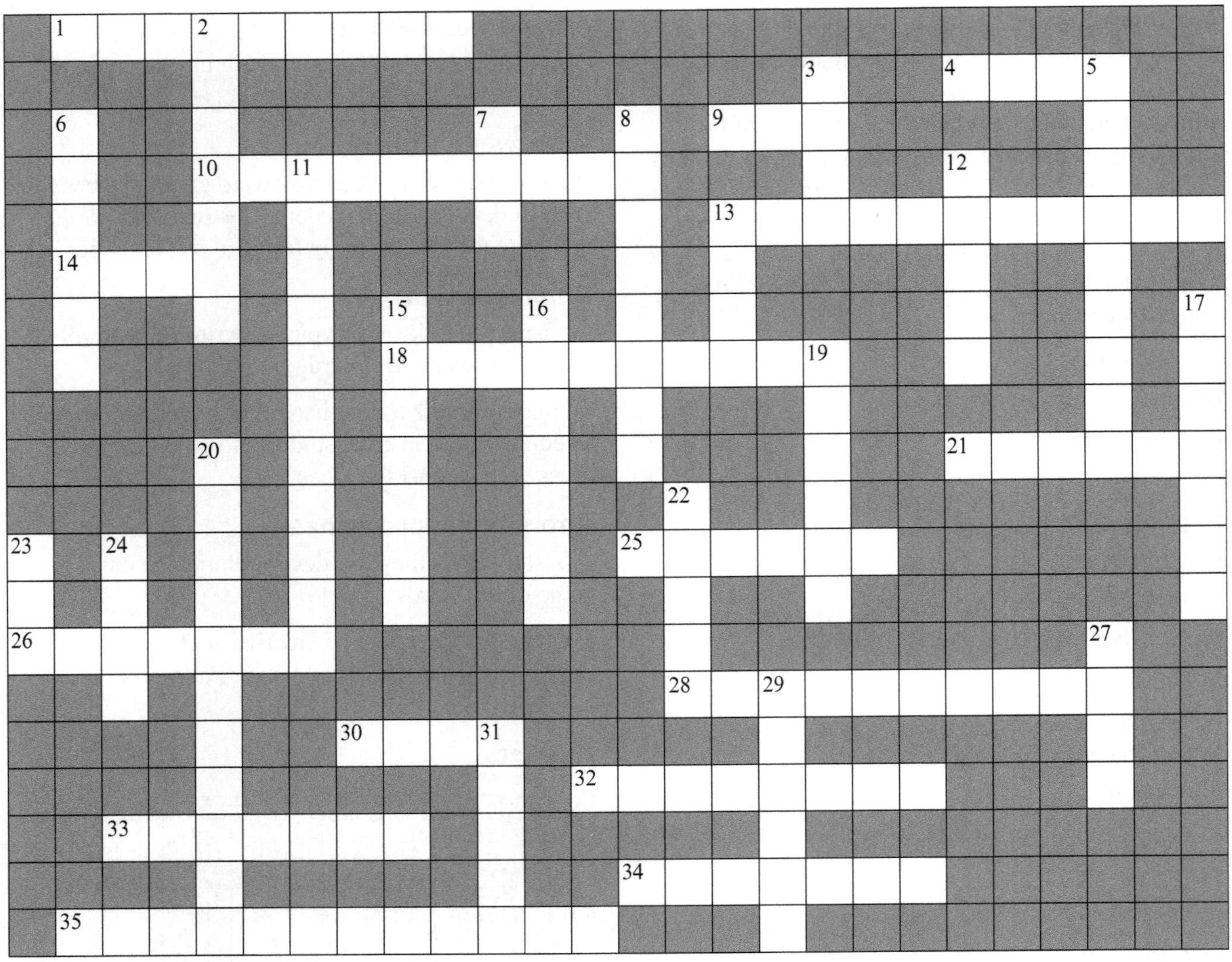

For clues, see the next page.

Probability crossword

Page 2 of 2

Across

1 Events that cannot happen at the same time are mutually .... (9)

4 Some people describe the probability of something that has an equal chance of happening or not happening as 'fifty-fifty'. A correct way to write this is as a .... (4)

9 The number of possible outcomes when a coin is tossed. (3)

10 An event is .... if it has the probability of 0. (10)

13 .... probability is calculated based on what you would expect to happen in theory, using number of successful outcomes divided by number of possible outcomes. (11)

14 Another word for information that is collected in a survey. It can be numerical or non-numerical. (4)

18 When you repeatedly roll dice and record the outcomes, you are conducting an .... (10)

21 Describes when some outcomes are favoured over others, or when a question asked in a survey steers the person answering towards a particular response. (6)

25 0.8 = ... per cent (6)

26 When you roll a normal, fair dice, the events 1, 2, 3, 4, 5 and 6 are all .... likely. (7)

28 .... events are events that happen one after the other. (10)

30 You roll a normal, fair dice 30 times. How many times would you expect to roll a 3? (4)

32 The sum of probabilities of all .... exclusive outcomes is one. (8)

33 A two-way table is a sample space .... (7)

34 The result of an experiment or other situation involving chance. (7)

35 .... probability is probability based on repeated trials, using number of successful trials divided by number of trials. (12)

Down

2 When listing all possible outcomes of an event it is important to list them in a .... (or systematic) order. (7)

3 The .... times an experiment is repeated the more reliable the results. (4)

5 Probabilities can be written as decimals, percentages or .... (9)

6 When something is chosen at .... , this means that all outcomes have a fair and equal chance of being selected. (6)

7 The number of possible outcomes when a normal, fair dice is rolled. (3)

8 Experimental probability is estimated by .... frequency. (8)

11 A number describing the chance of an event happening. (11)

12 0.07 = ... per cent (5)

13 A sample space diagram that uses branches to show outcomes is a .... diagram. (4)

15 An event is .... if it has a probability of 1. (7)

16 0.3 is an example of a probability written as a .... (7)

17 If the probability of an event happening is 25%, to find the probability of it not happening we need to subtract 25 from a .... per cent. (7)

19 You roll a normal, fair dice 60 times. The number of times you would expect to see an even number is .... (6)

20 The list of all the possible outcomes of an experiment. (6, 5)

22 The probability of an event not happening is 1 .... the probability of the event happening. (5)

23 Probabilities which are extremely likely are close to .... (3)

24 The number of possible outcomes when a tetrahedral dice is rolled. (4)

27 Probabilities which are extremely unlikely are close to .... (4)

29 Something that has a probability of $\frac{1}{2}$ has an even .... of happening. (6)

31 An .... is an outcome of a particular experiment. (5)

Different bases investigation

Teacher's notes

Different bases are not in the Secondary 1 framework. This investigation is useful for practicing thinking and working mathematically skills such as

- **Specializing** – questions 3 and 4 in the worksheet.
- **Generalizing, conjecturing** and **convincing** in the extension work.

The aim is to improve students' ability to apply a variety of mathematical skills in a context that is unfamiliar to them. The context is using different number base systems.

Introductory activity

It is a nice idea to give students some background information about different number bases and an example of where they are used.

Begin by explaining about the base 10 number system (the denary system) that we use. Explain that the column headings in the base 10 number system are powers of 10, and that the number 7458, for example, can be written in a table like this:

Thousands $10^3 = 1000$	**Hundreds** $10^2 = 100$	**Tens** $10^1 = 10$	**Units** $10^0 = 1$
7	4	5	8

So 7458, in base 10, can be partitioned (or expanded) to:
$7 \times 1000 + 4 \times 100 + 5 \times 10 + 8 \times 1$
$= 7000 + 400 + 50 + 8$

In base 10 we use all the single digits:
0, 1, 2, 3, 4, 5, 6, 7, 8, 9.

In base 5, the column headings are powers of 5 and the digits 0, 1, 2, 3 and 4 are used.

Example 1

Convert 3412 (base 5) into base 10.

$5^3 = 125$	$5^2 = 25$	$5^1 = 5$	$5^0 = 1$
3	4	1	2

So 3412 (base 5) $= 3 \times 125 + 4 \times 25 + 1 \times 5 + 2$
$= 375 + 100 + 5 + 2$
$= 482$

To change a base 10 number to a base 5 number you can look at the multiples of powers of 5 that add up to the number.

Example 2

Change 89 to base 5.

$89 = 75 + 10 + 4$
$= 3 \times 25 + 2 \times 5 + 4 \times 1$
$= 3 \times 5^2 + 2 \times 5^1 + 4 \times 5^0$
$=$ 324 (base 5)

An alternative method for Example 2 is to repeatedly divide by the number representing the base and then look at the remainders:

```
     0 r 3
   5)3 r 2
  5)17 r 4
5) 89
```

So 89 = 324 (base 5)

Talk to students about how the binary number system is used in computing. In the binary number system only the digits 1 and 0 are used and column headings are powers of 2. The processing unit in a computer is made up of millions of tiny switches that can be either ON or OFF. The digit 1 is used to represent ON, while 0 represents OFF.

Different bases investigation

Give students a copy of the investigation sheet, which gives two further examples on working with base 5. Instruct them to work through the conversion questions, 1 to 6, then they can choose one of the topics in the extension section to investigate if they get that far.

Differentiation

To make it harder don't give students the alternative method for Example 2. See if they can come up with it themselves. To make it easier go through some more examples, in particular for the binary number system. The extension section allows students to work at their own level.

Alternative approaches

The extension work provides opportunities to explore many different avenues.

Answers

1 **a** 21 **b** 18 **c** 42
d 71 **e** 228 **f** 468

2 **a** 241 **b** 31 **c** 132
d 310 **e** 441 **f** 2224

5 **a** 4 **b** 12 **c** 10
d 26 **e** 22

6 **a** 1000 **b** 1111 **c** 10001
d 10011 **e** 10111

Extension work answers

Square number	0^2	1^2	2^2	3^2	4^2
Base 5	0	1	4	14	31

Not possible to have a square number in base 5 ending with a 2 or 3.

Different bases investigation

Introduction

In maths we work with the base 10 (denary) number system. Other number systems can be used.

In base 5, for example, the column headings are powers of 5 and only the digits 0, 1, 2, 3 and 4 are used.

Example 1

Convert 1243 (base 5) to base 10.

$5^3 = 125$	$5^2 = 25$	$5^1 = 5$	$5^0 = 1$
1	2	4	3

$$\begin{aligned} 1243 \text{ (base 5)} &= 1 \times 125 + 2 \times 25 + 4 \times 5 + 3 \times 1 \\ &= 125 + 50 + 20 + 3 \\ &= 198 \end{aligned}$$

To change a base 10 number to a base 5 number, you can look at the powers of 5 that add up to that number.

Example 2

Change 66 to base 5.

$$\begin{aligned} 66 &= 50 + 15 + 1 \\ &= 2 \times 25 + 3 \times 5 + 1 \times 1 \\ &= 2 \times 5^2 + 3 \times 5^1 + 1 \times 5^0 \\ &= 231 \text{ (base 5)} \end{aligned}$$

An alternative method for Example 2 is to repeatedly divide by the number representing the base and then look at the remainders:

0 r 2
5)2 r 3
5)13 r 1
5) 66

So 66 = 231 (base 5)

Questions

1 Change these base 5 numbers to base 10.

a 41 b 33 c 132 d 241 e 1403 f 3333

2 Change these base 10 numbers to base 5.

a 71 b 16 c 42 d 80 e 121 f 314

3 Do you agree that 11 (base 2) = 3 (base 10)?

4 Do you agree that 13 (base 10) = 1101 (base 2)?

$2^3 = 8$	$2^2 = 4$	$2^1 = 2$	$2^0 = 1$
		1	1
1	1	0	1

5 Change these binary (base 2) numbers to base 10.

a 100 b 1100 c 1010 d 11010 e 10110

6 Change these base 10 numbers to binary.

a 8 b 15 c 17 d 19 e 23

Extension work

Choose one (or more) of the following areas to investigate:

- Use what you have learned about base 2 and base 5 to do some conversions with other bases of your choice. Check your answers using the number base calculator at www.cleavebooks.co.uk/scol/calnumba
- Write down the values in base 5 of 0^2, 1^2, 2^2, 3^2, 4^2. Is it possible to have a square number in base 5 ending with a 2 or 3? Why?
- Investigate adding, subtracting and multiplying numbers in different bases. In these examples, numbers are all in base 5: $23 \times 2 = 101$ (base 5) and $23 + 12 = 40$ (base 5)
- Find out about using decimals in different bases – for example, how to change 1.1 (base 2) to base 10 – at www.mathsisfun.com/binary-number-system
- Find out more about the binary system and its applications, and about the hexadecimal number system

Crossword puzzle

For use with Chapter 8, Exercise 8G, Question 7

Across

4. When you have found the value of one unknown you __________ it into one equation to find the other unknown.

5. To eliminate unknowns with the same sign you ________ the two equations.

6. Before adding two simultaneous equations to eliminate a variable you need the ____________ of the unknowns to be the same.

9. You need to solve two ____________ equations if there are two unknowns.

10. To eliminate unknowns with the opposite sign you ___ the two equations.

Down

1. The _________ to the simultaneous equations need to be the same for both equations.

2. Solving two equations ______________ means solving them at the same time.

3. If the coefficients of one variable are not the same you can ________ one of the equations by an appropriate number.

7. Always use the ________ two equations when substituting or checking.

8. When you have found both solutions it is a good idea to _____ by substitution that it works in the other equation.

Answer

1 solutions
2 simultaneously
3 multiply
4 substitute
5 subtract
6 coefficients
7 original
8 check
9 simultaneous
10 add

Tessellation cut-outs

For use with Chapter 9, Section 9.2 Activity

Copy this onto card for your students to cut out to use for the Activity in Chapter 9, section 9.2 (page 140).

For use with Chapter 9, Exercise 9C, Question 5

Copy this onto card for your students to cut out to use for Chapter 9, Exercise 9C, Question 5.

Multi-choice Exercise A

This exercise can be used as a resource for revising the work in Chapters 1 to 5 of the Student Book.

Note: Multi-choice questions are not part of Cambridge Checkpoint tests; they are included here as a useful revision tool.

Integers, powers and roots

1 Which of these is an irrational number?

a 1.7 **b** $\frac{2}{3}$ **c** $\sqrt{5}$ **d** $0.\dot{8}$

2 Here are some statements.

A rational number can be written as a fraction.

An irrational number can be written as an integer.

The square root of all odd integers under 6 will be irrational.

How many of these statements are correct?

a 0 **b** 1 **c** 2 **d** 3

3 Which of these statements is incorrect?

a $3^2 = 6$ **b** $4^{-2} = \frac{1}{4^2}$

c $\frac{1}{16} = 2^{-4}$ **d** $3^0 = 1$

4 Which of these is correct?

a $4^{10} \div 4^2 = 4^5$ **b** $7^{-2} = \frac{1}{14}$

c $3^{-3} = \frac{1}{27}$ **d** $\frac{1}{6} = {}^{-}6^1$

5 Which of these is correct?

a $\sqrt{79} < 9$ **b** $7 < \sqrt{47}$

c $5 < \sqrt{6} < 7$ **d** $\sqrt[3]{9} \geq 3$

6 Andy starts with a number and then squares it.
He ends up with the same number that he started with.
For how many starting numbers would this work?

a None **b** Two

c One **d** More than two

7 Which of these is a rational number?

a π **b** $\sqrt{25}$ **c** $\sqrt{8}$ **d** $\sqrt{13}$

8 Which of these estimates is incorrect?

a $\sqrt{88} \approx 9.4$ **b** $\sqrt{33} \approx 5.7$

c $\sqrt{22} \approx 5.3$ **d** $\sqrt{44} \approx 6.6$

9 Estimate $\sqrt[3]{60}$ to 1 d.p.

a 3.5 **b** 3.9 **c** 4.1 **d** 3.7

10 Estimate the side length of this square, to the nearest mm.

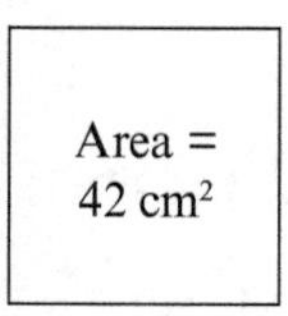

a 6.9 cm **b** 6.1 cm

c 6.5 cm **d** 7.1 cm

11 The fraction $\frac{1}{16}$ can be written as 2^x. What is the value of x?

a 3 **b** ⁻3 **c** 4 **d** ⁻4

12 Work out the number to go in the box to make this statement correct. $8^3 \times 8^2 = 8^6 \times 8^{\square}$

a ⁻1 **b** 0 **c** 1 **d** 2

13 Which of these statements is correct?

a $7^0 > 3^0$ **b** $6^{-3} > 6^{-2}$ **c** $5^{-3} < 0$ **d** $({}^{-}4)^{-1} < 0$

14 Simplify $\frac{(5^4 \times 5^{-1})^2}{5^3 \times 5}$

a 5^{-2} **b** 5^1 **c** 5^2 **d** 5^3

Expressions and formulae

15 Which of these statements is incorrect?

a $p^6 \div p^3 = p^2$ **b** $w^4 \times w = w^5$

c $4y^2 \times 3y^3 = 12y^5$ **d** $q^3 \div q^5 = q^{-2}$

16 Simplify $5x^4 \times 3x^5 \times 2x$

a $10x^{20}$ **b** $30x^9$ **c** $30x^{20}$ **d** $30x^{10}$

17 Using the formula $A = 100m - 7r^2$,
find A when $r = {}^{-}2$ and $m = 0.8$

a 108 **b** 52 **c** ⁻116 **d** ⁻20

18 Which of these statements is correct?

a $(x + 5)^2 = x^2 + 25$

b $(x + 2)(x + 5) = x^2 + 7$

c $(x - 3)(x + 5) = x^2 - 2x - 15$

d $(x - 2)(x - 3) = x^2 - 5x + 6$

19

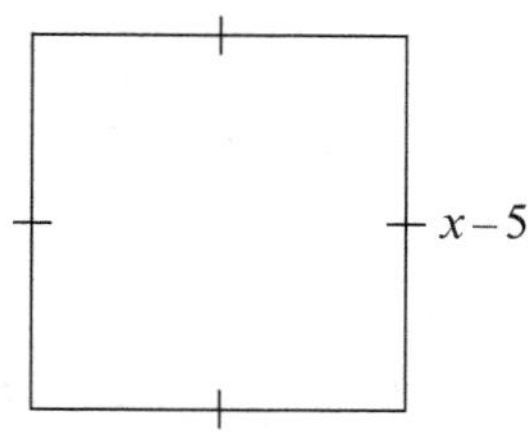

The area of this square is

a $x^2 - 25$ **b** $x^2 + 25$

c $x^2 - 10x + 25$ **d** $4x - 20$

20 $x^2 - 36$ is the product of which two expressions?

a x and $x - 36$ **b** $x - 6$ and $x - 6$

c $x + 6$ and $x - 6$ **d** $x + 6$ and $x + 6$

21 $H = 3(x - r)$ rearranged to make x the subject is

a $H - 3 + r = x$ **b** $\frac{H}{3} + r = x$

c $\frac{H+r}{3} = x$ **d** $\frac{H}{3} - r = xv$

22 Which of these statements is correct?

a $3n^2 - 2n^2 = 1$ **b** $(c^3)^2 = c^5$

c $x^{10} \div x^2 = x^5$ **d** $y^3 \times y^5 \times y^{-2} = y^6$

23 Simplify this algebraic fraction: $\frac{25x+15}{15}$

a $25x$ **b** $\frac{5x+15}{3}$ **c** $\frac{25x+1}{1}$ **d** $\frac{5x+3}{3}$

24 When $(x + 8)(x - 7)$ is expanded and simplified, what is the coefficient of the term in x?

a ⁻15 **b** ⁻1 **c** 1 **d** 15

25 What two values correctly complete this statement?

$x^2 + \square x + 81 = (x + \square)^2$

a 18 and 9 **b** 9 and 9 **c** 18 and 81 **d** 9 and 81

26 I think of a number, multiply it by 3, square and then subtract 5. Which expression correctly shows this?

a $3n^2 - 5$ **b** $n \times 3^2 - 5$ **c** $(3n - 5)^2$ **d** $(3n)^2 - 5$

27 Find the value of $2x^4 - x^3 - 10x$ when $x = 3$

a 105 **b** 125 **c** 123 **d** 1239

28 $\frac{2d}{3} + \frac{3d}{4}$ as a single fraction is

a $\frac{17d}{12}$ **b** $\frac{5d}{7}$ **c** $1\frac{5}{12}$ **d** $\frac{6d^2}{12}$

29 $2+\frac{3}{h}$ as a single fraction is

a $\frac{2h+3}{h}$ b $\frac{5}{h}$ c $\frac{5h}{h}$ d $\frac{3}{2+h}$

30 James is rearranging $t=\frac{7y-5}{2}$ to make y the subject of the formula.
Choose the answer that is not a correct rearrangement or a correct step towards rearranging.

a $2t=7y-5$ b $2t+5=7y$
c $2t+5\div 7=y$ d $(2t+5)\div 7=y$

Shapes and mathematical drawings

31 The diagram shows a map of six towns A to F.

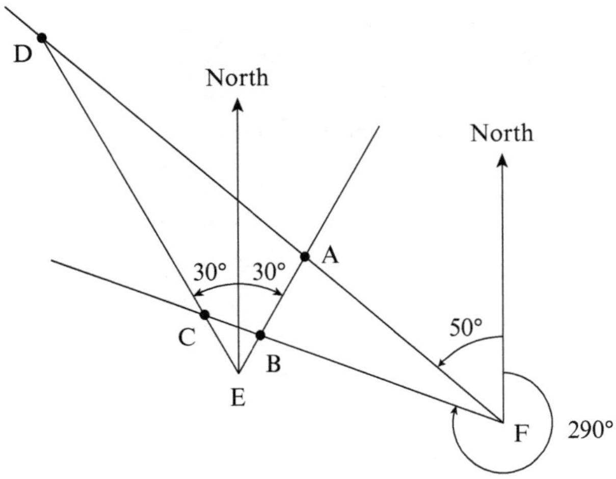

Which town is on a bearing of 030° from E and 310° from F?

a A b B c C d D

32 In this scale drawing a town is on a bearing of 270° and 12 km from city X.

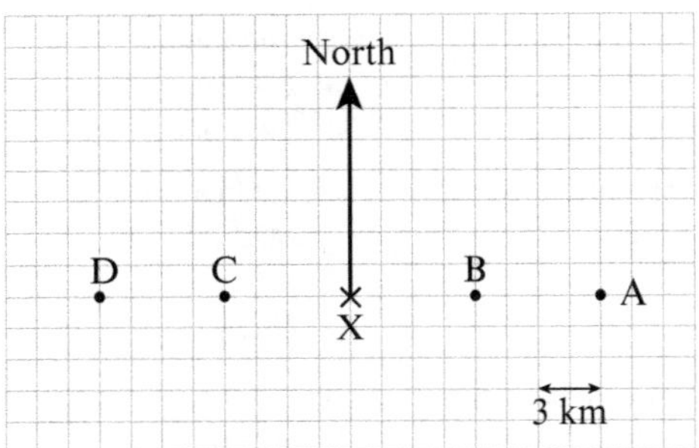

Which of these is the position of the town?

a A b B c C d D

33 How many planes of symmetry does this regular pentagonal prism have?

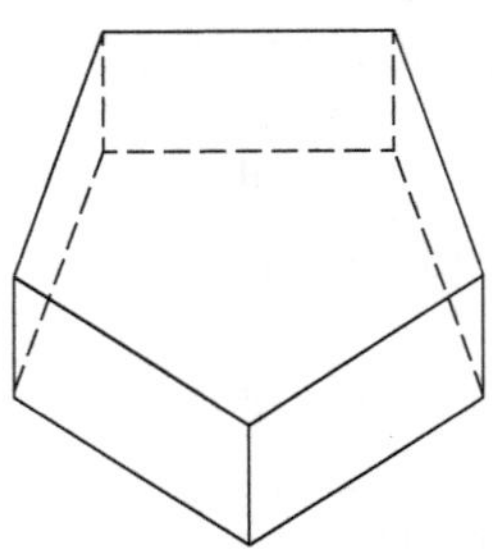

a 5 b 6 c None d 3

34 A hexagon inscribed inside a circle is

a a hexagon inside a circle that doesn't touch the circumference
b a hexagon inside a circle that touches the circumference at some of its vertices
c a hexagon inside a circle that touches the circumference at all of its vertices
d a circle inside a hexagon

35 Which diagram shows the correct starting point for constructing a perpendicular line to AB from C?

a
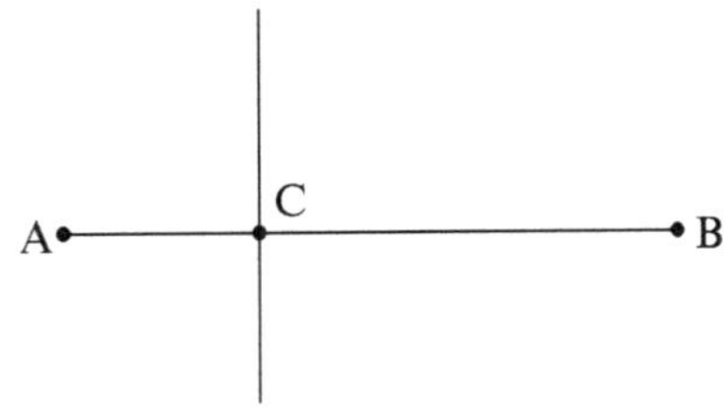

b
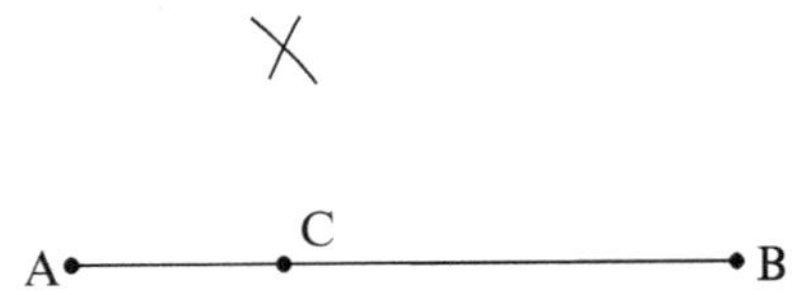

c
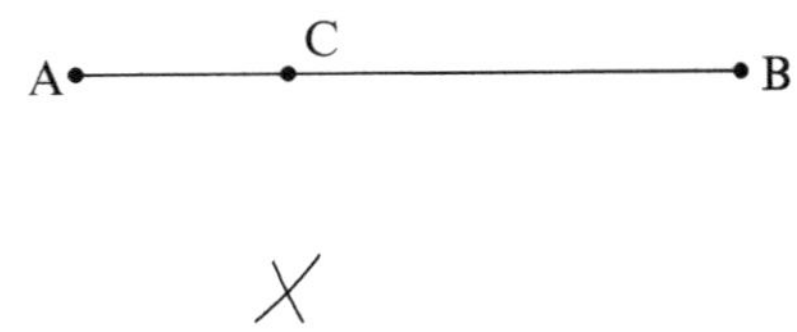

d
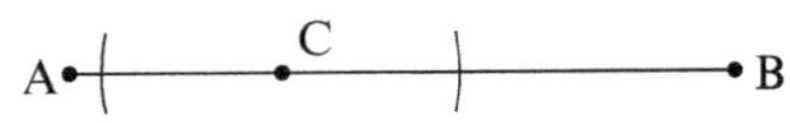

36 A shape with 4 planes of symmetry could be a

a triangular prism b cube
c cuboid d cylinder

Place value

37 Work out $82\div 0.01$

a 0.082 b 8200 c 820 d 0.82

38 $0.00023\div 10^{\square}=2.3$
What is the missing number?

a 4 b 3 c $^-4$ d $^-5$

39 Which of these is incorrect?

a $7000\times 10^{-3}=7$
b $1\,700\,000\times 10^{-6}=1.7$
c $0.0031\div 10^{-2}=0.31$
d $0.0009\div 10^{-4}=0.9$

40 Which of these is correct?

a $0.1=\frac{1}{10}=10^{1}$ b $10^{-3}=\frac{1}{100}=0.001$
c $\frac{1}{100}=10^{-2}=0.01$ d $\frac{1}{1000}=0.001=10^{3}$

41 Which of these numbers is not written correctly in standard form?

a 6.5×10^{3} b 33×10^{5}
c 1.25×10^{-2} d 8×10^{4}

42 Write this number, that is in standard form, as an ordinary number: 1.03×10^{4}

a 0.000103 b 10300
c 103000 d 1030000

43 Which of these is correct standard form for $4\div 1000$?

a $4\div 10^{3}$ b $\frac{4}{1000}$ c 4×10^{3} d 4×10^{-3}

44 Use your calculator to work out $(4 \times 10^4) \times (5 \times 10^2)$.
Give your answer in standard form.

a 20×10^6 b 2×10^5 c 2×10^7 d 2×10^9

45 A number rounded correct to the nearest ten is 320.
What is the lower bound?

a 319.5 b 315 c 319.95 d 325

46 A number, n, rounded correct to 2 d.p. is 2.48. What is the correct inequality showing possible values for the number?

a $2.475 \le n < 2.485$ b $2.475 \le n \le 2.485$
c $2.475 < n < 2.485$ d $2.475 < n \le 2.485$

Grouped data and sampling

47 The table shows the mass of 25 animals in a zoo.

Mass (m kg)	f
$20 \le m < 30$	1
$30 \le m < 40$	6
$40 \le m < 50$	8
$50 \le m < 60$	5
$60 \le m < 70$	3
$70 \le m < 80$	2

What is the estimate of the mean of the mass of the animals?

a 48.6 kg b 52.6 c 48.1 kg d 43.6 kg

48 For the data from question **47** what is the modal class?

a $30 \le m < 40$ b $40 \le m < 50$
c $50 \le m < 60$ d $60 \le m < 70$

49 For the data from question **47** what is an estimate of the range?

a 20 b 50 c 60 d 70

50 For the data from question **47** what interval is the median in?

a $30 \le m < 40$ b $40 \le m < 50$
c $50 \le m < 60$ d $60 \le m < 70$

51 Here are some pieces of data:
mass of a person, height of a person, hair colour of a person, number of children in the person's family.
How many of these pieces of data are quantitative?

a 1 b 2 c 3 d 4

52 Enrique is investigating whether adults' books have more lines of writing on a page on average than children's books.
He has one adults' book and one children's book.
Then he chooses one of these methods.
Method 1: count the number of lines of writing on every page of both books.
Method 2: measure the length of writing on every page in cm.
Method 3: count the number of lines of writing on a sample of pages of both books.
Complete these sentences with the correct numbers.
Method … will be time consuming but will give the most reliable results.
Method … is the least reliable method to answer the investigation.

a 1 and 2 b 1 and 3 c 2 and 3 d 3 and 1

53 Stephan is investigating whether or not students in his school exercise regularly.
He writes four questions on his questionnaire.
Choose the question that is most likely to be a source of bias in his investigation.

a How many times do you run in a week?
b Do you agree that exercising regularly is good for you? Yes/No
c How many hours exercise do you do each week on average?
d How many times do you go swimming each week?

54 Here are some sampling methods. Choose the method that is least likely to be biased.

a standing outside a library and asking people if they like reading
b a population has 50 adults and 60 children; choosing a sample of 5 adults and 5 children to survey
c collecting data from a book published in 2015
d putting the names of 200 children in a hat and drawing out a sample of 20 to survey

Multi-choice Exercise B

This exercise can be used as a resource for revising the work in Chapters 6 to 10 of the Student Book.

Note: Multi-choice questions are not part of Cambridge Checkpoint tests; they are included here as a useful revision tool.

Area, perimeter and volume

1 Which of these is incorrect?

a 45 000 ml = 45 ℓ
b 0.0008 t = 80 g
c 740 000 cm = 7.4 km
d 0.09 ℓ = 9 cl

2 0.3 t + 800 kg + 72 000 g =

a 72 800.3 g **b** 872.3 kg
c 1172 kg **d** 1.82 t

3 This cuboid has a volume of 210 cm^3. Calculate the missing side length.

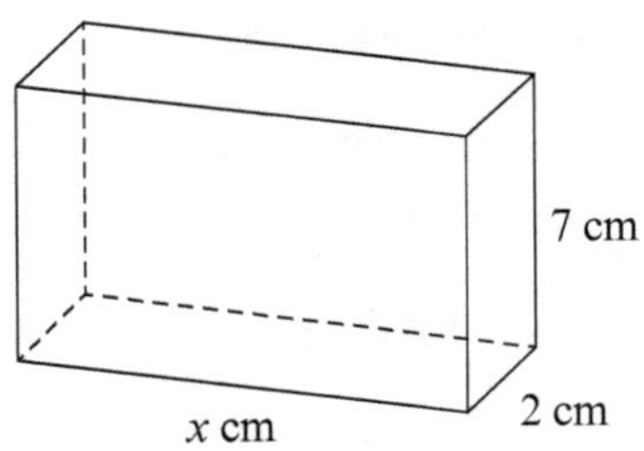

a 15 mm **b** 14 cm **c** 1.5 cm **d** 15 cm

4 Find the area of a circle with radius 4.5 cm. Give your answer correct to 1 d.p.

a 15.9 cm^2 **b** 28.3 cm^2 **c** 63.6 cm^2 **d** 254.5 cm^2

5 The area of a circle is 25 cm^2. The radius in centimetres is approximately

a 1.8 **b** 2.8 **c** 3.2 **d** 5.6

6

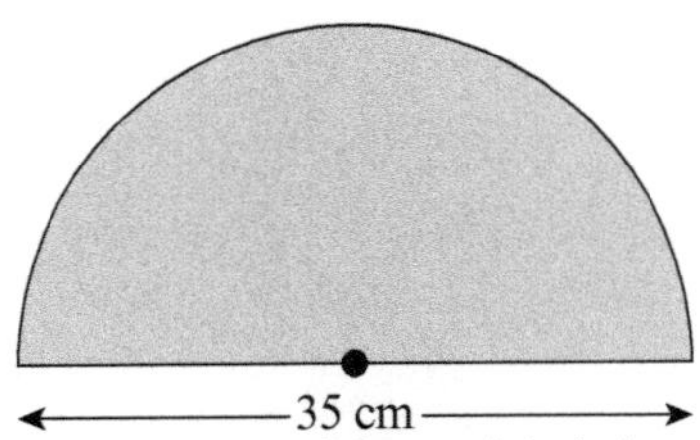

The perimeter of the semicircle is approximately

a 110 cm **b** 90 cm **c** 55 cm **d** 52 cm

7 A circle has diameter 8 cm. What is its area? Use $\pi = 3.14$.

a 12.56 cm^2 **b** 25.12 cm^2
c 50.24 cm^2 **d** 200.96 cm^2

8 Find the side length x.

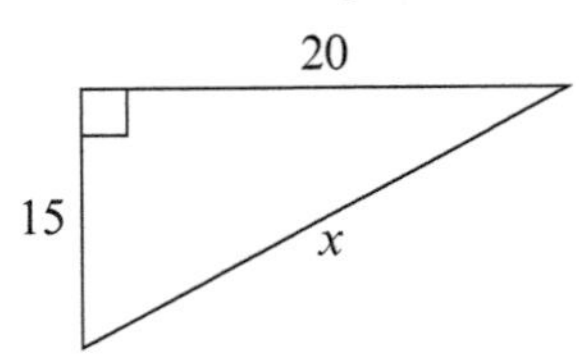

a 35 **b** 25 **c** 13.2 **d** 300

9 Find the side length x.

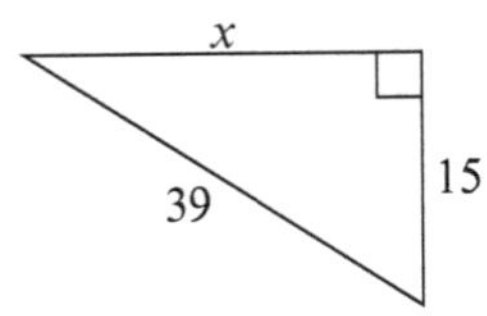

a 41.8 **b** 24 **c** 15 **d** 36

10 Find the volume of this shape.

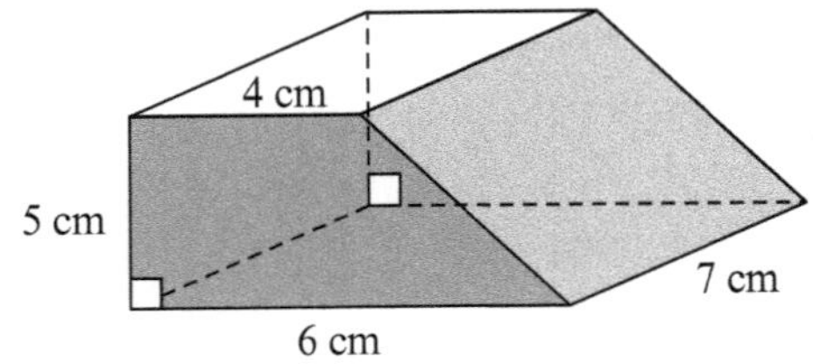

a 210 cm^3 **b** 175 cm^3
c 350 cm^3 **d** 840 cm^3

11 A cylindrical tank has base area 2 m^2. How much water can it hold if its height is 1 m?

a 2 litres **b** 20 litres
c 200 litres **d** 2000 litres

12 What is the perimeter of this semicircle? Give your answer to 1 d.p.

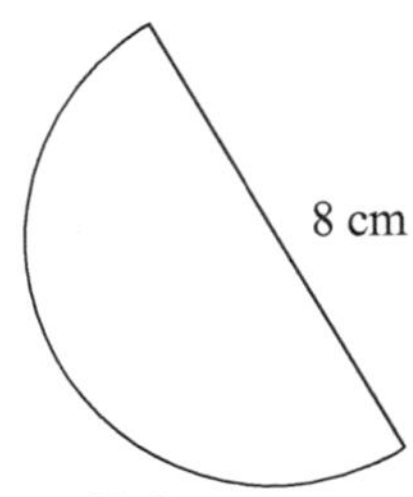

a 33.1 cm **b** 12.6 cm
c 20.6 cm **d** 25.1 cm

13 Find the volume of a cylinder with radius 3 cm and length 8 cm.

a 12 cm^3 **b** 24 cm^3
c 72 cm^3 **d** 226 cm^3

14 Find the area of this shape made up of a rectangle and a semicircle.

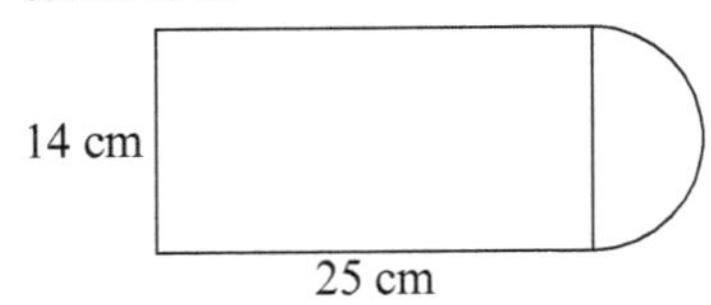

a 504 cm^2 **b** 658 cm^2
c 427 cm^2 **d** 86 cm^2

15 The solid with the smallest volume is

a

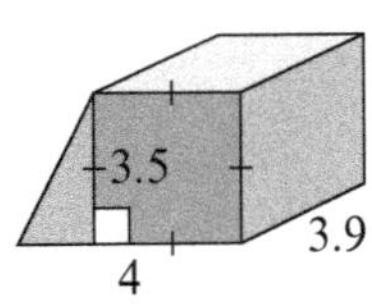

b

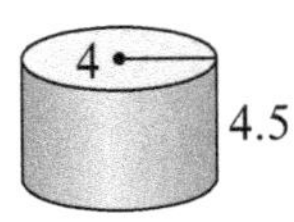

c

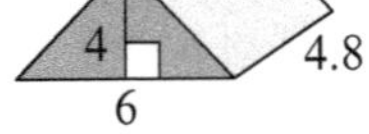

d

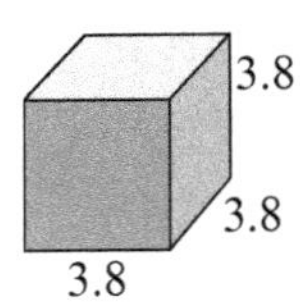

16 Find the surface area of this cylinder.

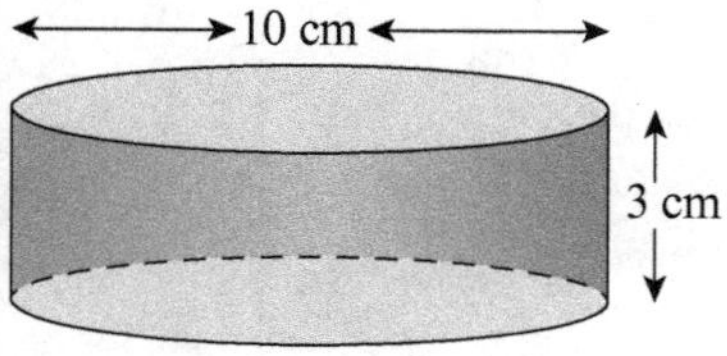

a $173\ cm^2$ b $722\ cm^2$
c $298\ cm^2$ d $251\ cm^2$

17 Find the area of this triangle.

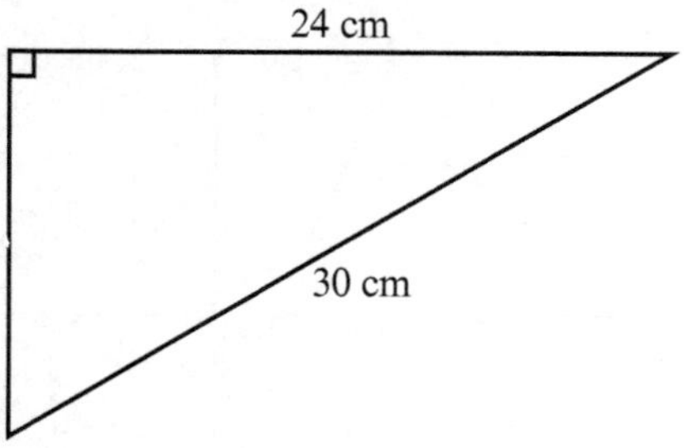

a $72 cm^2$ b $216 cm^2$ c $360 cm^2$ d $432 cm^2$

18 Find the surface area of this cylinder. Give your answer correct to the nearest cm^2.

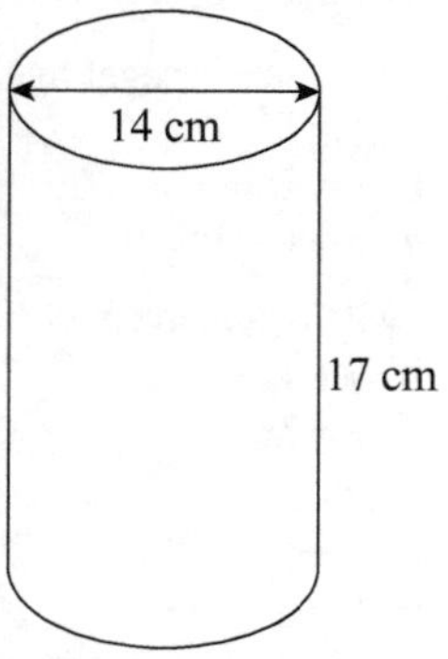

a $308 cm^2$ b $748 cm^2$ c $902 cm^2$ d $1056 cm^2$

19 Find the surface area of this triangular prism.

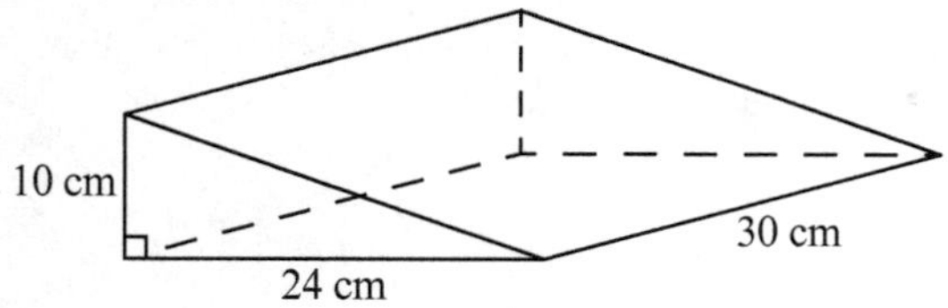

a $240 cm^2$ b $1260 cm^2$ c $1800 cm^2$ d $2040 cm^2$

20 Which of these is a correct conversion?
a 2000 MB = 2 KB b 5 GB = 5000 MB
c 7 TB = 700 GB d 900 KB = 9 MB

21 Which of these is the shortest distance?
a 10 μm b 10 Mm c 10 mm d 10 ly

Fractions and decimals

22 Without working out the answer, which statement will have an answer less than 219?
a 219×1.36 b $219 \div 0.8$
c $219 \div 1.42$ d $219 \times 1\frac{4}{9}$

23 Without working out the answer, which statement will have an answer greater than 48.5?
a $48.5 \div 0.63$ b $48.5 \div 2.7$
c 48.5×0.98 d $48.5 \times \frac{3}{7}$

24 Two pieces of wood are both $2\frac{9}{16}$ m long. What is the total length of wood, in its simplest form?
a 5.125 cm b $4\frac{18}{16}$ m c $4\frac{18}{32}$ m d $5\frac{1}{8}$ m

25 The answer to $1\frac{3}{5} \times 3\frac{2}{11}$, in its simplest form, is
a $5\frac{1}{11}$ b $3\frac{6}{55}$ c $5\frac{5}{55}$ d $4\frac{5}{16}$

26 The answer to $2\frac{5}{8} \div 2\frac{3}{16}$, in its simplest form, is
a $5\frac{95}{128}$ b $\frac{236}{280}$ c $1\frac{1}{5}$ d $1\frac{7}{35}$

27 Before running a marathon, Jan has a mass of $62\frac{1}{2}$ kg. After, her mass has decreased by $\frac{1}{50}$ of her mass at the start. What is her mass after running the marathon?
a $1\frac{1}{4}$ kg b $62\frac{12}{25}$ kg c $61\frac{1}{4}$ kg d 62.5 kg

28 The answer to $2\frac{1}{7} \div 1\frac{2}{3}$ as a mixed number in its simplest form is
a $1\frac{2}{7}$ b $\frac{7}{25}$ c $1\frac{3}{14}$ d $1\frac{10}{35}$

29 Calculate $\left(1\frac{3}{4} + 0.25\right)^3 + \frac{5}{8}$
a $6\frac{5}{8}$ b $8\frac{5}{8}$ c $2\frac{25}{64}$ d $2\frac{5}{8}$

30 This rectangle has side lengths $4\frac{3}{5}$ cm and $2\frac{1}{4}$ cm. What is its area?

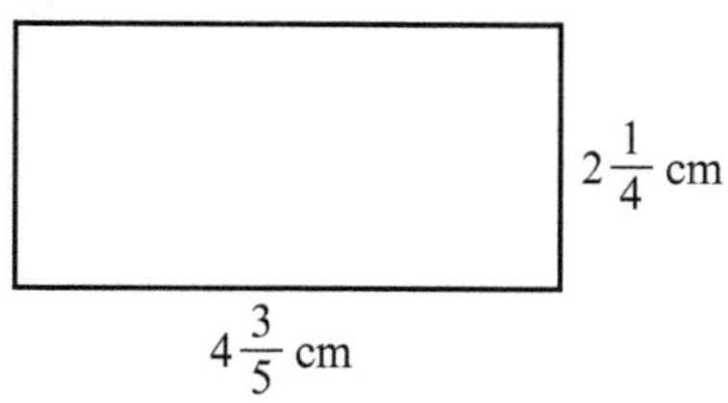

a $8\frac{3}{20}\ cm^2$ b $10\frac{7}{20}\ cm^2$
c $6\frac{4}{9}\ cm^2$ d $13\frac{7}{10}\ cm^2$

31 Work out $\dfrac{4\frac{3}{4}}{1\frac{1}{4} + \frac{1}{3}}$
a $4\frac{2}{15}$ b $3\frac{25}{36}$ c $\frac{1}{3}$ d 3

32 Which of these fractions is equivalent to a recurring decimal?
a $\frac{7}{28}$ b $\frac{5}{7}$ c $3\frac{7}{14}$ d $\frac{525}{700}$

33 Which of these is an incorrect conversion?
a $\frac{7}{9} = 0.\dot{7}$ b $\frac{4}{9} = 0.\dot{4}$ c $1\frac{2}{9} = 1.\dot{2}$ d $\frac{2}{3} = 0.\dot{2}$

34 Choose the correct statement about the equivalent decimal of the described fraction.
a If the denominator of a fraction is 7 and the numerator is a prime number the decimal will be terminating.
b If the denominator of a fraction has a factor of 3 the decimal will be recurring.
c If the denominator of a fraction is 12 the decimal will sometimes be recurring and sometimes be terminating depending what the numerator is.
d If the numerator of a fraction is 9 the decimal will always be recurring.

Equations and inequalities

35 Solve $4(x + 3) = 2(5x - 3)$
a $x = 1.5$ b $x = {}^{-}3$ c $x = 3$ d $x = 4$

36 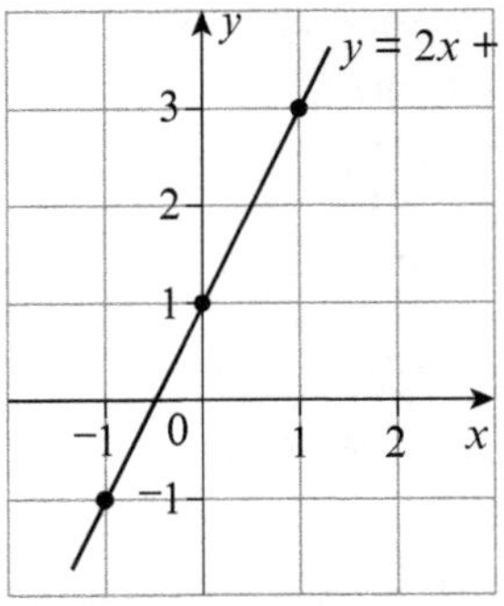

The graph shows the line $y = 2x + 1$. Where does it intercept the y-axis?

a (⁻1,⁻1) **b** (1, 0) **c** (0, 1) **d** (2, 3)

37 Solve $2(5x - 2) = 4(2x - 3)$

a $x = 4$ **b** $x = {}^-8$ **c** $x = {}^-4$ **d** $x = 8$

38 Jane is 3 years older than her sister Sue. Their combined age is 19. An equation to show this is

a $2x + 3 = 19$ **b** $x + 3 = 19$
c $2x = 19$ **d** $x + 3 - x = 19$

39 If $\frac{15}{x+1} = 5$ the value of $2x + 4$ is

a 2 **b** 8 **c** 10 **d** 12

40 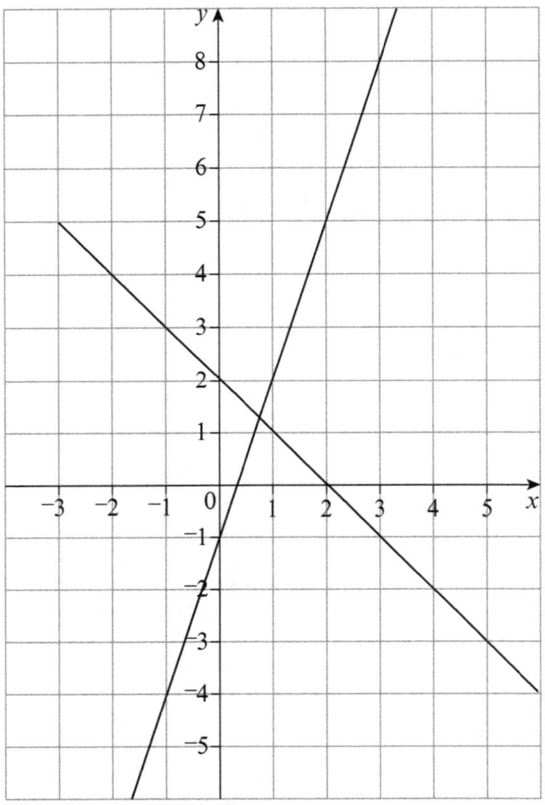

The gradients of these lines are

a ⁻1 and 3 **b** 1 and 3
c 1 and ⁻3 **d** ⁻1 and 2

41 A plant starts with a height of 15 cm. The plant grows 3 cm every week. The plant is expected to grow to a height of at least 43 cm. Using w to represent the number of weeks the plant grows, which of these inequalities shows this information?

a $15 + 3w < 43$ **b** $15 + 3w > 43$
c $15 + 3w \leq 43$ **d** $15 + 3w \geq 43$

42 Which of these lines could be the graph of $y - 3 = 2x$?

a 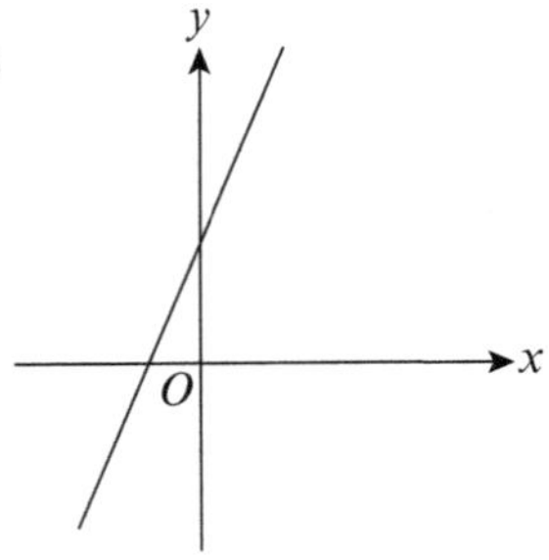

b 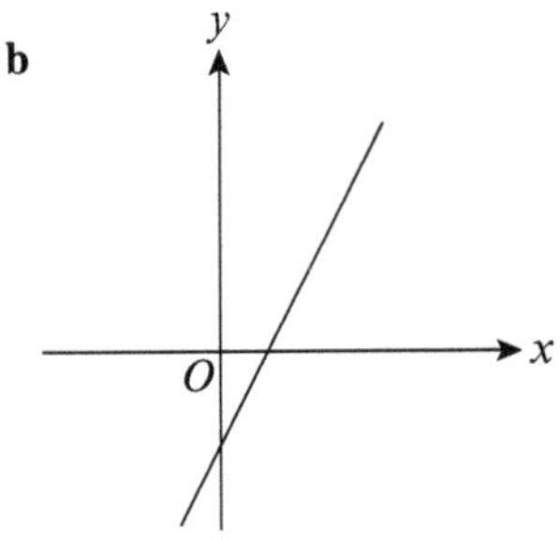

c 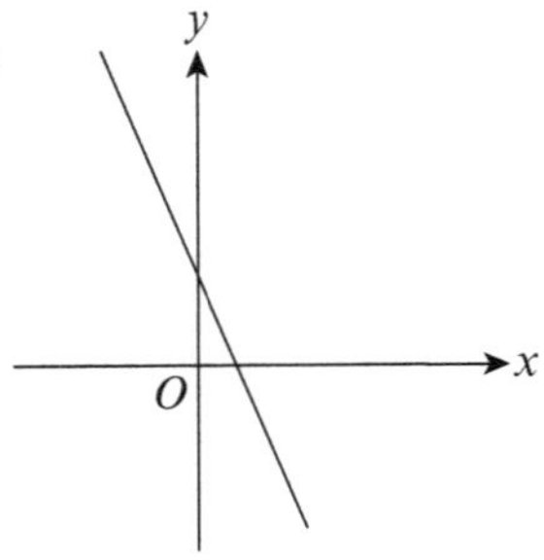

d 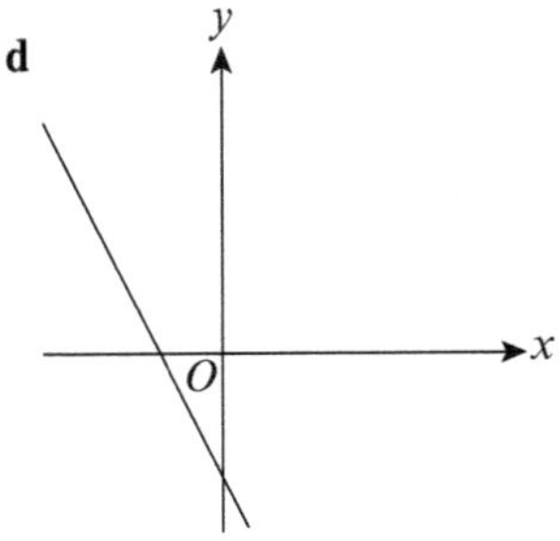

43 Rearrange $4y + 12x = 8$ to make y the subject.

a $y = 2 - 12x$ **b** $y = 2 - 3x$
c $y = 3x + 2$ **d** $y = 8 - 3x$

44 Which equation does not represent the same straight line graph as the other three?

a $2y + 2 = 6x$ **b** $y = 3x + 1$
c $15x - 5y = 5$ **d** $y + 1 - 3x = 0$

45 Which line is parallel to $y = 4x - 5$ with a y-intercept of (0, 7)?

a $y = 0x + 7$ **b** $y = 4x + 0$
c $y = 4x + 7$ **d** $y = 7x - 5$

46 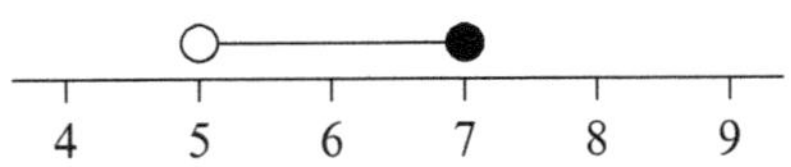

The number line shows the inequality

a $5 \leq x < 7$ **b** $5 < x < 7$
c $x < 5$ and $x \geq 7$ **d** $5 < x \leq 7$

47 The solution to $3x > 12$ is shown on which number line?

a 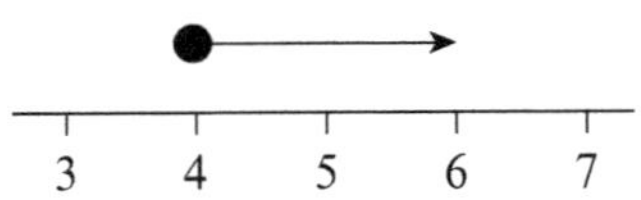

b 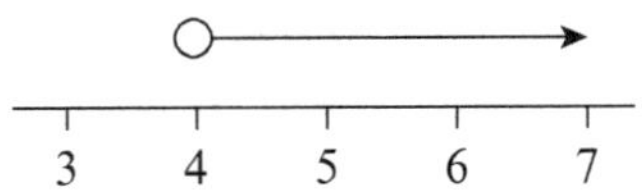

c 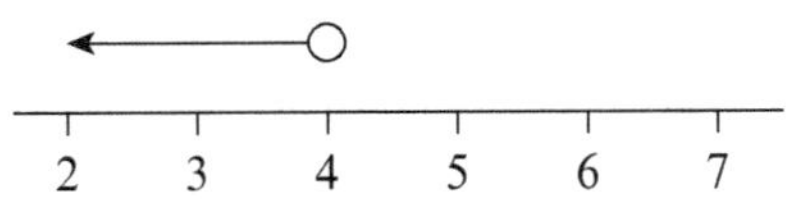

d 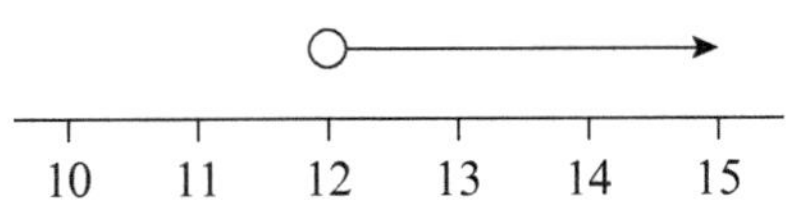

48 Solve $8x - 5 > 6x - 13$

a $x = {}^-4$ **b** $x > 4$ **c** $x < 4$ **d** $x > {}^-4$

49 $x = 3$ and $y = 5$ is the solution to which simultaneous equations?

a $9x + 5y = 34$
$9x + 2y = 19$

b $7x + 3y = 27$
$4x + 3y = 18$

c $3x + 5y = 16$
$3x + 2y = 1$

d $7x + 2y = 31$
$4x + 2y = 22$

50 The length of this rectangle is greater than its width.

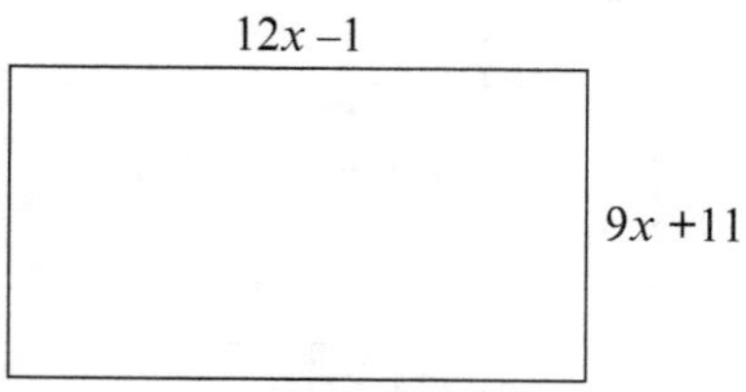

A correct inequality for this information is

a $12x - 1 \geq 9x + 11$

b $12x > 9x + 12$

c $12x - 1 < 9x + 11$

d $x \geq 4$

51 Solve $^-3x \leq 9$

a $x \leq {}^-3$ **b** $x < {}^-3$
c $x \geq {}^-3$ **d** $x \leq 3$

52 Solve these simultaneous equations
$6x + 2y = 10$
$2x + 3y = 1$

a $x = 2$ and $y = {}^-1$

b $x = 2.25$ and $y = {}^-1.75$

c $x = 2$ and $y = 1$

d $x = 2.25$ and $y = {}^-9$

53 The perimeter of this rectangle is 41 cm.

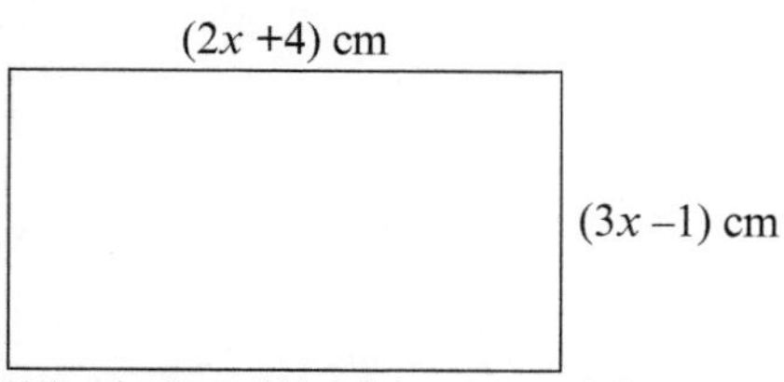

What is the width of the rectangle?

a $x = 3.5$ **b** 9.5 cm
c 11 cm **d** 104.5 cm^2

54 I am thinking of a number. I subtract 13 from my number. Then I divide 350 by this result. The answer is 14. An equation to show this is

a $\frac{13-x}{350} = 14$ **b** $\frac{x-13}{350} = 14$
c $\frac{350}{13-x} = 14$ **d** $\frac{350}{x-13} = 14$

Geometry

55 The exterior angle of a regular octagon is

a 8° **b** 22.5° **c** 135° **d** 45°

56 The interior angle of a regular pentagon is

a 60° **b** 72° **c** 108° **d** 120°

57 Which expression does not find the sum of the interior angles of a polygon with n sides?

a $180(n - 2)$ **b** $90(2n - 4)$
c $180n - 2$ **d** $180n - 360$

58 Which of these regular polygons will not tessellate?

a hexagon **b** triangle
c quadrilateral **d** pentagon

59 The diagram shows a hexagon.
Find the value of x.

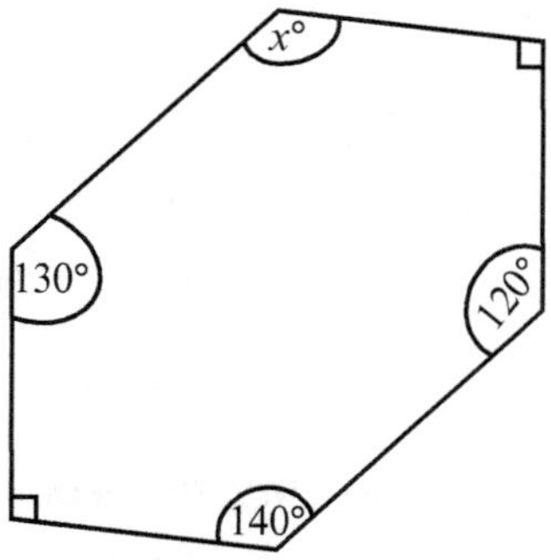

a 140 **b** 150 **c** 160 **d** 170

60 The diagram shows a pentagon with exterior angles marked.

Find the value of y.

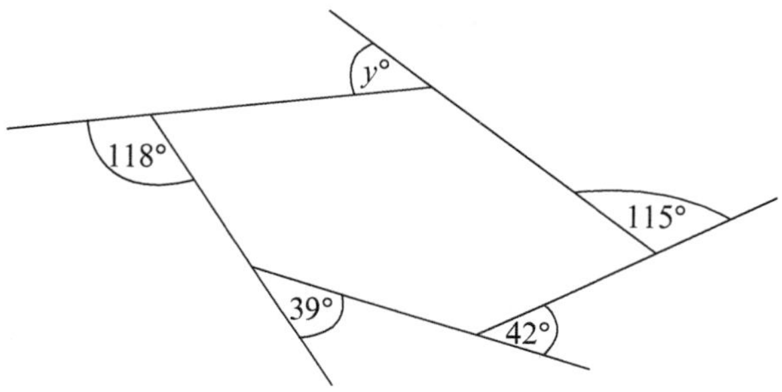

a 36 **b** 46 **c** 56 **d** 66

61 The diagram is made from three parallel lines and two straight lines crossing these lines.

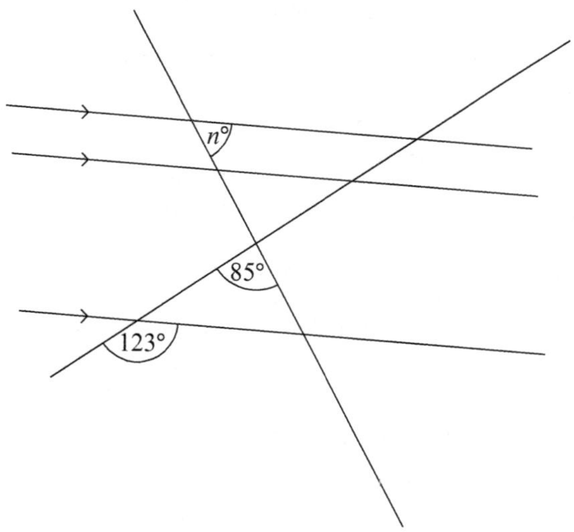

Find the value of n.

a 38 **b** 57 **c** 85 **d** 95

62 The diagram is made from two parallel lines and two straight lines.
Find the value of h.

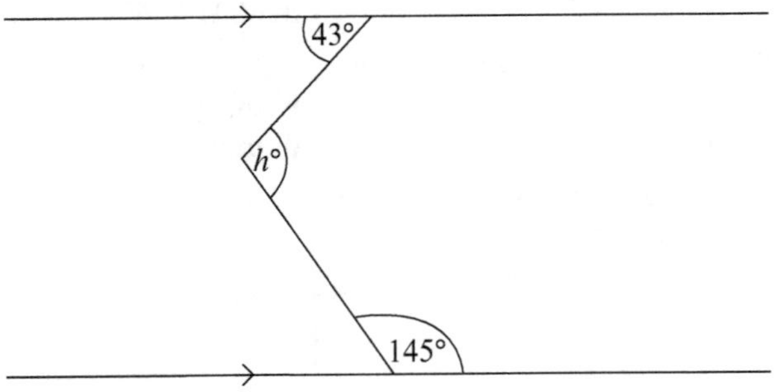

a 35 b 43 c 78 d 145

63 Find the coordinates of the point that is $\frac{2}{3}$ of the way along AB from A, where A is (4, 2) and B is (22, ⁻13).
a (6, ⁻5) b (10, ⁻3) c (12, ⁻10) d (16, ⁻8)

Presenting data and interpreting results

64 Which of these is not a feature of a stem-and-leaf diagram?
- a Numbers written in order, smallest to largest
- b A key
- c Numbers evenly spaced
- d The median point circled

65

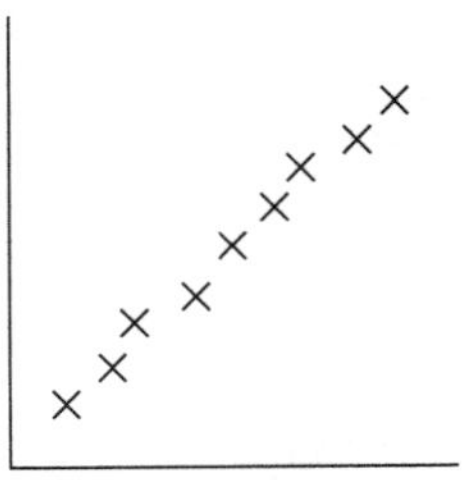

This graph shows
- a A strong positive correlation
- b A weak positive correlation
- c No correlation
- d A strong negative correlation

66 Which of these statements about scatter graphs is not true?
- a The line of best fit should be ruled.
- b A scatter graph will show positive correlation, negative correlation or no correlation.
- c The line of best fit must pass through the origin.
- d The line of best fit can be used to estimate the value of one variable when you know the value of the other.

67 The back-to-back stem-and-leaf diagram shows the length, in millimetres, of two samples of 25 screws produced by two different machines.

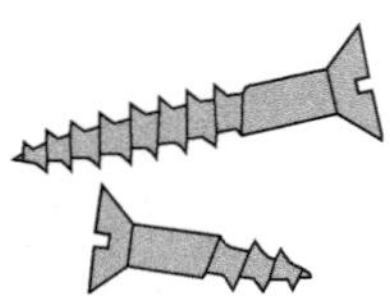

Machine A		Machine B
9 8 8 6	1	5 6 7 7 8 8 9
9 5 5 4 0	2	1 2 4 4 6 7 7 9 9 9 9 9
7 7 7 6 6 5 5 4 3 2 1	3	0 0 2 4 5
7 5 4 1 0	4	6

Key
| 1 | 5 = 15 mm and 6 | 1 | = 16 mm

The median screw length for Machine B is
a 34 mm b 43 mm c 27 mm d 26 mm

68 Using the back-to-back stem-and-leaf diagram from question **67**, the largest screw made by Machine A is
a 74 mm b 47 mm
c 40 mm d 92 mm

69 Using the back-to-back stem-and-leaf diagram from question **67**, the range of the screw lengths for **all** screws made on machine A **and** machine B is
a 24 b 31 c 32 d 15 to 47

70 The histogram shows the mass of some animals.

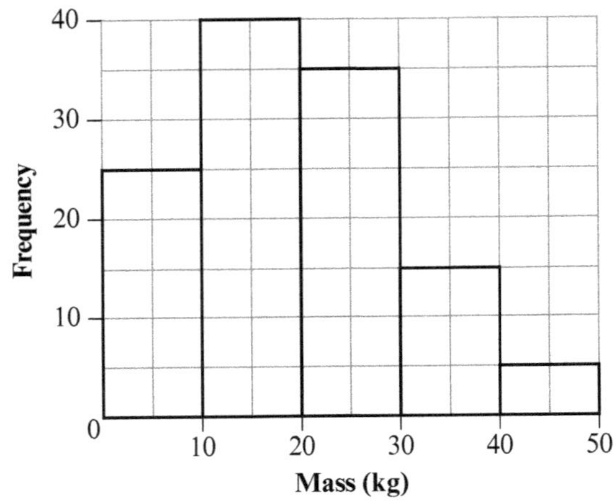

Which of these is definitely true?
- a The smallest animal has a mass of 4 kg
- b There are 110 animals in total
- c The modal mass is 10–20 kg
- d The largest animal has a mass of 50 kg

71 Weak negative correlation is shown in which graph?

a

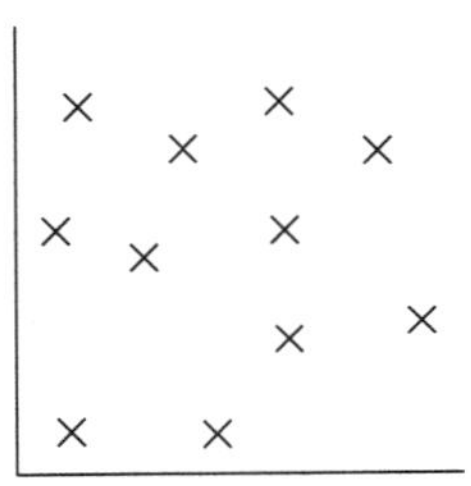

b

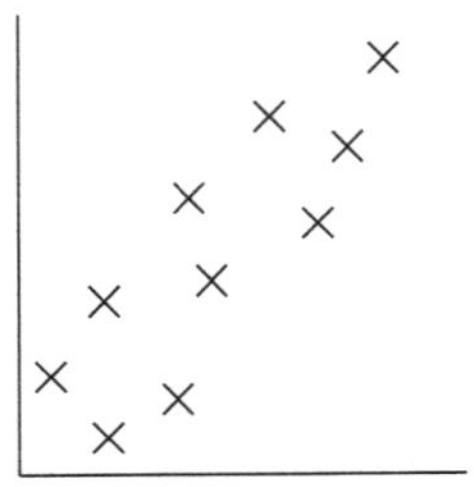

c

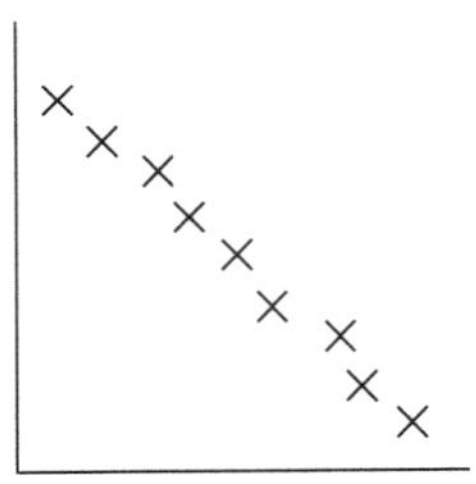

d

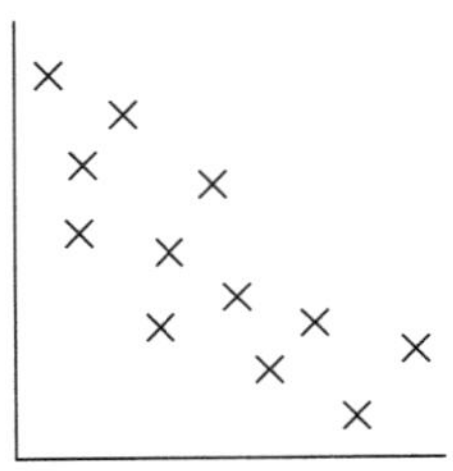

72 Some information about the mass, in kilograms, of each of 25 boxes is shown in the table.

Mass (m, kg)	Frequency
$10 \le m < 15$	1
$15 \le m < 20$	3
$20 \le m < 25$	6
$25 \le m < 30$	9
$30 \le m < 35$	4
$35 \le m < 40$	2

Johannes draws a frequency polygon to represent this data. He starts by plotting all the points ready to join them up with straight lines.

Which of these is **not** a correct point for him to plot?

a (12.5, 1) **b** (22.5, 3) **c** (32.5, 4) **d** (27.5, 9)

73 The frequency polygon shows information about the time, in minutes, each of 50 students spent on their homework one evening.

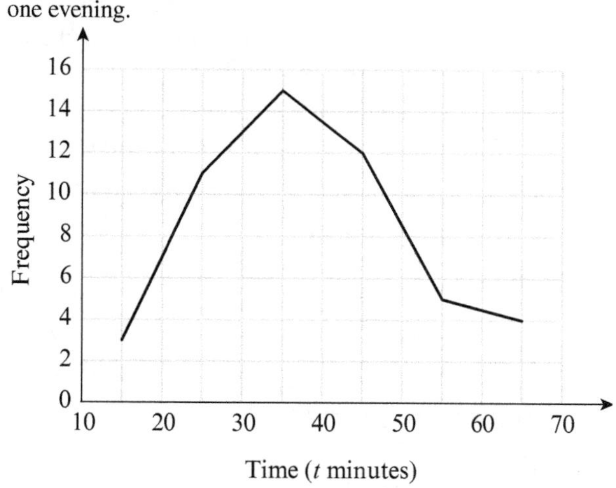

Choose the statement that could be true but is not definitely true.

a 15 students spent between 30 and 40 minutes on their homework.

b All students spent at least 10 minutes on their homework.

c 5 students spent 55 minutes on their homework.

d 9 students spent 50 minutes or more on their homework.

Multi-choice Exercise C

This exercise can be used as a resource for revising the work in Chapters 11 to 15 of the Student Book.

Note: Multi-choice questions are not part of Cambridge Checkpoint tests; they are included here as a useful revision tool.

Ratio and proportion

1 Leela and Parvati share some money in the ratio 5 : 9. Parvati gets $252 more than Leela. How much money does Leela get?

a $567 **b** $90 **c** $140 **d** $315

2 x and y are directly proportional.

x	3		10
y	51	68	

The missing numbers are

a 4 and 170 **b** 20 and 58
c 6.5 and 85 **d** Not enough information to say

3 Write $60\% : 2.4 : 1\frac{4}{5}$ as a ratio in its simplest whole number form.

a 3 : 12 : 9 **b** 60 : 24 : 18
c 1 : 4 : 3 **d** 6 : 24 : 18

4 A green paint is a mix of blue and yellow in the ratio 4 : 5.

How many litres of yellow paint need to be mixed with 3800 ml of blue paint to make the green paint?

a 4.75 ℓ **b** 2.1 ℓ
c 19 ℓ **d** 5 ℓ

5 I exchange 200 US dollars ($) for 3010 South African rand (R). At the same rate of exchange, how many dollars would I get for R5418?

a $350.00 **b** $2408.00
c $81540.90 **d** $360.00

6 x and y are directly proportional.

x	4	8	16
y	10	20	40

The equation connecting x and y is

a $y = 2x$ **b** $y = 0.4x$
c $y = 2.5x$ **d** $x = 2.5y$

7 A box contains pink, red and green sweets. The ratio of pink to red sweets is 2 : 3. The ratio of red to green sweets is 5 : 14. Find the ratio of pink to green sweets in its simplest form.

a 10 : 42 **b** 2 : 14 **c** 1 : 7 **d** 5 : 21

8 It takes 10 machines 6 minutes to complete a job. How many minutes will it take 4 machines?

a 15 **b** 24 **c** 40 **d** 60

9 m is inversely proportional to b. Use the table to find the value of x.

m	20	15
b	36	x

a 27 **b** 31 **c** 48 **d** 720

10 In a survey the ratio of people who preferred chocolate ice cream to the people who preferred vanilla ice cream was 8 : 5. 48 more people chose chocolate than vanilla ice cream. How many people chose vanilla ice cream?

a 30 **b** 80 **c** 128 **d** 208

11 Some dollars are shared in the ratio 5 : 4 : 3 between Anna, Ben, and Carlos. Ben gets $4.20. How many dollars were shared in total?

a $10.08 **b** $12.60 **c** $16.80 **d** $50.40

12 Which equation represents inverse proportion between x and y?

a $x = y \times 10$ **b** $x = \frac{y}{10}$
c $xy = 10$ **d** $x = y + 10$

Sequences, functions and graphs

13 What is the next number in this sequence? 1, 4, 9, 16, 25, ...

a 41 **b** 34 **c** 36 **d** 38

14 Peter cycled from home to Matthew's house, stayed for a while and cycled back home, stopping at a shop on the way. The distance–time graph is shown below.

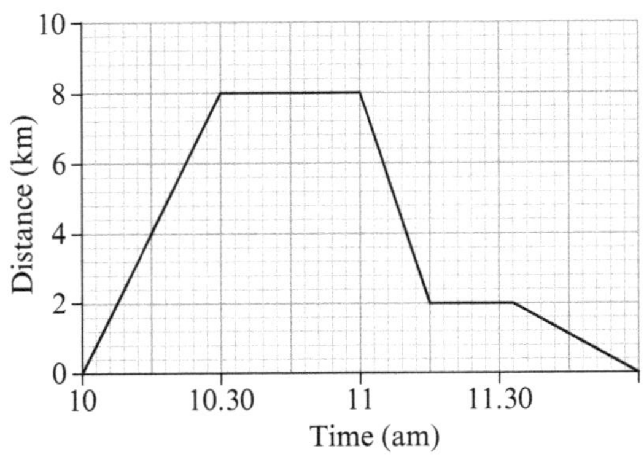

What was Peter's average speed when he cycled from home to Matthew's house?

a 8 km/h **b** 4 km/h
c $\frac{8}{30}$ km/h **d** 16 km/h

15 Find the average speed of an object that travels 180 km in 2 hours and 15 minutes.

a 83.7 km/h **b** 80 km/h
c 405 km/h **d** 2.4 km/h

16 The graph shows the fees charged by a plumber.

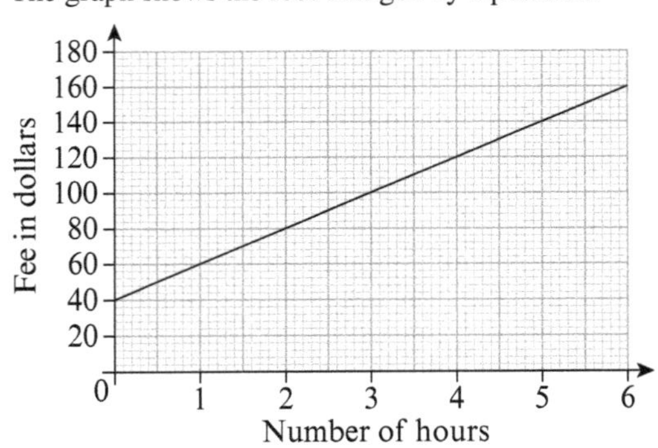

The intercept on the fee axis shows

a The call-out fee
b The hourly fee
c The cost of parts
d The discount if paying by cash

17 Using the graph in question **16**, the gradient of the line is

a 40 **b** 60 **c** 0 **d** 20

18 The simultaneous equations

$3x + 2y = 31$
$-x + y = -2$

are shown on the graph.

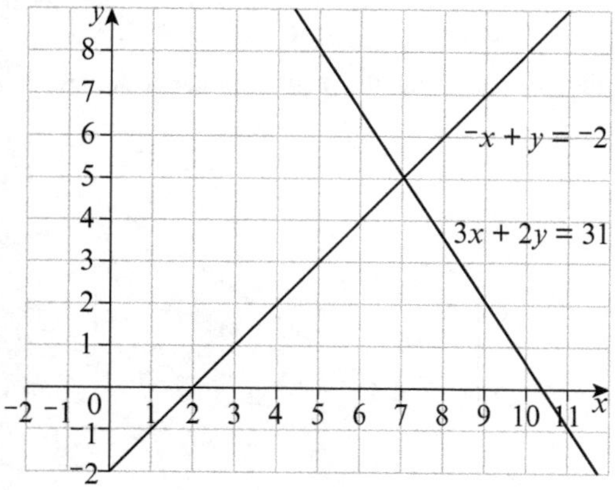

The solution is

a 7 **b** $x = 7, y = 5$
c $x = 7$ **d** $x = 5, y = 7$

19 The diagram shows the speed–time graph over a period of 10 seconds, for a car that starts from rest.

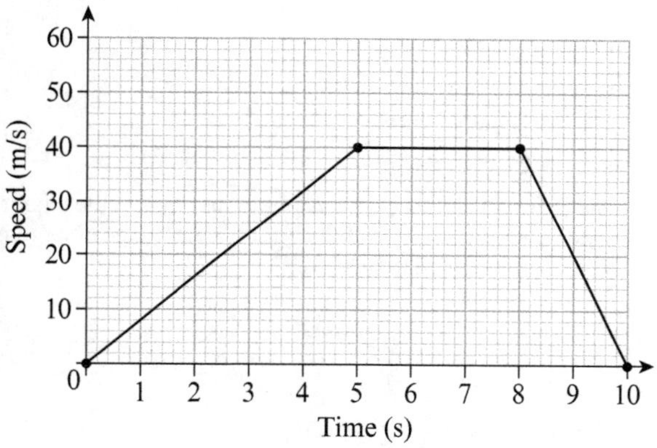

How far has the car travelled in the 10 seconds?

a 40 m **b** 10 m **c** 260 m **d** 400 m

20 Using the graph in question **19**, what part of the journey shows acceleration?

a Between 5 and 8 seconds
b Between 0 and 5 seconds
c Between 8 and 10 seconds
d Can't tell from this graph

21 The term-to-term rule of a sequence is square then subtract 3. The first term of the sequence is 2. Which of these is **not** a term in the sequence?

a -2 **b** -1 **c** 1 **d** 2

22 What is the *n*th term of the sequence $-3, 0, 5, 12, 21, 32, \dots$?

a $-4n^2$ **b** $(n-4)^2$ **c** $n-4$ **d** n^2-4

23 The position-to-term rule is square then add 1. Which of these is **not** a term in the sequence?

a 5 **b** 45 **c** 65 **d** 145

24 Here is a function machine.

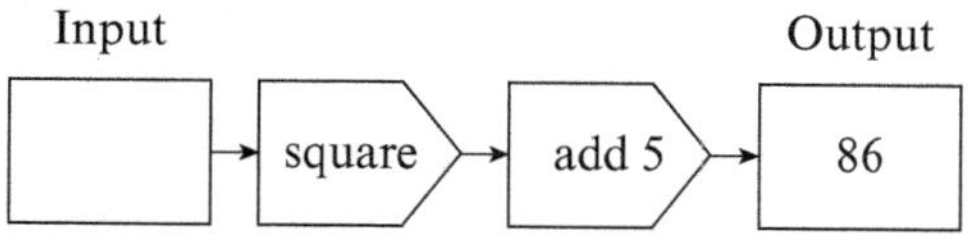

How many input values would give the output of 86 shown in the function machine?

a 0 **b** 1 **c** 2 **d** more than 2

25 Which of these values is a term in the sequence $\frac{n}{7}$?

a $\frac{1}{14}$ **b** $\frac{3}{77}$ **c** $\frac{7}{2}$ **d** 5

26 What is the *n*th term of the sequence 1, 8, 27, 64, 125, …?

a $3n$ **b** n^2 **c** n^3 **d** $2n$

27 Which of these graphs of $y = x^2 - 1$ has been drawn correctly?

a

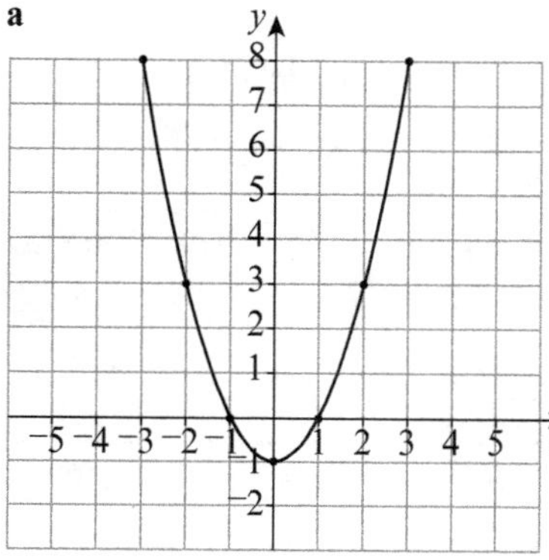

b

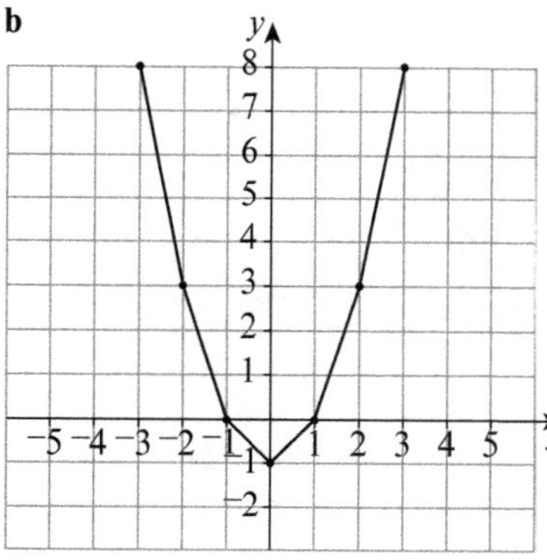

c

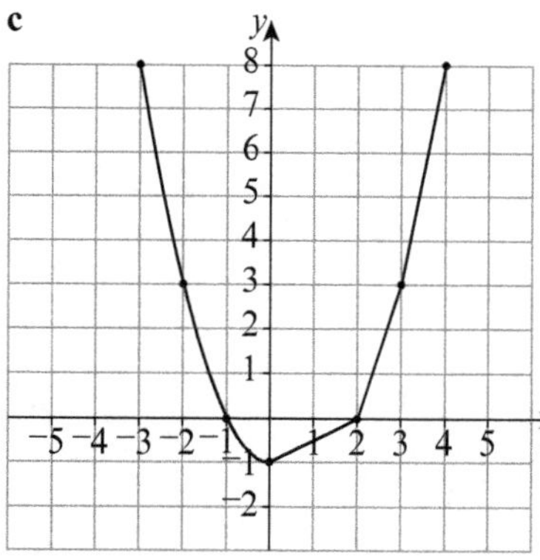

d

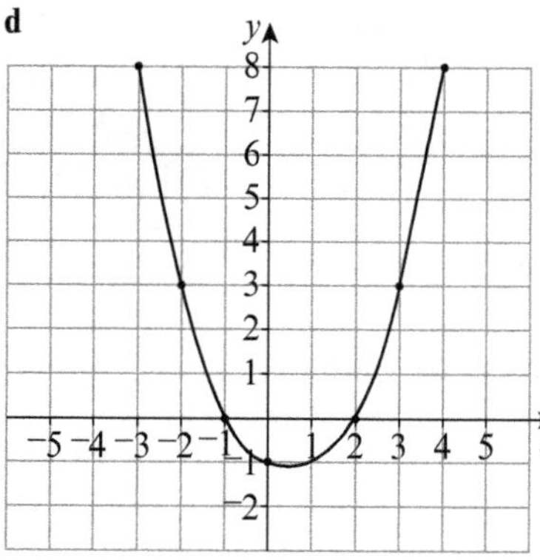

28 Rearrange the equation $3y - 4x = 12$ to make y the subject.

a $y = 4 + 4x$ **b** $y = 4 + \frac{4x}{3}$

c $y = 4 - \frac{4x}{3}$ **d** $y = \frac{12 - 4x}{3}$

29 What is the gradient of the graph of $5y = 4 - 10x$?

a $^{-}10$ **b** 4 **c** $\frac{4}{5}$ **d** $^{-}2$

30 The line $5y + 2x = 10$ crosses the y-axis at the point

a (0, 5) **b** (0, 2) **c** (5, 0) **d** (2, 0)

31 How long does a car take to make a 125 km journey if it travels at an average speed of 100 km/h?

a 1 hour 25 minutes **b** 1 hour 15 minutes

c 12 500 minutes **d** 48 minutes

32 An aeroplane is travelling at an average speed of 840 km/h. How far does it travel in 4 hours 10 minutes?

a 3444 km **b** 201.6 km

c 3370 km **d** 3500 km

33 Which coordinates are in the wrong place in the Carroll diagram?

	On the line $2y - 3x = 12$	**Not** on the line $2y - 3x = 12$
Above the x-axis	($^{-}2$, 3)	($^{-}3$, 2)
Below the x-axis	($^{-}6$, $^{-}3$)	($^{-}8$, $^{-}6$)

a ($^{-}2$, 3) **b** ($^{-}3$, 2) **c** ($^{-}6$, $^{-}3$) **d** ($^{-}8$, $^{-}6$)

Transformations

34

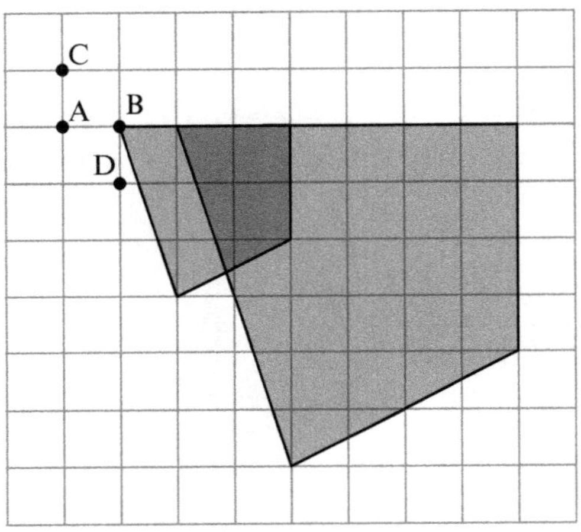

The centre of enlargement is

a A **b** B **c** C **d** D

35

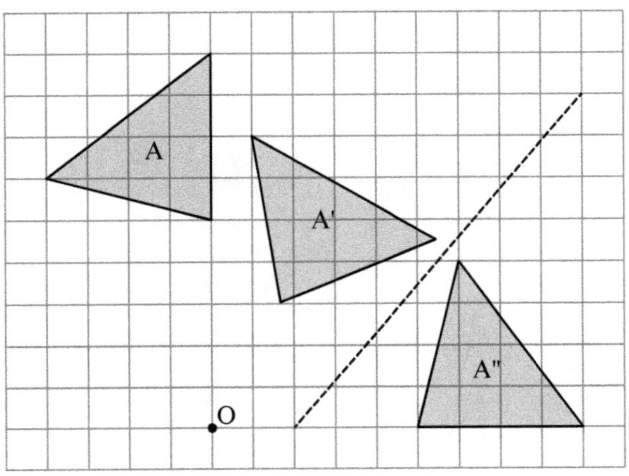

Shape A has undergone two transformations, firstly to A′ and secondly to A″. These transformations are

a translation then reflection

b reflection then rotation

c rotation then reflection

d reflection then enlargement

36 Write down the coordinates of the reflection of the point (1, 2) in the line $y = {}^{-}x$

a ($^{-}2$, 1) **b** (1, $^{-}2$) **c** ($^{-}2$, $^{-}1$) **d** ($^{-}1$, $^{-}2$)

37 Describe the transformation mapping triangle A onto triangle A′.

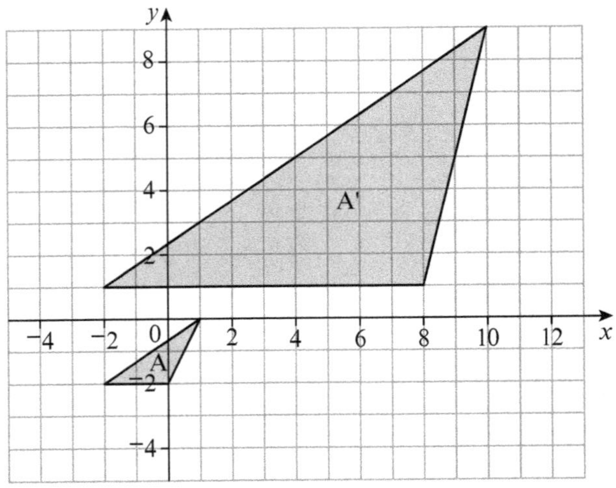

a Enlargement

b Enlargement scale factor 4

c Enlargement centre ($^{-}2$, $^{-}3$)

d Enlargement scale factor 4, centre ($^{-}2$, $^{-}3$)

38 The diagram shows Daryl's attempt to enlarge the smaller quadrilateral using a scale factor of 3, centre of enlargement X, to make quadrilateral ABCD.

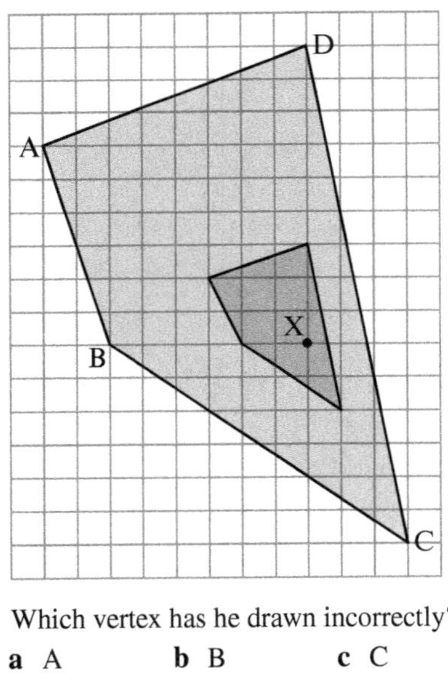

Which vertex has he drawn incorrectly?

a A **b** B **c** C **d** D

39 A shape has area 40 cm². It is enlarged by a scale factor of 2. What is the area of the new shape?

a 40 cm² **b** 80 cm² **c** 160 cm² **d** 320 cm²

40 A quadrilateral has area 23 mm². It is enlarged by scale factor k. The image has area 828 mm². Find the value of k.

a 6 **b** 18 **c** 36 **d** 805

41 Shape A is transformed to shape A′ by the first transformation shown in the table. Then shape A′ is transformed to shape A″ using the second transformation shown in the table.

First transformation, A to A′	Second transformation, A′ to A″
Reflection in $y = -x$	Translation through $\begin{pmatrix} 0 \\ ^-2 \end{pmatrix}$
Rotation 90° clockwise about (1, 3)	Reflection in the line $y = 8$
Translation through $\begin{pmatrix} 2 \\ ^-5 \end{pmatrix}$	Rotation 90° anticlockwise about the origin

For how many of these three combinations of transformations will the shape A″ be congruent to shape A?

a 0 **b** 1 **c** 2 **d** 3

42 A point A is translated by the vector $\begin{pmatrix} c \\ d \end{pmatrix}$ to make point B.

Then point B is translated by the vector $\begin{pmatrix} e \\ f \end{pmatrix}$ to get back to point A.
None of the values in the vectors are 0.
Sanjay writes these statements.

$c = e$

$c = {}^-e$

$d = f$

$d = {}^-f$

How many of Sanjay's statements are true?

a 0 **b** 1 **c** 2 **d** 3

43 A is the point ($^-2$, 1).
Point A is reflected in the line $x = {}^-1$ to make point B.
Point B is reflected in the line $x = 3$ to make point C.
The single transformation that maps point A to point C is a

a reflection in the line $x = 0$
b reflection in the line $x = 1$
c reflection in the line $x = 2$
d reflection in the line $x = 4$

44 Line segment AB, where A = (2, 3) and B = (5, 8), is reflected in the line $y = x$, then rotated 90° anticlockwise about the origin. The gradient of the image of AB is

a $\frac{5}{3}$ **b** $\frac{3}{5}$ **c** $\frac{^-5}{3}$ **d** $\frac{^-3}{5}$

Decimals and percentages

45 Given that $56 \times 435 = 24\,360$ the answer to $^-5.6 \times 4.35$ is

a $^-24.36$ **b** $^-2.436$
c $^-243.6$ **d** $^-0.002436$

46 Work out $28 \div 0.001$

a 28 **b** 0.0028
c 2800 **d** 28 000

47 Work out $5.6 \div 0.08$

a 7 **b** 0.7 **c** 70 **d** 700

48 How many 0.125 kg food portions can be made up from a 50 kg sack?

a 40 **b** 400 **c** 4000 **d** 6.25

49 A gas stove bought for $2000 depreciates by 10% each year. What is its value after two years?

a $200 **b** $400 **c** $1600 **d** $1620

50 The table shows the percentage profit made on selling four items.

Cost price	Selling price	Percentage profit
$230.00	$322.00	40%
$2.40	$2.64	10%
$464.00	$835.20	80%
$32.00	$64.00	50%

Which figure in the percentage profit column is incorrect?

a 40% **b** 10% **c** 80% **d** 50%

51 Work out $0.204 \div 0.06$

a 34 **b** 0.34 **c** 0.034 **d** 3.4

52 Guntur bought a phone for $260. What is his percentage loss if he sells the phone for $228.80?

a 31.2% **b** 12% **c** 88% **d** 14%

53 What is the selling price of a $400 television if a sales tax of 20% is charged?

a $480 **b** $320 **c** $333.33 **d** $80

54 A bank has a simple interest rate of 4%. What will be the value of a $300 investment after 5 years?

a $360 **b** $60 **c** $240 **d** $320

55 Last week there were 1400 visitors to a museum. This week there are 1470. What is the percentage increase in the number of visitors?

a 4.8% **b** 7% **c** 70% **d** 5%

56 What principal balance will earn $127.50 in 5 years at 3% simple interest?

a $85 000 **b** $255 **c** $850 **d** $425

57 A bag had an original price of $110.
In the sale it is reduced to $82.50.
What percentage discount was given on the bag?

a 75% **b** 25% **c** 33% **d** 27.5%

58 Stephan earns $35 000 a year. He has to pay tax on his earnings. He has a tax-free allowance of $12 000, then pays tax at 30% on the rest of his earnings.
What are his net (take-home) earnings?

a $16 100 **b** $6900 **c** $23 000 **d** $28 100

59 Without working out the answer, which of these calculations will give an answer less than 415?

a $415 \div 0.281$ **b** $415 \times {}^-2.35$
c $415 \times 2.4 \div 1.2$ **d** $415 \div \frac{1}{25}$

60 Without working out exact answers, which of these statements is correct?

a $765 \times 0.352 > 765$ **b** $819 \div 0.0329 < 819$
c $220 \times 2.32 < 440$ **d** $376 \div 1.84 < 300$

61 Work out the volume of this cuboid in cm³.

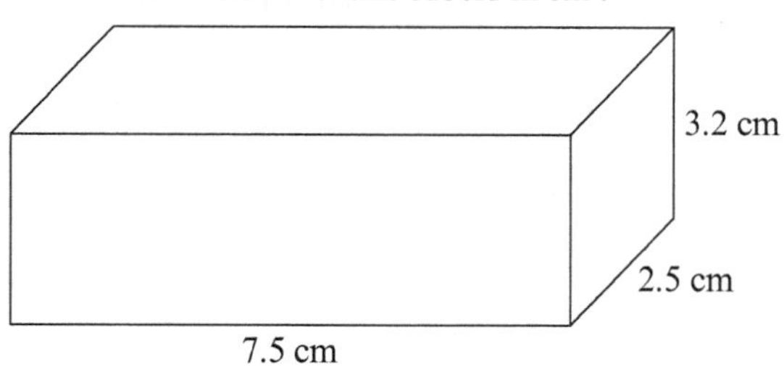

a 13.2 **b** 52.8 **c** 60 **d** 101.5

62 Which of these statements is not correct?
- **a** to increase by 40% the multiplier is 1.4
- **b** to decrease by 30% the multiplier is $^{-}1.3$
- **c** to decrease by 80% the multiplier is 0.2
- **d** to increase by 8% the multiplier is 1.08

63 Jenny buys a diamond necklace for $2400. The necklace increases in value by 5% each year.
What is the value of the necklace after 3 years?

a $378.30 **b** $2760 **c** $2778.30 **d** $10 800

64 In a shop all sale items are reduced by 20%. The following week the sale items are reduced by a further 25%. What percentage have sale items been reduced by overall?

a 40 **b** 45 **c** 50 **d** 55

65 Harry invests $9500 in a bank paying compound interest of 3% each year for 4 years. Which calculation works out the value of his investment after 4 years?

a 9500×0.03^4 **b** 9500×1.03^4
c 9500×1.04^3 **d** $\frac{9600 \times 3 \times 4}{100}$

Probability

66 A biased 4-sided spinner is spun. The results are shown in the table below.

Colour on spinner	Red	Blue	Yellow	Green
Probability	0.2		0.15	0.45

What is the missing probability for getting the colour blue?

a 0.38 **b** 0.2 **c** 0.8 **d** 1

67 Using the table from question **66**, what is the probability of spinning a red or a yellow?

a 0.17 **b** 0.55 **c** 0.35 **d** 0

68 Using the table from question **66**, what is the probability of not spinning a red or a yellow?

a 1 **b** 0.83 **c** 0.45 **d** 0.65

69 The spinner from question **66** is spun and a dice is rolled. What are the missing entries labelled X and M in this sample space diagram?

		Spinner			
		Red	**Blue**	**Yellow**	**Green**
Dice	1				G, 1
	2			Y, 2	
	3				
	4		B, 4		
	5	M			
	6			X	

a R,5 and Y,6 **b** R,5 and Y,5
c B,5 and Y,6 **d** M,5 and X,6

70 The length of time two different models of cell phone worked for before the battery needed charging are shown in the table.

Cell phone working time (hours)	1–10	11–20	21–30	31–40	More than 41
Cell phone A	3	15	27	4	1
Cell phone B	18	63	91	18	10

The estimated probability of cell phone A lasting between 21 and 30 hours is

a 27% **b** 0.54 **c** 27 : 50 **d** 91%

71 Using the table from question **70**, the estimated probability of cell phone B lasting between 1 and 10 hours is

a 18% **b** 0.9 **c** 0.18 **d** 9%

72 A box contains sweets of different colours. The sweets are red, orange, pink or yellow. One sweet is chosen at random. The probability of choosing a red sweet is 0.22, and the probability of choosing an orange sweet is 0.3. There are twice as many yellow sweets as pink sweets. What is the probability of choosing a yellow sweet?

a 0.16 **b** 0.25 **c** 0.32 **d** 0.48

73 Two fair 6-sided dice are rolled. What is the probability of the score on the first dice being higher than the score on the second dice?

a $\frac{5}{12}$ **b** $\frac{1}{2}$ **c** $\frac{7}{12}$ **d** $\frac{7}{18}$

74 Shannon rolls a fair 6-sided blue dice and a fair 6-sided red dice. Here are some events.
Event A – Shannon rolls an even number on the red dice.
Event B – Shannon rolls an odd number on the blue dice.
Event C – Shannon rolls a 6 on the red dice.
Event D – Shannon rolls a prime number on the blue dice.
Which pair of events are not independent?

a A and B **b** B and C **c** B and D **d** A and D

75 Gemma rolls a fair 6-sided dice and writes down the score. She then spins a fair spinner with five sections numbered 1, 2, 3, 4 and 5 and writes down that score too.
Work out the probability that both scores are odd.

a $\frac{1}{2}$ **b** $\frac{3}{10}$ **c** $\frac{1}{4}$ **d** $\frac{1}{5}$

76 The probability of Mohammed spinning a 10 on a spinner is 0.8
Work out the number of times he is expected to spin a 10 if he spins the spinner 60 times.

a 8 **b** 10 **c** 12 **d** 48

77 A black card is chosen at random from the black cards, then a white card is chosen at random from the white cards.

Which is the correct tree diagram?

a

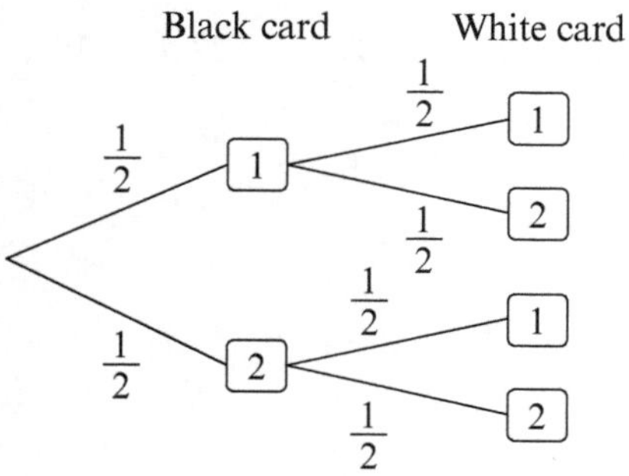

b

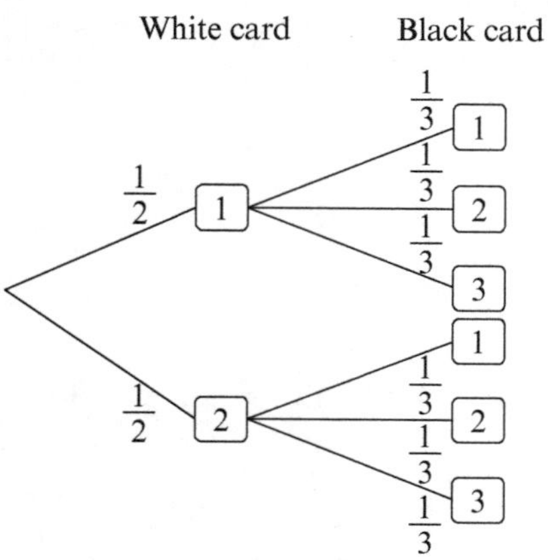

c

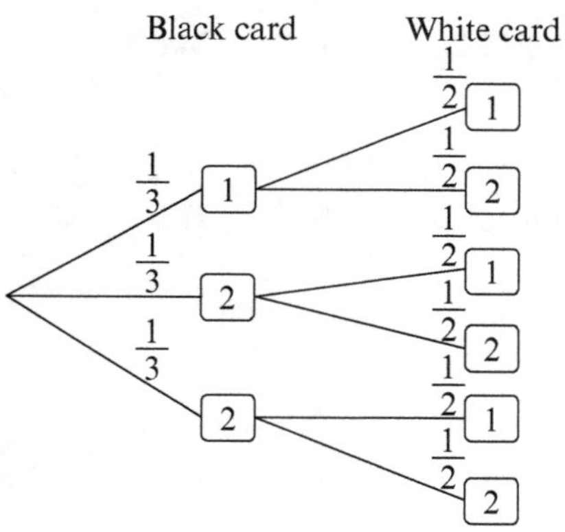

d

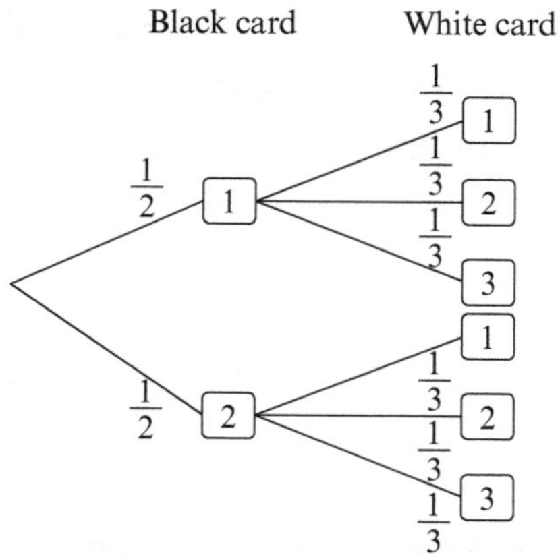

78 Franca has a large box of counters that are either black or red. She draws this graph showing the relative frequency of picking a black counter from the box after every 20 counters picked.

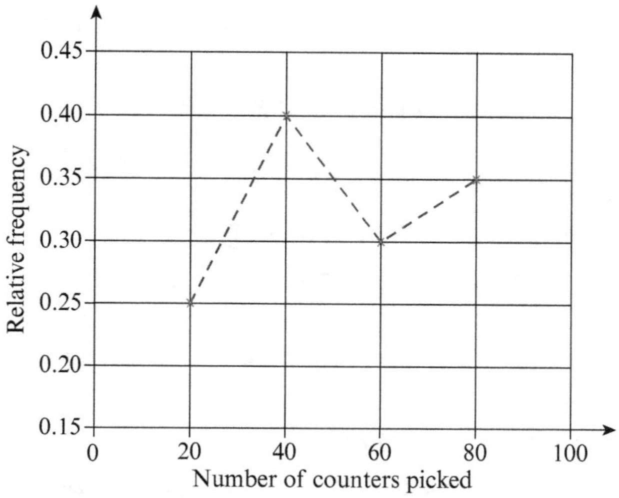

What is the best estimate of the probability of picking a black counter from the box?

a 0.25 **b** 0.3 **c** 0.35 **d** 0.4

Answers to multi-choice Exercises A, B and C

Multi-choice Exercise A

Integers, powers and roots

1 c 2 b 3 a 4 c 5 a 6 b 7 b 8 c 9 b 10 c 11 d 12 a 13 d 14 c

Expressions and formulae

15 a 16 d 17 b 18 d 19 c 20 c 21 b 22 d 23 d 24 c 25 a 26 d 27 a 28 a
29 a 30 c

Shapes and mathematical drawings

31 a 32 d 33 b 34 c 35 d 36 a

Place value

37 b 38 c 39 d 40 c 41 b 42 b 43 d 44 c 45 b 46 a

Grouped data and sampling

47 a 48 b 49 c 50 b 51 c 52 b 53 b 54 d

Multi-choice Exercise B

Area, perimeter and volume

1 b 2 c 3 d 4 c 5 b 6 b 7 c 8 b 9 d 10 b 11 d 12 c 13 d 14 c
15 a 16 d 17 b 18 d 19 d 20 b 21 a

Fractions and decimals

22 c 23 a 24 d 25 a 26 c 27 c 28 a 29 b 30 b 31 d 32 b 33 d 34 c

Equations and inequalities

35 c 36 c 37 c 38 a 39 b 40 a 41 d 42 a 43 b 44 b 45 c 46 d 47 b 48 d
49 d 50 b 51 c 52 a 53 b 54 d

Geometry

55 d 56 c 57 c 58 d 59 b 60 b 61 a 62 c 63 d

Presenting data and interpreting results

64 d 65 a 66 c 67 c 68 b 69 c 70 c 71 d 72 b 73 c

Multi-choice Exercise C

Ratio and proportion

1 d **2** a **3** c **4** a **5** d **6** c **7** d **8** a **9** c **10** b **11** b **12** c

Sequences, functions and graphs

13 c **14** d **15** b **16** a **17** d **18** b **19** c **20** b **21** b **22** d **23** b **24** c **25** d **26** c
27 a **28** b **29** d **30** b **31** b **32** d **33** d

Transformations

34 a **35** c **36** c **37** d **38** a **39** c **40** a **41** d **42** c **43** c **44** c

Decimals and percentages

45 a **46** d **47** c **48** b **49** d **50** d **51** d **52** b **53** a **54** a **55** d **56** c **57** b **58** d
59 b **60** d **61** c **62** b **63** c **64** a **65** b

Probability

66 b **67** c **68** d **69** a **70** b **71** d **72** c **73** a **74** c **75** b **76** d **77** d **78** c

Topic Test – Stage 9

Chapters 1–5 (non-calculator)

50 minutes

Name ....

Additional materials:

- ruler
- protractor
- compasses

Calculators are not allowed

READ THESE INSTRUCTIONS FIRST

Answer all questions in the spaces provided on the question paper.

You should show all your working on the question paper.

The number of marks is given in brackets [ ] at the end of each question or part question.

The total number of marks for this paper is 40.

Note: This Topic test is not a Cambridge Checkpoint test, but it is included here to help assess the work covered in Chapters 1–5 of the Student Book.

For Teacher's Use

1 Work out the values of *x* and *y* that make these statements correct.

(a) $0.428 \times 10^x = 4280$

$x =$ .... [1]

(b) $39 \div 10^y = 3900$

$y =$........................... [1]

2 Expand the brackets and simplify

$(x - 3)(x + 8)$

............................[2]

For Teacher's Use

3 Here is a list of numbers.

$^-3$ 9 $\sqrt{25}$ 1.4 $\sqrt{7}$ $\frac{2}{3}$

Draw a ring around the irrational number. [1]

4 Simplify

(a) $(x^4)^3$

.... [1]

(b) $5x^2 - x^2$

.... [1]

5 Gemma is 20 years old.
She wants to investigate how her resting heart rate is affected by exercise over a 20-week swimming training period.

She considers these two methods for finding how her resting heart rate has changed.

Method 1: measure her resting heart rate once before starting the training and once at the end of the training.

Method 2: measure her resting heart rate every Friday for the whole 20 weeks of training.

(a) Give one reason why Method 1 might be the best choice.

....

....

[1]

(b) Give one reason why Method 2 might be the best choice.

....

....

[1]

(c) Gemma also wants to compare her results with those for other women aged 20 doing the same training.

She decides to sample 20 people at random from those at her local swimming pool.

Explain why her sampling method may not give her reliable data to compare.

....

....

[1]

For Teacher's Use

6 Write each of these numbers in standard form.

(a) 420 000 000

.... [1]

(b) 0.00713

.... [1]

7 Write as a single fraction

$\frac{7}{8m} - \frac{1}{2m}$

.... [2]

8 The diagram shows a prism drawn on isometric paper.

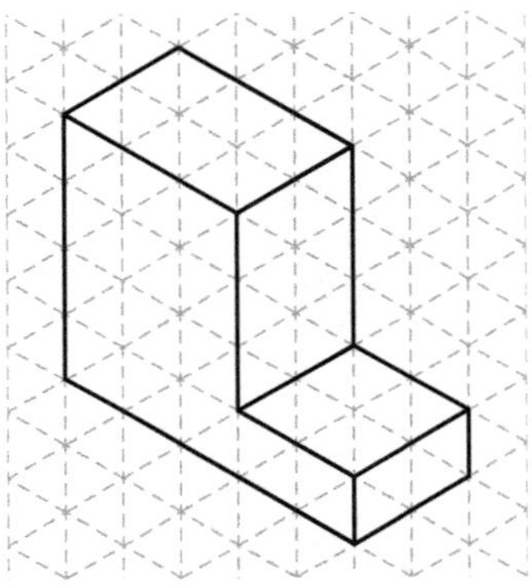

Write down the number of planes of symmetry of this prism.

.... [1]

9 Write a number on the line to complete this statement correctly.

$x^2 + 8x + 16 = (x + \_\_\_\_\_\_)^2$

[1]

10 Ahmed records the time, in seconds, it takes for each of 20 students to solve a maths problem.

Some information about the data he collects is shown in the table.

Time (t, seconds)	Frequency
$2 \leq t < 4$	8
$4 \leq t < 6$	5
$6 \leq t < 8$	4
$8 \leq t < 10$	2
$10 \leq t < 12$	1

(a) Find an estimate for the mean.

..........................s [3]

(b) Find the interval where the median time lies.

.... [1]

(c) Find the modal class interval.

.... [1]

(d) Find an estimate for the range of the times.

..........................s [1]

For Teacher's Use

For Teacher's Use

11 James thinks of a number, n, multiplies it by 5, squares the result and then adds 2.

Write down an expression, in terms of n, for the number James ends up with.

.... [1]

12 A number rounded correct to the nearest 10 is 240.

Kulwinder writes this inequality to show the bounds of the original number, x.

$$239.5 \leq x \leq 245$$

Write down **two** mistakes that Kulwinder has made.

....

....

.... [2]

13 Using the formula $F = 100c - 2h^2$,
find F when $h = {}^{-}5$ and $c = 0.7$

$F =$ .... [2]

14 Look at the cards below. Put a tick (✔) under all the correct statements:

$3^4 \times 3^2 = 9^6$	$7^{-4} \div 7^4 = 7^{-1}$	$6p^6 \div 3p^{-2} = 2p^8$
☐	☐	☐
$2t^{-2} \times 5t^5 = 10t^3$	$5^0 = 0$	$\frac{6^2}{6^4} = 6^{0.5}$
☐	☐	☐

[2]

For Teacher's Use

15 Simplify

$$\frac{15x+3}{3}$$

.... [1]

16 Construct a 60° angle.

[1]

17 Here are some numbers.

$\sqrt{11}$ 2.1 3.9 $\sqrt{7}$

Write the numbers in order, starting with the smallest.

....

smallest largest

[1]

18 Here is a formula.

$$h = \frac{2p-5}{m}$$

Rearrange the formula to make p the subject.

p = ...[2]

For Teacher's Use

19 The diagram shows the position of two cities, A and B.

A third city, C, is on a bearing of 075° from A and on a bearing of 315° from B.

Work out the position of C and mark it on the diagram.

[2]

20 **(a)** I start with any two **consecutive integers**.

I square each of them, then add the two squares together.

Construct an algebraic expression to show this.

Simplify this expression.

.... [3]

(b) Show that the expression in part **(a)** must be an odd number.

For Teacher's Use

[1]

Topic Test – Stage 9

Chapters 6–10 (non-calculator)

50 minutes

Name ....

Additional material:
- ruler
- protractor

Calculators are not allowed

READ THESE INSTRUCTIONS FIRST

Answer all questions in the spaces provided on the question paper.

You should show all your working on the question paper.

The number of marks is given in brackets [ ] at the end of each question or part question.

The total number of marks for this paper is 40.

Note: This Topic test is not a Cambridge Checkpoint test, but it is included here to help assess the work covered in Chapters 6-10 of the Student Book.

For Teacher's Use

1 Here are some measurements.

5 cm   5 nm   5 mm   5 μm

Write the measurements in order starting with the smallest.

....

smallest   largest

[1]

2 Use the fact that $\frac{1}{9} = 0.\dot{1}$ to write $\frac{7}{9}$ as a decimal.

.........................[1]

For Teacher's Use

3 Solve $\frac{24}{x} = 6$

$x =$ ....[1]

4 A is the point (6, 4) and B is the point (24, $^{-}11$).

Find the coordinates of the point that is $\frac{1}{3}$ of the way along AB from A.

(.............. , ...) [2]

5 The diagram shows an isosceles triangle and a pair of parallel lines.

Find the value of x.

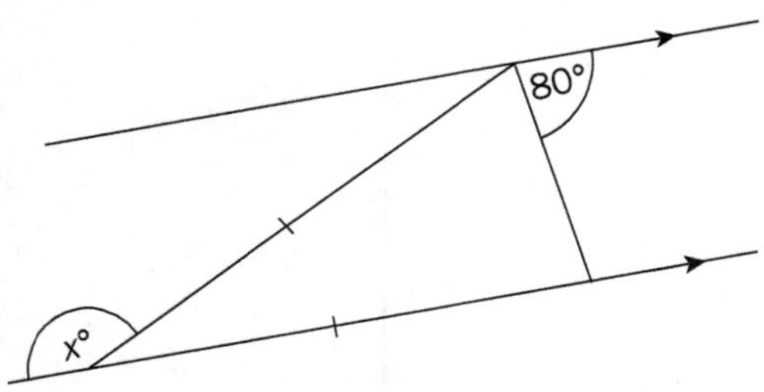

NOT TO SCALE

$x =$ ....[2]

For Teacher's Use

6 Here is an equation of a straight line.

$2y = 5x + 8$

Find the coordinates of the point where the line intercepts the *y*-axis.

(.............. , ...) [1]

7 Write a fraction in the box to make this calculation correct.

$2\frac{3}{5} \times \square = 4\frac{1}{3}$

[2]

8 Work out the perimeter of this semicircle.

Use 3.14 for π.

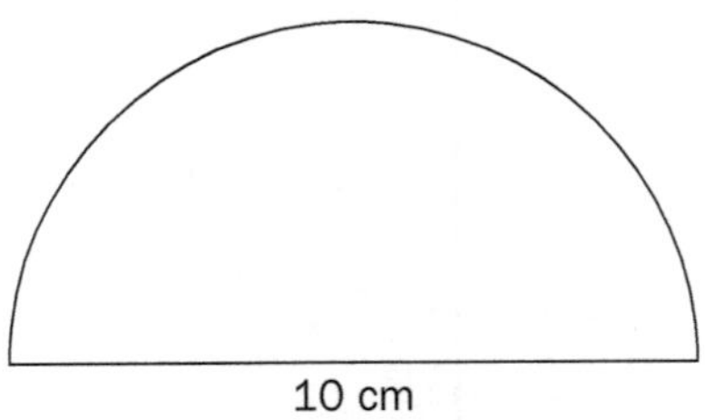

.... cm [2]

9 Solve the simultaneous equations

$3x + y = 19$

$2x - y = 6$

Show your working.

For Teacher's Use

$x =$ ....

$y =$ ....[2]

10 Work out the length of side x.

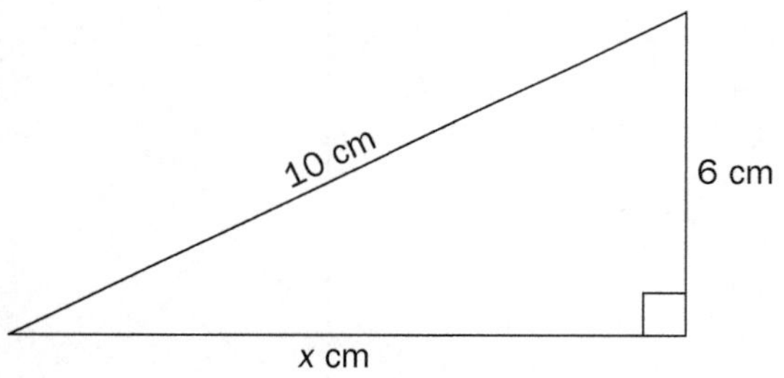

$x =$ .... cm [2]

For Teacher's Use

11 15 students sit a maths test and a history test.

The back-to-back stem-and-leaf diagram shows the results.

Both tests were marked out of 100.

Maths		History
7 3 1	40	
7 5 3 1 9	50	8
9 3 6	60	6 2 5
1 0	70	2 5 3 1
2	80	0 9 8 4 2
1	90	1 7

(a) Make two criticisms of the diagram.

....

....

....

....

[2]

(b) Which test do you think is easier?

Give a reason for your answer.

....

....

....

[1]

For Teacher's Use

12 (a) A group of students took a test.
The marks they scored are shown in the pie chart below.

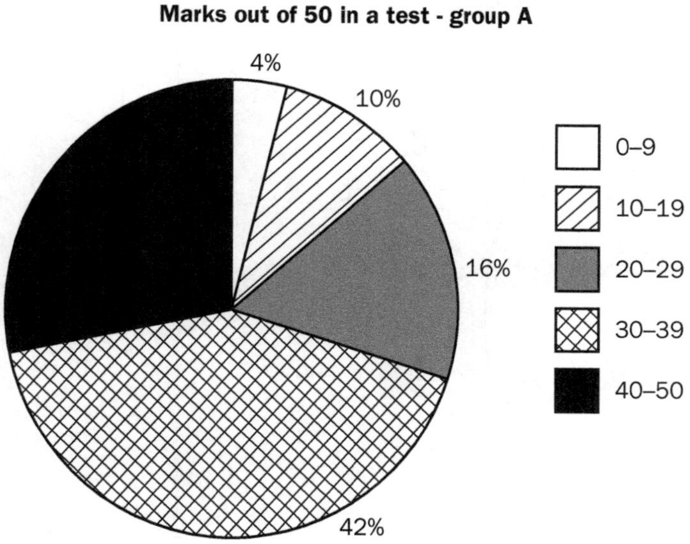

If 21 students had a mark from 30 to 39, how many students had a mark from 40 to 50?

.... [1]

(b) Another group of students took the same test. Their results are shown in the table below. Complete the table by working out the angles for the pie chart for Group B.

Group B		
Mark out of 50	**Number of students**	**Angle on pie chart**
0–9	9	
10–19	8	
20–29	5	
30–39	4	
40–50	4	
Total	30	

[2]

For Teacher's Use

(c) Complete the pie chart below for Group B using your answer to part **(b)**.

Marks out of 50 in a test – group B

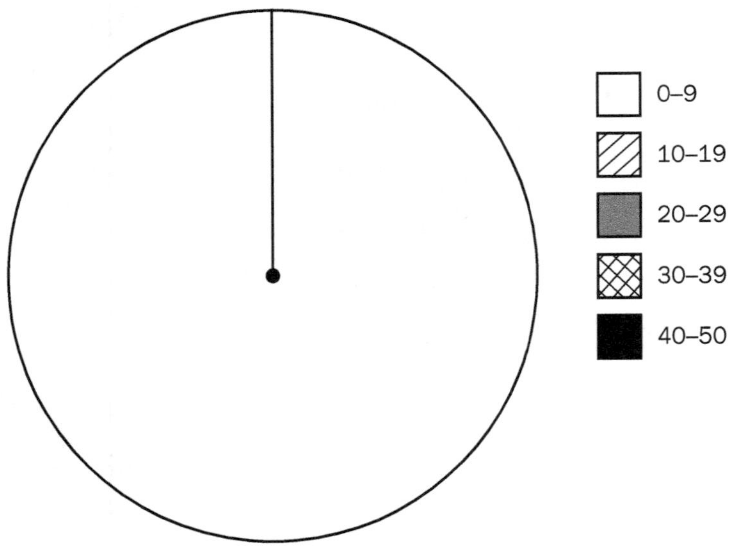

[2]

(d) Out of Group A and Group B, only one group did a lot of revision work. The other group did no revision work.
Which group do you think did the revision work?
Give a reason for your answer.

Group .... did revision work because ....

....

.... [1]

13 **(a)** Solve the inequality

$4x + 10 \geq x + 25$

.... [2]

(b) Show your answer to part **(a)** on the number line.

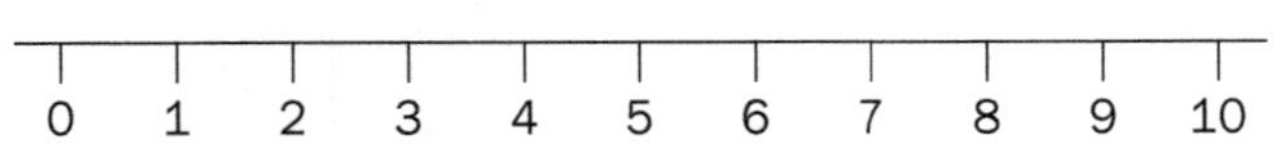

[1]

(c) Using your answer to part **(a)**, write down the smallest number that x can be.

.... [1]

For Teacher's Use

14 Ten people take part in a competition.

Two judges each give a mark out of 20.

The results for 7 of these people are shown on the scatter graph.

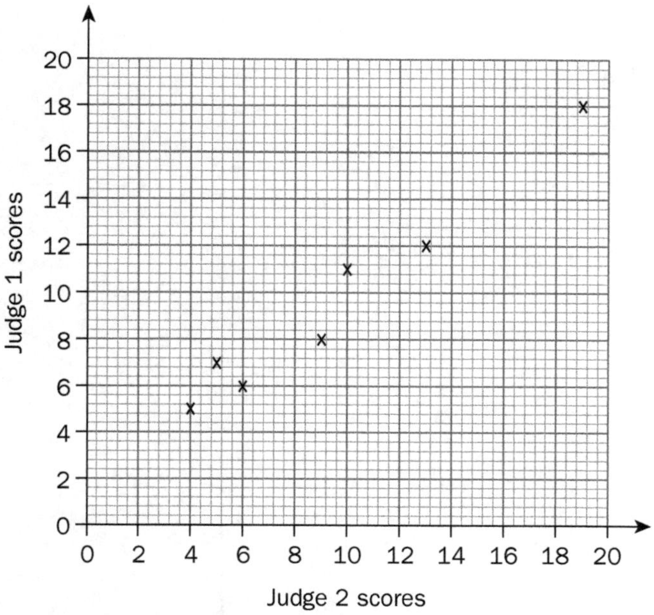

The other three results are shown in the table.

Judge 1	12	17	7
Judge 2	14	15	8

(a) Plot the other three points on the scatter graph. [1]

(b) Make two statements about the correlation between the judges' scores.

....

....

....

....[2]

For Teacher's Use

15 **(a)** The diagram shows a pentagon.

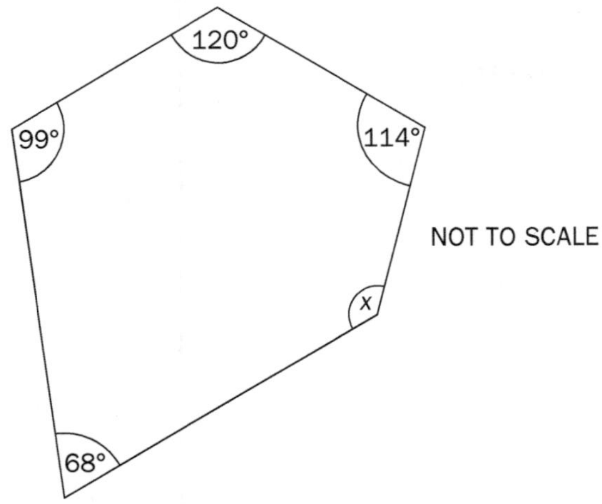

Work out the value of x.

x = ... [2]

(b) Work out the exterior angle of a regular pentagon.

.... [1]

16 Tick (✔) the regular polygons that will tessellate.

Hexagon	Quadrilateral	Heptagon	Triangle
☐	☐	☐	☐

[1]

For Teacher's Use

17 Here is a cylinder and a cube.

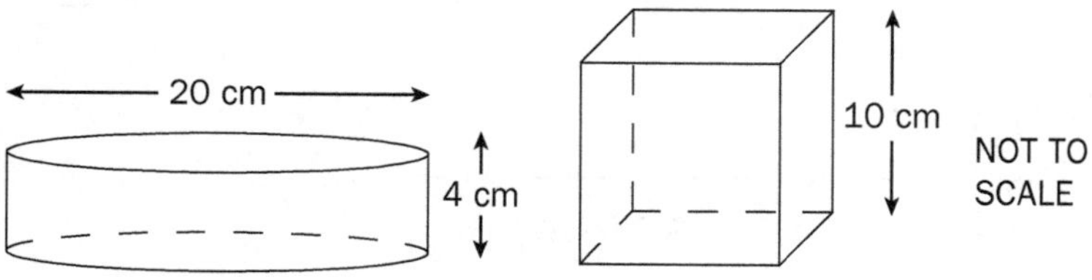

The cylinder has a diameter of 20 cm and a height of 4 cm.
The cube has a side length of 10 cm.

By rounding π to 1 decimal place, show that the volume of the cylinder is greater than the volume of the cube.

[3]

18 Draw lines to match all the pairs of parallel lines.

$y = 3x + 4$	$^{-}5x = y$
$y - 5x = 3$	$y = 10x - 7$
$2y + 10x = 1$	$8 + 3x = y$
$y - 10x + 4 = 0$	$y = 5x + 4$

[1]

Topic Test – Stage 9

Chapters 11–15 (non-calculator)

50 minutes

Name ....

Additional material:

- ruler
- tracing paper (optional)

Calculators are not allowed

READ THESE INSTRUCTIONS FIRST

Answer all questions in the spaces provided on the question paper.

You should show all your working on the question paper.

The number of marks is given in brackets [ ] at the end of each question or part question.

The total number of marks for this paper is 40.

Note: This Topic test is not a Cambridge Checkpoint test, but it is included here to help assess the work covered in Chapters 11–15 of the Student Book.

For Teacher's Use

1 A sequence has a first term 2 and a term-to-term rule of square and subtract 1.
Find the 3rd term in the sequence.

.... [1]

2 A car travels 210 km at a speed of 70 km/h.
Work out how long this takes.

.... hours [1]

3 A sequence has nth term $\frac{n}{13}$
Find the 4th term in the sequence.

.... [1]

For Teacher's Use

4 A car cost $11 000. It decreases in value by 7% each year for 3 years.

Draw a ring around the calculation that gives the value of the car after the 3 years.

11000×0.7^3 11000×0.07^3 11000×0.93^3 11000×1.07^3

[1]

5 Amir and Yasmina share some money in the ratio 3 : 7

Yasmina gets $720 more than Amir.

How much money does Amir get?

$ .... [2]

6 Complete these sentences.

The graph $3x + 7y = 21$ intercepts the y-axis at the point

(............ , ...)

The graph $3x + 7y = 21$ intercepts the x-axis at the point

(............ , ...)

[2]

7 It takes 12 builders 30 days to complete a building.

(a) How long would it take 4 builders to complete the building?

.... days [1]

(b) How many builders would it take to complete the building in 20 days?

.... [1]

8 The diagram shows triangle A and triangle A' drawn on a grid.

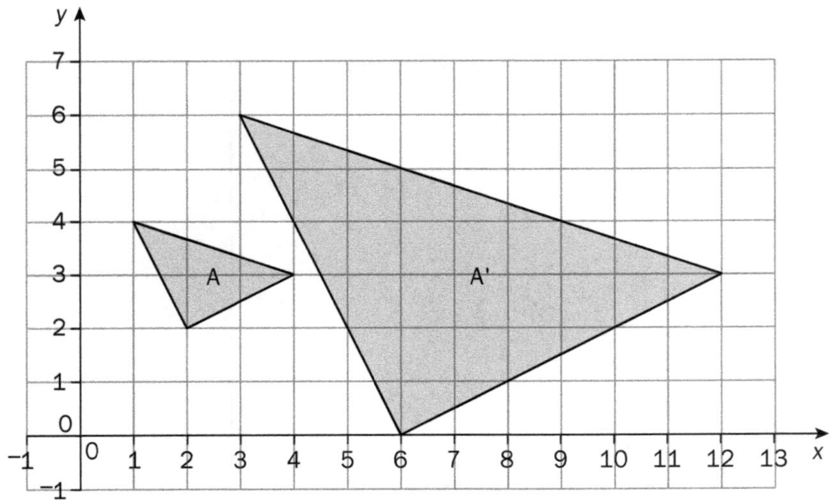

Describe fully the **single** transformation that maps triangle A to triangle A'.

....

....[3]

9 Look at this calculation.

$28 \times 549 = 15\ 372$

Work out

(a) $^{-}2.8 \times 54.9$

.... [1]

(b) $15.372 \div 0.28$

.... [1]

For Teacher's Use

For Teacher's Use

10 A bag contains some blue, red and yellow coloured balls.

One ball is taken from the bag at random, the colour is noted and then the ball is replaced in the bag.

The results are shown in the table.

Colour	Blue	Red	Yellow
Frequency	60	40	100

(a) Estimate the probability that the next ball chosen is red.

.... [1]

(b) There are 150 balls in the bag altogether.

How many of each colour ball would you expect there to be?

Blue .... Red .... Yellow .... [2]

11 In a shop all sale items are reduced by 20% off the full price.
The following week the sale items are reduced by a further 20% off the reduced price.
Farook says, "All sale items have now been reduced by 40% off the full price."
Show that Farook is not correct by working out the correct percentage reduction off the full price.

............................% [2]

For Teacher's Use

12 The diagram shows three lines drawn on a grid.

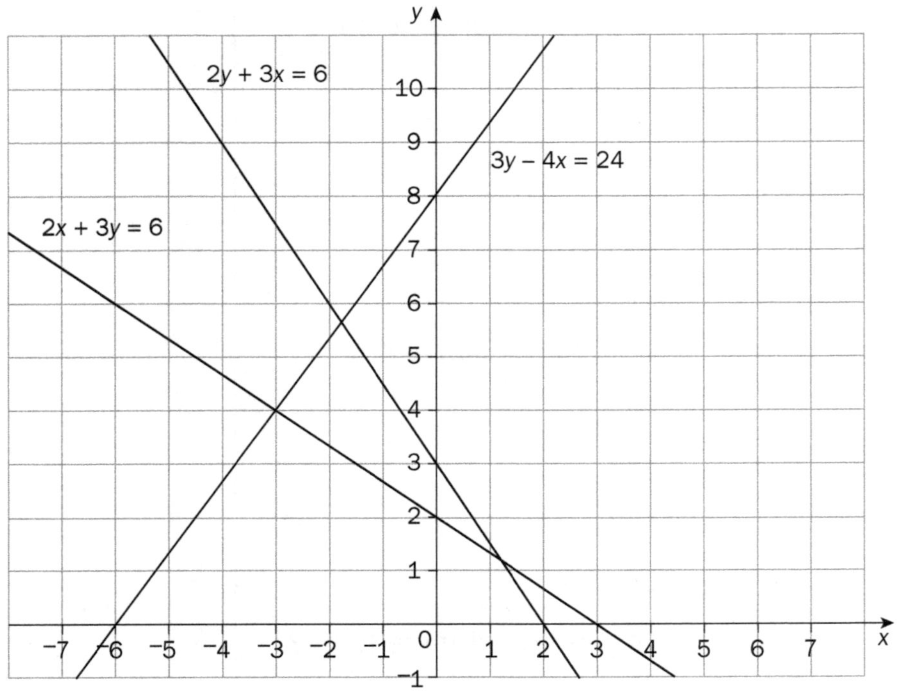

Use the diagram to solve the simultaneous equations.

$2x + 3y = 6$

$3y - 4x = 24$

$x =$ ...., $y =$ .... [1]

13 A biased 4-sided dice is rolled.

The probability of the dice landing on 1, 2 or 4 is shown in the table.

Score on dice	1	2	3	4
Probability	0.25	0.3		0.35

(a) Complete the table by filling in the missing probability of rolling a 3. [1]

(b) Find the probability of rolling a 1 or a 2.

........................[1]

For Teacher's Use

14 Reflect quadrilateral ABCD in the mirror line EF.
Label the reflected shape X.
Then rotate Shape X 90° anticlockwise about the centre O.
Label the rotated shape Y.

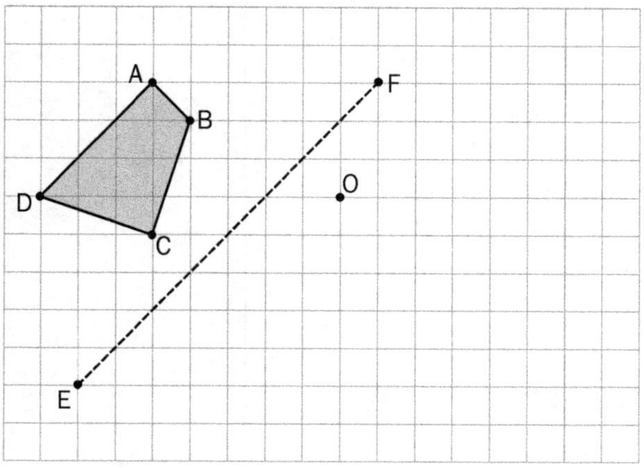

[2]

15 The distance–time graph shows the journey of a car.

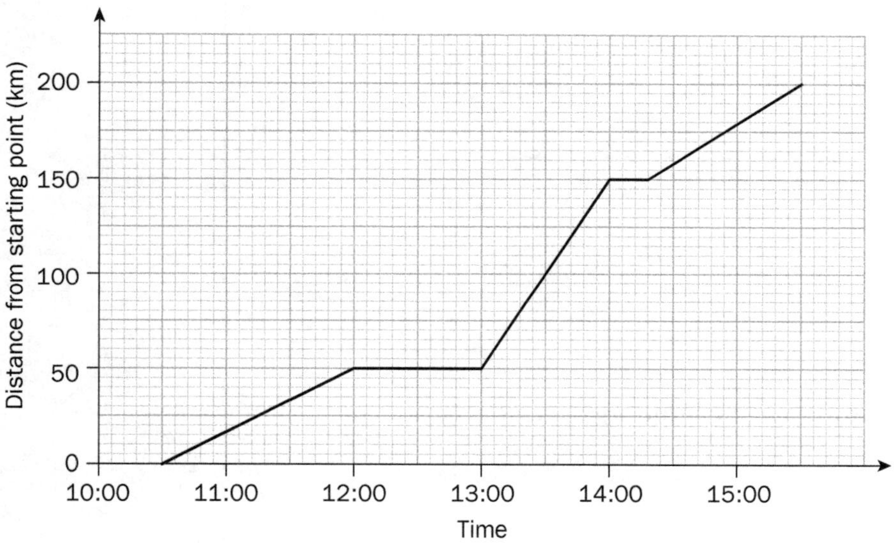

(a) Tick (✔) the section of the graph when the car is travelling the fastest. [1]

(b) Work out the average speed for the fastest part of the journey.

.... km/h [1]

For Teacher's Use

16 **(a)** Complete this table of values for the graph of $2y + x = 4$

x		0	1		$^-2$
y	$^-1$			0	3

[2]

(b) Use your table of values to draw the graph of $2y + x = 4$

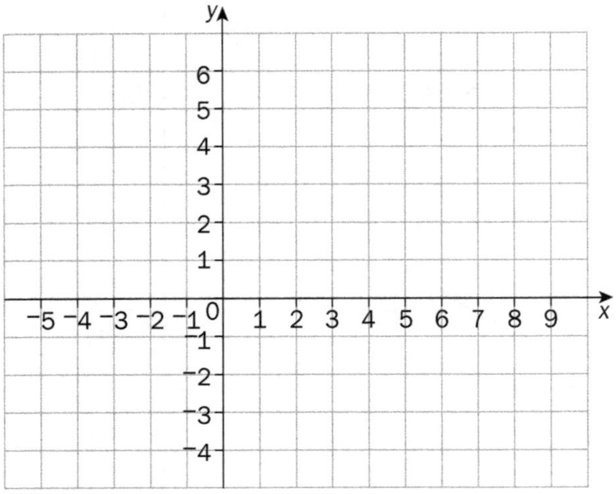

[1]

17 Here is a mapping diagram for the function $y = x^2 + 3$

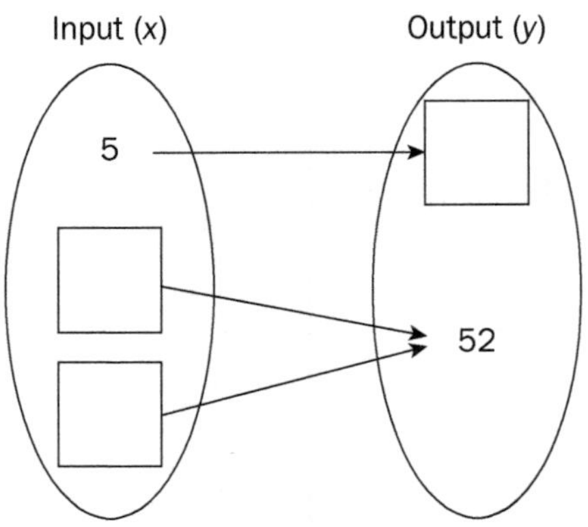

Use three **different** numbers to complete the mapping diagram.

[2]

For Teacher's Use

18 Here are two spinners.

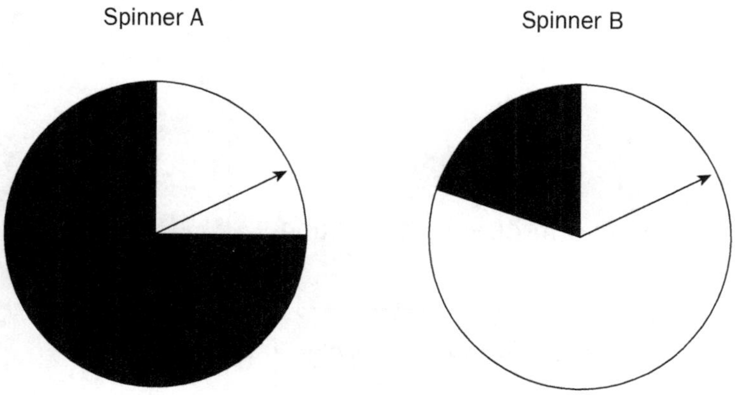

The probability of spinning black on spinner A is $\frac{3}{4}$

The probability of spinning black on spinner B is $\frac{1}{5}$

(a) Complete the tree diagram.

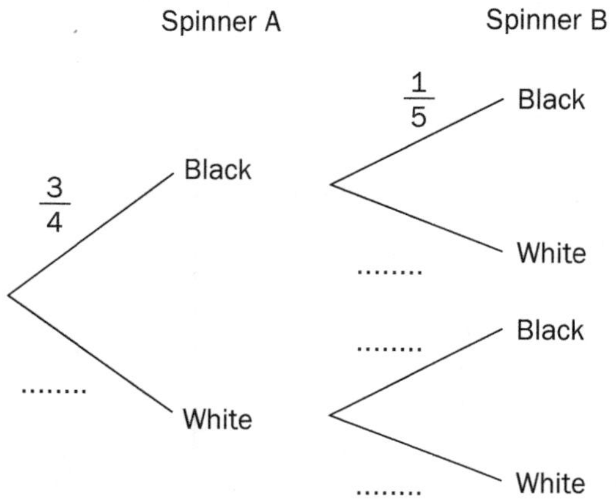

[2]

(b) Calculate the probability of spinning white on spinner A and black on spinner B.

.... [2]

19 The *n*th term of a sequence is $2n + b$

The third term of the sequence is 11

(a) Work out the value of *b*.

$b =$........................... [1]

(b) Write down the first four terms of the sequence.

.... [1]

For Teacher's Use

20 Enrique has a large bag of coloured counters.
Counters are either red or blue.
Enrique picks a counter from the bag at random, notes the colour and then replaces the counter.
He repeats this 20 times.
He continues to repeat this in batches of 20.
He draws this table to show how the relative frequency of picking a red changes every 20 picks.

Total number of picks	20	40	60	80	100
Number of reds	10	16	27	32	42
Relative frequency of red	0.5	0.4	0.45	0.4	

(a) Complete the table.

[1]

(b) Enrique draws this graph to show how the relative frequency of picking a red changes every 20 picks.

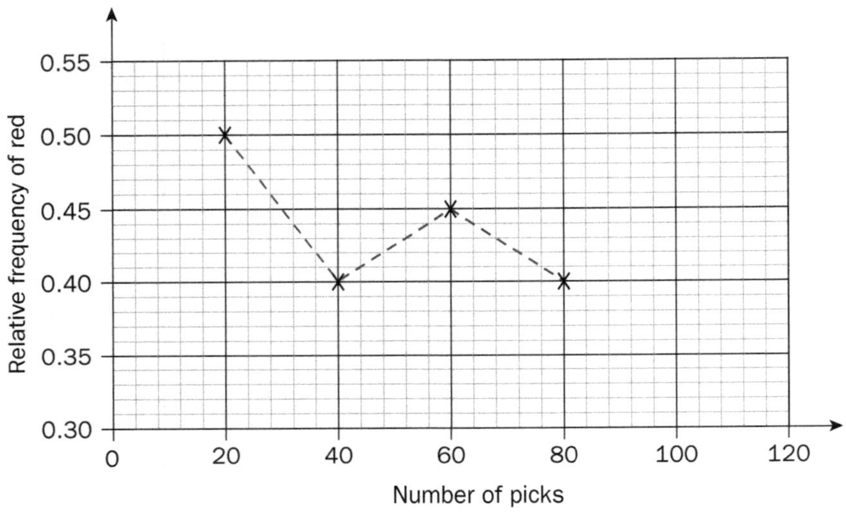

Complete the graph.

[1]

Answers to topic tests

Cambridge Assessment International Education bears no responsibility for the example answers to test-style questions contained in this publication which have been written by the author.

Topic test – Chapters 1–5

1. **(a)** $x = 4$ [1]
(b) $y = {}^{-}2$ [1]

2. $x^2 + 5x - 24$ [2]
If not simplified, 1 mark for
$x^2 - 3x + 8x - 24$

3. $\sqrt{7}$ [1]

4. **(a)** x^{12} [1]
(b) $4x^2$ [1]

5. **(a)** A correct reason, e.g. she only has to take two measurements so this will be quicker [1]
(b) A correct reason, e.g. she can see how her resting heart rate changes over time [1]
(c) A correct reason, e.g. her sample may not be all women aged 20; her sample may not have all followed the same training. [1]

6. **(a)** 4.2×10^8 [1]
(b) 7.13×10^{-3} [1]

7. $\frac{3}{8m}$ [2]
If answer incorrect, 1 mark for $\frac{7}{8m} - \frac{4}{8m}$

8. 1 [1]

9. 4 [1]

10. **(a)** 5.3 [3]
1 mark for correct midpoints
(3, 5, 7, 9, 11) implied by 24, 25, 28, 18 and 11 seen or 106 seen
1 mark for (sum of their midpoints $\times f$) $\div$ 20
(midpoints must be within range including each end of ranges)
(b) $4 \leq t < 6$ [1]
(c) $2 \leq t < 4$ [1]
(d) 10 [1]

11. $(5n)^2 + 2$ or $25n^2 + 2$ [1]

12. She has used one wrong inequality sign (it should be $\leq x <$) [2]
The lower bound is incorrect (it should be 235)
1 mark for one mistake found

13. 20 [2]
1 mark for $100 \times 0.7 = 70$ **and** $({}^{-}5)^2 = 25$
(implied by – 50)

14.

$3^4 \times 3^2 = 9^6$	$7^{-4} \div 7^4 = 7^{-1}$	$6p^6 \div 3p^{-2} = 2p^8$
		✔
$2t^{-2} \times 5t^5 = 10t^3$	$5^0 = 0$	$\frac{6^2}{6^4} = 6^{0.5}$
✔		

[2]
2 marks if two correct ticks and no extra ticks
1 mark if one correct (even if one incorrect tick)

15. $5x + 1$ [1]

16. Correct construction with arcs [1]

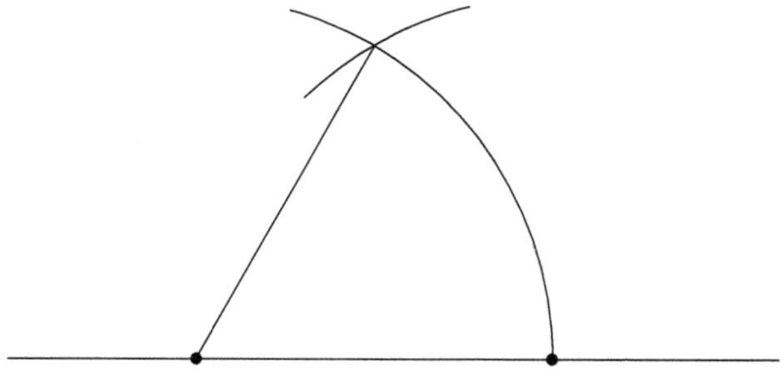

17. 2.1, $\sqrt{7}$, $\sqrt{11}$, 3.9 [1]

18. $p = \frac{hm+5}{2}$ [2]
1 mark for correct first step
(or better) e.g. $hm = 2p - 5$

19. Correct position of city C [2]
Tolerance ± 2° on each bearing.

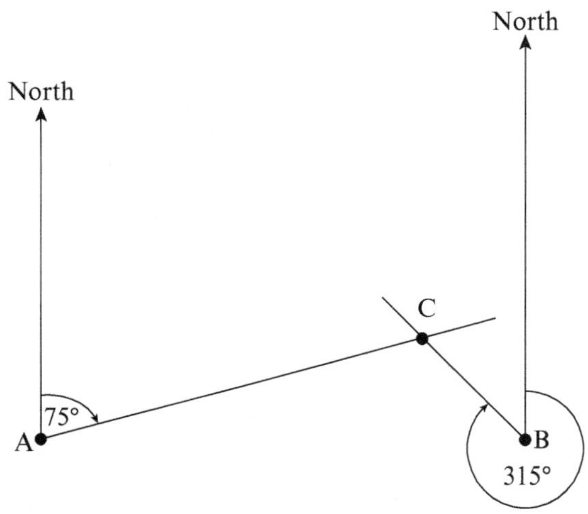

1 mark for one correct bearing in tolerance.

20. **(a)** $2n^2 + 2n + 1$ [3]
1 mark for n^2 and $(n + 1)^2$ seen added
1 mark for a correct unsimplified expression
(b) The expression can be factorised to $2(n^2 + n) + 1$.
$2(n^2 + n)$ is a multiple of 2 and so must be even. Adding 1 to an even number makes it odd. [1]

Topic test – Chapters 6–10

1. 5 nm   5 μm   5 mm   5 cm [1]

2. $0.\dot{7}$ [1]

3. 4 [1]

4. (12, ⁻1) [2]
1 mark for 1 correct coordinate

5. 160 [2]
1 mark for any correct interior angle found for the triangle (may be marked on diagram)

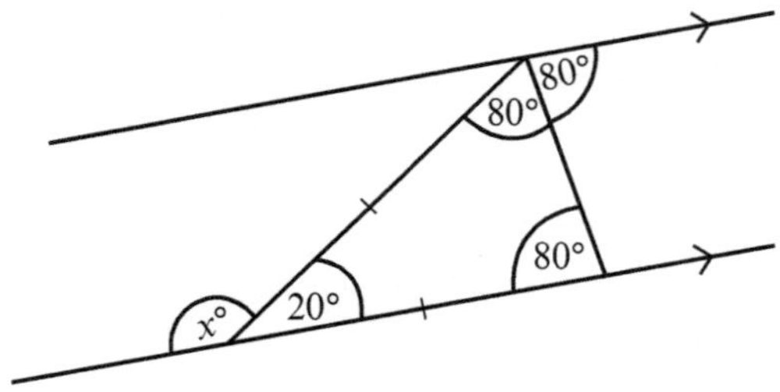

6. (0, 4) [1]

7. $1\frac{2}{3}$ [2]
If answer incorrect, 1 mark for $4\frac{1}{3} \div 2\frac{3}{5}$ seen

8. 25.7 [2]
1 mark for $3.14 \times 10 \div 2$ or 15.7 seen
Allow π or more accurate value in place of 3.14

9. $x = 5, y = 4$ [2]

10. $x^2 = 10^2 - 6^2$ for 1 method mark
$x = \sqrt{64} = 8$ [2]

11. (a) The numbers are not in order; there is no key. [2]
(b) History is easier: more students got higher marks. [1]

12. (a) 14 [1]
(b)

Group B		
Mark out of 50	Number of students	Angle on pie chart
0–9	9	108°
10–19	8	96°
20–29	5	60°
30–39	4	48°
40–50	4	48°
Total	30	360°

[2]

(c)

Marks out of 50 in a test - group B

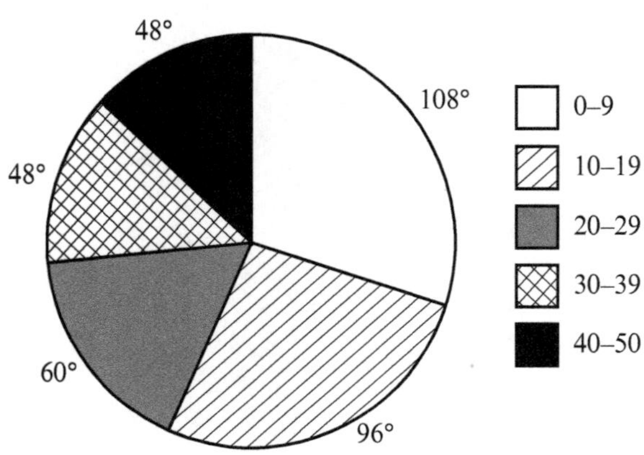

[2]

(d) Group A did revision work because a higher percentage of students got higher marks. For example, nearly three quarters got a mark between 30 and 50, whereas in Group B only just over one quarter got a mark between 30 and 50. [1]

13. (a) $4x + 10 \geq x + 25$
$3x + 10 \geq 25$ 1 method mark for correct first step (may be subtracting 10)
$3x \geq 15$
$x \geq 5$ [2]

(b)

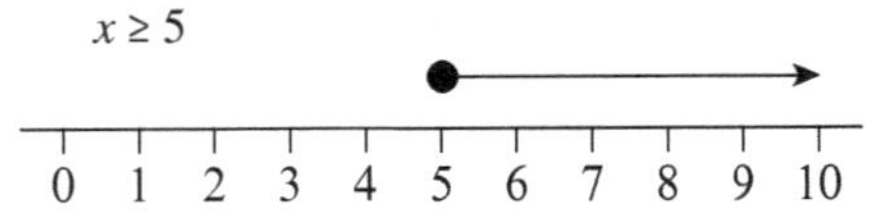

[1]

(c) 5 [1]

14. (a)

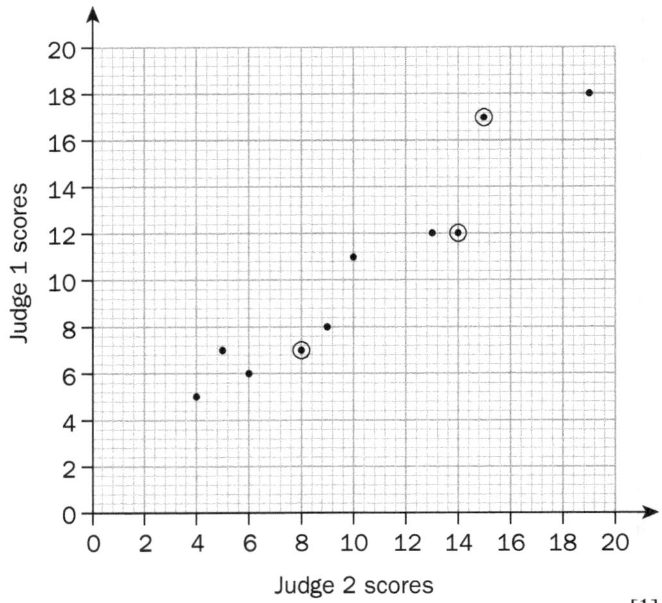

[1]

(b) There is a positive correlation.
The correlation is strong.
Judges are in agreement (contextual interpretation).
Any two statements [2]

15. (a) Angles in a pentagon sum to $180 \times 3 = 540°$ for 1 method mark
139° [2]
(b) 72° [1]

16.

Hexagon	Quadrilateral	Heptagon	Triangle
✔	✔		✔

[1]

17. 1 mark for $10 \times 10 \times 10$ [3]
1 mark for $3.1 \times 10^2 \times 4$
1 mark for 1240 and 1000

18.

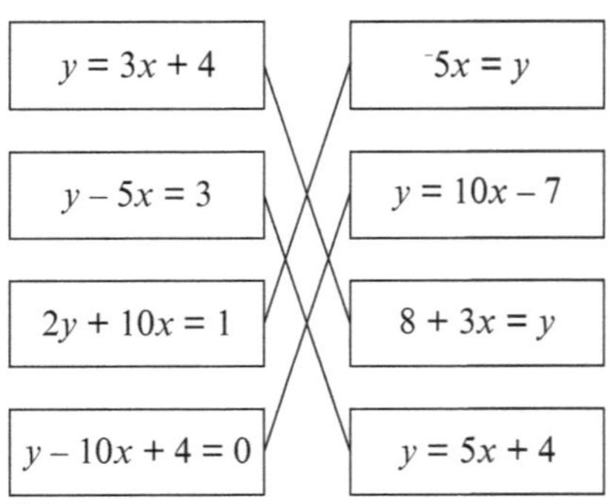

[1]

Topic test – Chapters 11–15

1. 8 [1]

2. 3 [1]

3. $\frac{4}{13}$ [1]

4. 11000×0.93^3 [1]

5. \$540 [2]
1 method mark for $720 \div 4 = 180$

6. y-axis (0, 3) [2]
x-axis (7, 0)
1 mark for one correct or correct intercepts but on incorrect answer lines

7. **(a)** 90 [1]
(b) 18 [1]

8. Enlargement, scale factor 3, centre (0, 3) [3]
0 marks if more than one transformation mentioned.
1 mark for enlargement
1 mark for scale factor 3
1 mark for centre (0, 3)

9. **(a)** 153.72 [1]
(b) 54.9 [1]

10. **(a)** $\frac{40}{200}$ or any equivalent e.g. 0.2, $\frac{4}{20}$, 20% [1]
(b) Blue 45 Red 30 Yellow 75 [2]
1 mark for two correct values

11. 36% [2]
1 mark for 0.8×0.8 or 0.64 seen
Allow *their invented value* $\times 0.8 \times 0.8$ to score 1 mark

12. $x = {}^{-}3$, $y = 4$ both correct for the mark [1]

13. **(a)**

Score on dice	1	2	3	4
Probability	0.25	0.3	**0.1**	0.35

[1]

(b) 0.55 [1]

14.

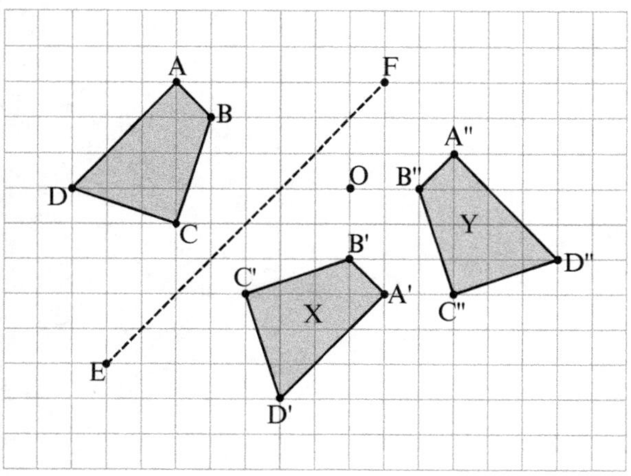

[2]

15. **(a)**

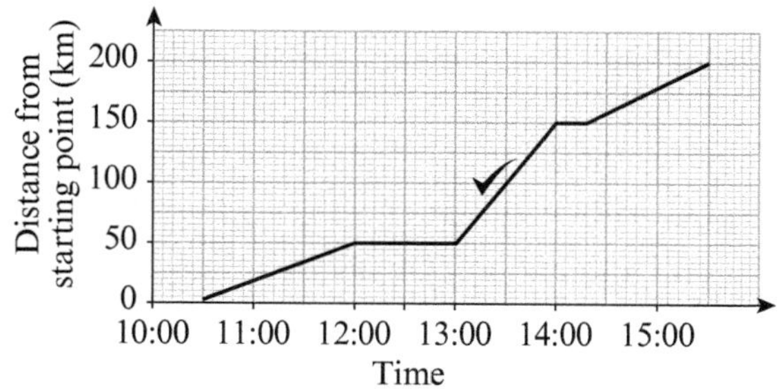

[1]

(b) 100 km/h [1]

16. **(a)**

x	**6**	0	1	**4**	−2
y	−1	**2**	**1.5**	0	3

[2]

1 mark for two out of four correct

(b)

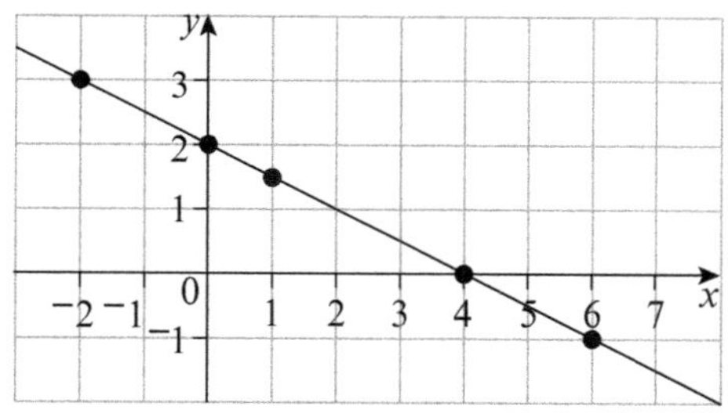

[1]

17.

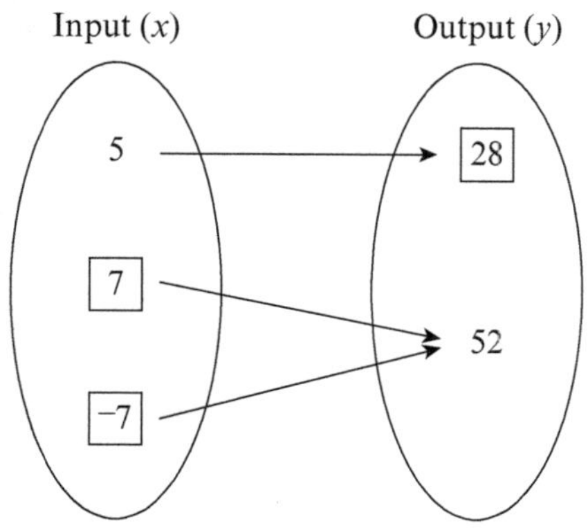

1 mark for 2 correct values [2]

18. **(a)**

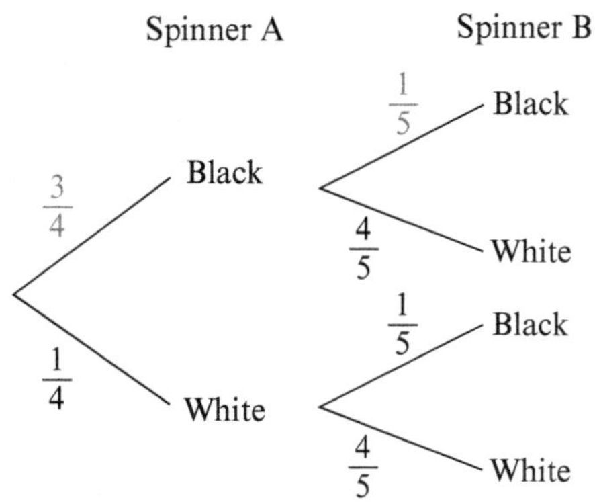

[2]

1 mark for two correct probabilities added to the diagram

(b) $\frac{1}{20}$ or equivalent e.g. 0.05, 5% [2]
1 mark for $\frac{1}{4} \times \frac{1}{5}$

19. **(a)** $b = 5$ [1]
(b) 7, 9, 11, 13 [1]

20. **(a)** 0.42 [1]

(b) point correctly plotted and joined to the previous point [1]

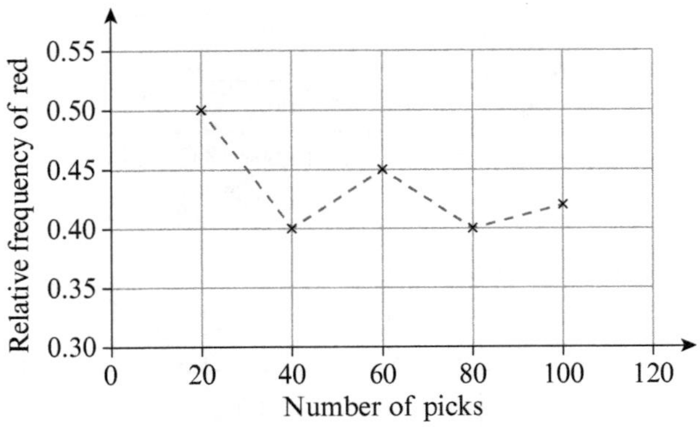

Follow through their 0.42 from the table provided it will fit on the graph.

Progression test-style paper

Practice test (non calculator)

Mathematics Paper 1 test-style questions

Stage 9

Name ....

Additional material:

- ruler
- compasses

Calculators are not allowed

READ THESE INSTRUCTIONS FIRST

Answer all questions in the spaces provided on the question paper.

You should show all your working on the question paper.

The number of marks is given in brackets [ ] at the end of each question or part question.

The total number of marks for this paper is 50.

Note: this test has been developed by the author, therefore does not necessarily match the Cambridge Progression tests.

For Teacher's Use

1 Solve

$\frac{36}{t} = 9$

$t =$............................[1]

2 The diagram shows a prism.

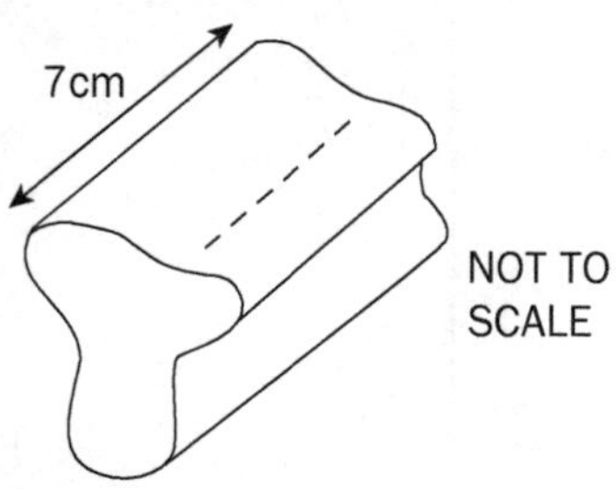

The area of the cross-section of this prism is 20 cm^2.
Work out the volume of the prism.

.... cm^3 [1]

For Teacher's Use

3 Draw a ring around **all** the rational numbers in this list.

π $0.\dot{6}$ $\sqrt{7}$ $\frac{3}{5}$ $\sqrt{125}$ 0.52

[2]

4 $2^{-3} = \frac{1}{k}$ where k is an integer.

Find the value of k.

k = ... [1]

5 Tick (✔) the calculations that have an answer bigger than 5.

$5 \div 0.06$	$5 \div 2.47$	5×0.86	5×1.43
☐	☐	☐	☐

[1]

6 Complete this calculation.

$$1.25 \times 4 \times 12 \times (1 + 0.25) = \frac{\square}{4} \times 48 \times \frac{\square}{4}$$

$$= \square \times 3 \times \square$$

$$= \square$$

[2]

7 Jenny thinks all fractions with a denominator of 12 are equivalent to recurring decimals.
She gives the example $\frac{5}{12} = 0.41\dot{6}$.
Write down an example to show that Jenny is not correct.

.... [1]

8 Will a regular pentagon tessellate? Yes ☐ No ☐

Explain how you know.

....

....[1]

9 Look at the graph showing the line $y = 4 - x$

For Teacher's Use

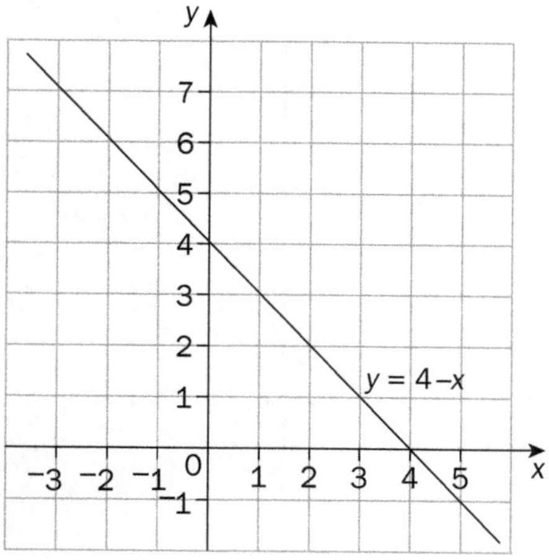

(a) Complete this table of values for the equation $y - 2x = 1$

x	0	1	2	3
y	1			7

[1]

(b) On the graph above draw the line $y - 2x = 1$ [1]

(c) Using your graph, solve the simultaneous equations

$y - 2x = 1$

$y = 4 - x$

$x =$.......................

$y =$....................... [1]

(d) What is the gradient of the line $y = 4 - x$?

.... [1]

For Teacher's Use

10 Here is a formula.

$a^2 + b^2 = c^2$

Rearrange the formula to make b the subject.

$b =$ .... [2]

11 Look at the diagram showing the transformation of shape P onto P′.

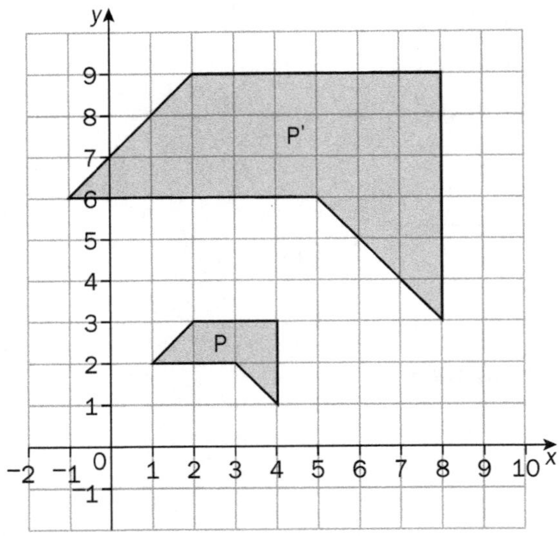

Describe fully the single transformation mapping shape P onto P′.

....

....[3]

12 Look at this calculation.

$27 \times 463 = 12\,501$

Work out $1250.1 \div 4.63$

.... [1]

For Teacher's Use

13 Tick (✔) which expressions are equivalent to $8x^6$.

$(2x^2)^3$ ☐

$16x^{12} \div 2x^2$ ☐

$4x^3 \times 4x^3$ ☐

$\frac{80x^{10}}{10x^4}$ ☐

[1]

14 Work out

(a) 2380×10^{-3}

.... [1]

(b) $245 \div 10^{-2}$

.... [1]

15 It takes 12 machines 10 minutes to make 400 cakes.

Find the number of minutes it would take 15 of the machines to make the 400 cakes.

.... min [1]

For Teacher's Use

16 Calculate the following, giving each answer as a mixed number in its simplest form.

(a) $2\frac{2}{5} \div \frac{24}{25}$

..............................[2]

(b) $\left(1-\frac{1}{4}\right)\times\left(1+\frac{5}{6}\right)$

..............................[3]

17 Match the equivalent expressions by drawing lines.
Some values have been rounded to 1 d.p.
One has been done for you.

$\sqrt[3]{68}$	2
$\sqrt[3]{8}$	3.1
$\sqrt[3]{15}$	5.2
$\sqrt[3]{30}$	2.5
$\sqrt[3]{140}$	4.1

[1]

For Teacher's Use

18 Stefan collects this information about the heights jumped by a group of children from each of two different clubs.

	Club A	Club B
Mean height	1.5 metres	1.1 metres
Range of heights	25 cm	35 cm

Stefan writes these comparisons.

1. The children in club A have a larger mean than those in club B.

2. The children in club B have a larger range than those in club A.

Write improved comparisons of the heights jumped by the children in each club.

Give your answers in context.

1. ....

....

2. ....

....[2]

19 Write each of these numbers in standard form.

350 000 = ...

0.0021 = ...

[2]

20 Tick (✔) to show whether each of the statements about the graph of $y = x^2 - 4$ is true or false.

	True	False
The graph of $y = x^2 - 4$ is a curve.	☐	☐
The graph of $y = x^2 - 4$ passes through the point $(^-4, 0)$	☐	☐
The graph of $y = x^2 - 4$ passes through the point $(^-3, ^-13)$	☐	☐

[1]

For Teacher's Use

21 Here is quadrilateral A.

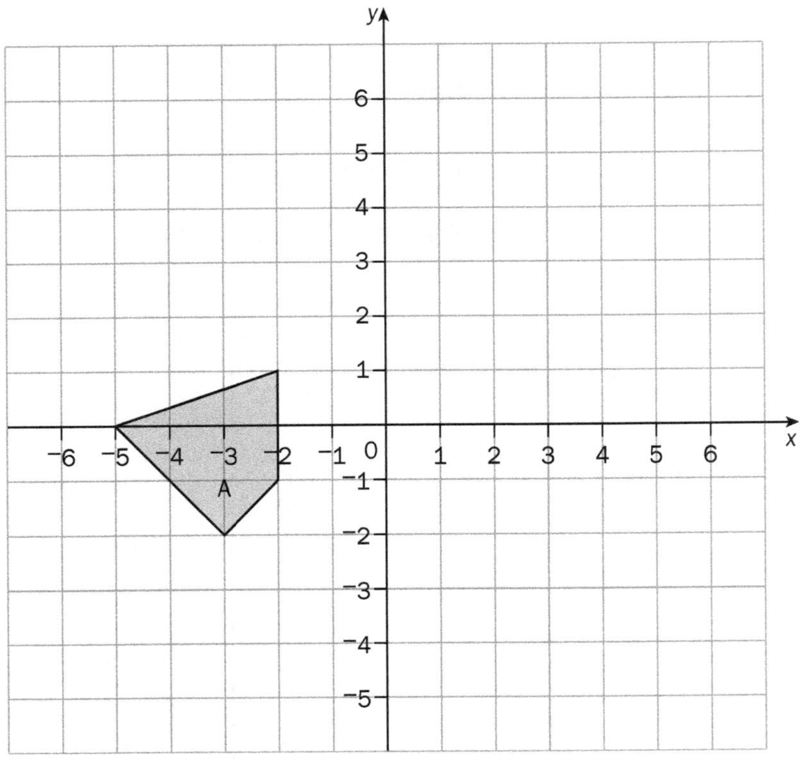

(a) Reflect quadrilateral A in the line $y = x$. Label it B.

Then reflect quadrilateral B in the line $y = {}^{-}x$. Label it C. [2]

(b) Describe fully the **single** transformation that maps quadrilateral A onto quadrilateral C.

....

.... [2]

22 A bag contains cards, each card with A or B or C or D written on it.

The probabilities of choosing at random a card marked with a particular letter are:

P(A) = 0.2, P(B) = 0.15, P(C) = 0.3

Find the probability of choosing a card with C or D written on it.

..................................[2]

For Teacher's Use

23 Expand the brackets and simplify

$(x - 4)(x - 3)$

..................................[2]

24 A is the point (3, 2) and B is the point (19, $^-18$).

Find the coordinates of the point that is $\frac{1}{4}$ of the way along AB from A.

(.............. , ...) [2]

25 Construct an inscribed regular hexagon with side length 5 cm inside a circle.

The radius of your circle should be 5 cm.

[2]

For Teacher's Use

26 Look at the scatter graphs below.
Some of the labels are missing.

(a) Complete this scatter graph by choosing a possible label for the *y*-axis.

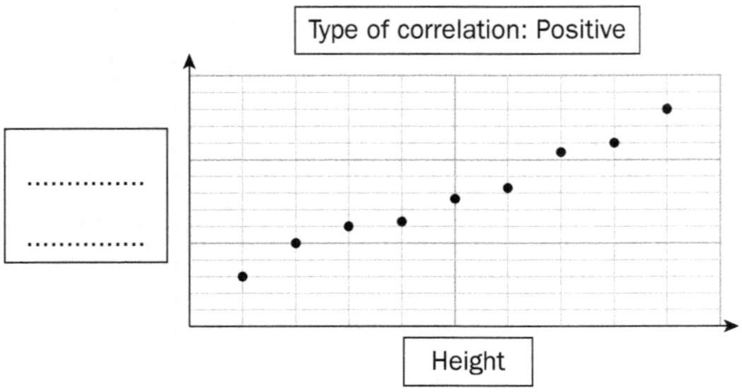

[1]

(b) Complete this scatter graph by describing the correlation.

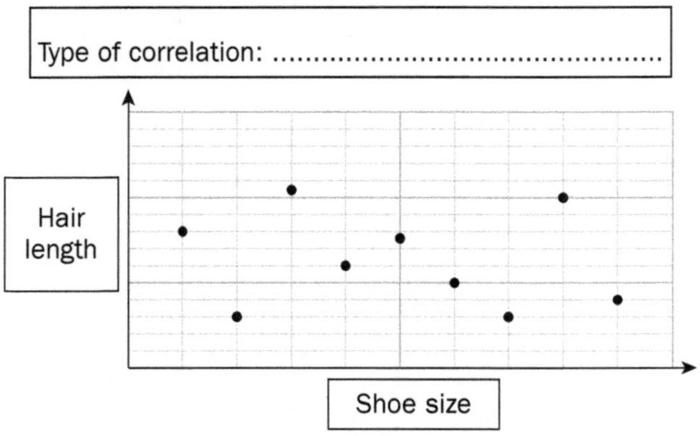

[1]

(c) Complete this scatter graph by drawing some possible points on the graph.

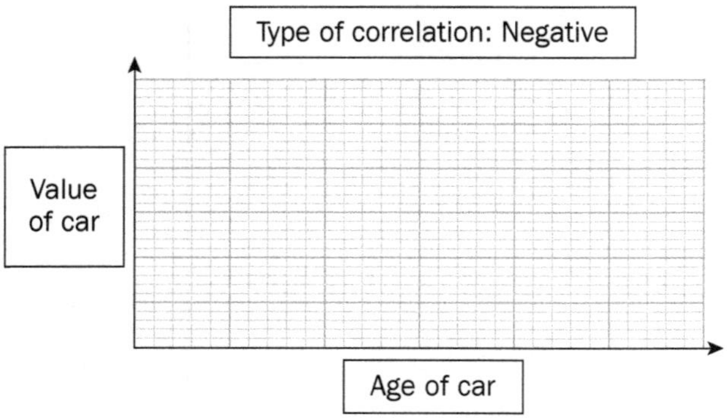

[1]

Mark scheme for paper 1

Cambridge Assessment International Education bears no responsibility for the example answers to test-style questions contained in this publication, which have been developed by the author.

M marks are method marks

A marks are accuracy/answer marks

B marks are independent marks

Question	Marks	Answer	Notes
1	1	4 [B1]	
2	1	140 [B1]	
3	2	$0.\dot{6}$, $\frac{3}{5}$ and 0.52 [B2]	Accept any clear indication. B1 for 2 correct and no incorrect or 3 correct with 1 incorrect.
4	1	8 [B1]	
5	1	5 ÷ 0.06 and 5 × 1.43 [B1]	Must have both; any extra ticked no marks.
6	2	$\frac{5}{4} \times 48 \times \frac{5}{4}$ $5 \times 3 \times 5$ [M1] **75** [A1]	
7	1	Any fraction with denominator 12 and numerator that is a multiple of 3. [B1]	E.g. $\frac{3}{12}, \frac{6}{12}, \frac{15}{12}$, etc. Accept mixed numbers, e.g. $7\frac{9}{12}$
8	1	No, because the interior angle of a regular pentagon is 108° and 360° is not a multiple of 108°. [B1]	Require decision 'no' and a correct reason. Must see 108° as part of the reason.
9(a)	1	<table><tr><td>x</td><td>0</td><td>1</td><td>2</td><td>3</td></tr><tr><td>y</td><td>1</td><td>**3**</td><td>**5**</td><td>7</td></tr></table> [B1]	Both need to be correct for the mark.
9(b)	1	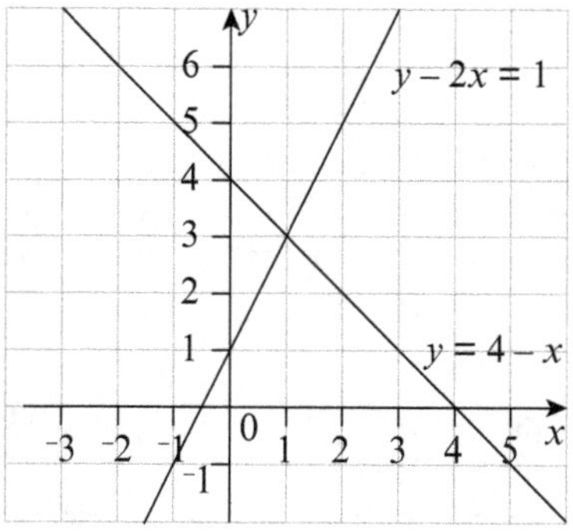 [B1]	Line must be ruled.
9(c)	1	$x = 1, y = 3$ [B1]	Both need to be correct for the mark.
9(d)	1	$^-1$ [B1]	
10	2	$b^2 = c^2 - a^2$ [M1] $b = \sqrt{c^2 - a^2}$ [A1]	M1 implied by the correct answer.

11	3	Enlargement [B1] Scale factor 3 [B1] Centre (2, 0) [B1]	
12	1	270 [B1]	
13	1	$(2x^2)^3$ and $\dfrac{80x^{10}}{10x^4}$ [B1]	No marks if more than these two ticked.
14(a)	1	2.38 [B1]	
14(b)	1	24 500 [B1]	
15	1	8 [B1]	
16(a)	2	$\dfrac{12}{5} \times \dfrac{25}{24}$ [M1] $2\dfrac{1}{2}$ [A1]	Need conversion to improper fraction **and** changing to multiply **and** the reciprocal fraction for the method mark. M1 implied by the correct answer. For the A1 mark, must be a mixed number in its simplest form.
16(b)	3	$\dfrac{3}{4} \times \dfrac{11}{6}$ or equivalent [M1] $\dfrac{1}{4} \times \dfrac{11}{2}$ or $\dfrac{33}{24}$ or equivalent [M1] $1\dfrac{3}{8}$ [A1]	Accept $\dfrac{3}{4} \times 1\dfrac{5}{6}$ with an attempt to convert $1\dfrac{5}{6}$ to an improper fraction For the A1 mark, must be a mixed number in its simplest form.
17	1	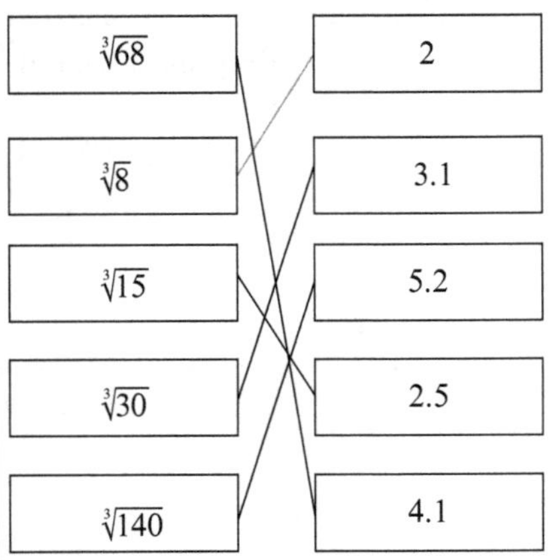 [B1]	All four matches need to be correct for the mark.
18	2	The children in club A jump higher on average than those in club B. [B1] The children in club B are less consistent in their jumping heights than the children in club A. [B1]	Must be correct comparisons in context for the mark.
19	2	3.5×10^5 [B1] 2.1×10^{-3} [B1]	
20	1	True False False [B1]	All correct for the mark.

21(a)	2	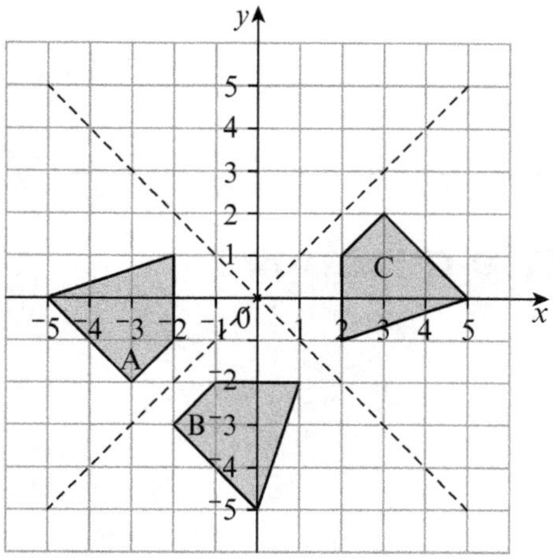 Correct position of B [B1] Correct position of C or correct reflection of their B in $y = -x$ [B1]	
21(b)	2	Rotation [B1] 180° about the origin [B1]	Accept origin or (0, 0) Follow through their quadrilateral C provided a single transformation is possible.
22	2	$1 - (0.2 + 0.15)$ [M1] 0.65 [A1]	Accept alternative equivalent methods.
23	2	$x^2 - 7x + 12$ [B2]	1 mark for 3 terms correct out of these four $x^2 - 3x - 4x + 12$, where $-7x$ counts as two terms correct.
24	2	$(7, {}^-3)$ [B2]	B1 for one correct coordinate.
25	2	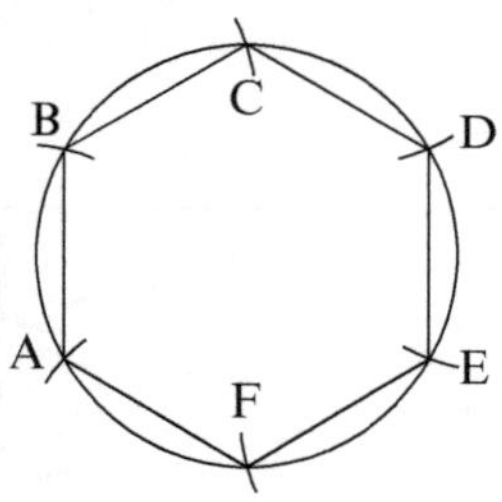 Construction arcs [M1] Accurate inscribed hexagon [A1]	M1 for an attempt at arcs at equal intervals at the points A, B, C, D, E, F (A, B, C, D, E, F do not need labelling). For this mark do not penalise an incorrect radius or the accuracy. A1 for an accurate regular hexagon side length 5cm ± 2 mm and angles all 120° ± 2°.
26(a)	1	Anything that might increase as height increases, e.g. mass, arm length, age [B1]	Must be numerical measure where positive correlation is reasonable.
26(b)	1	No (correlation) or none [B1]	
26(c)	1	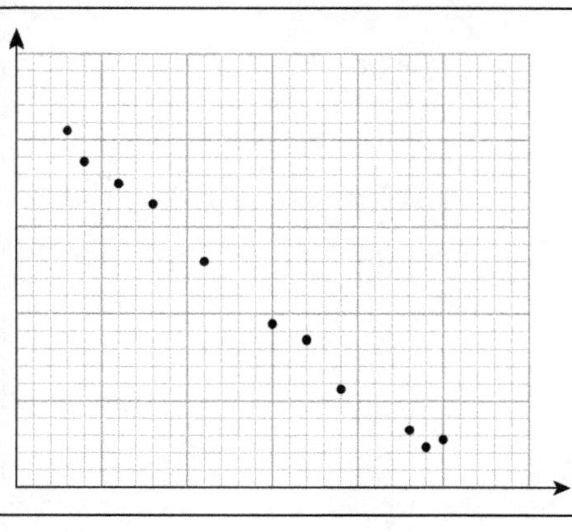 [B1]	This is just an example. Any points showing negative correlation, strength of correlation unimportant. Must be more than 4 points.

Progression test-style paper

Practice test (calculator)

Mathematics Paper 2 test-style questions

Stage 9

Name ....

Additional material:

- ruler
- compasses
- calculator

READ THESE INSTRUCTIONS FIRST

Answer all questions in the spaces provided on the question paper.

You should show all your working on the question paper.

The number of marks is given in brackets [ ] at the end of each question or part question.

The total number of marks for this paper is 50.

Note: this test has been developed by the author, therefore does not necessarily match the Cambridge Progression tests.

For Teacher's Use

1 Work out the value of $\frac{n^4}{24} + 5$ when $n = 6$

..........................[1]

2 Calculate $\sqrt{6} \times \sqrt{54}$

..........................[1]

3 A sequence has nth term $n^2 - 3$

Work out the value of the 5th term.

.... [1]

For Teacher's Use

4 The diagram shows a 60° angle.

Use this angle to construct a 30° angle.

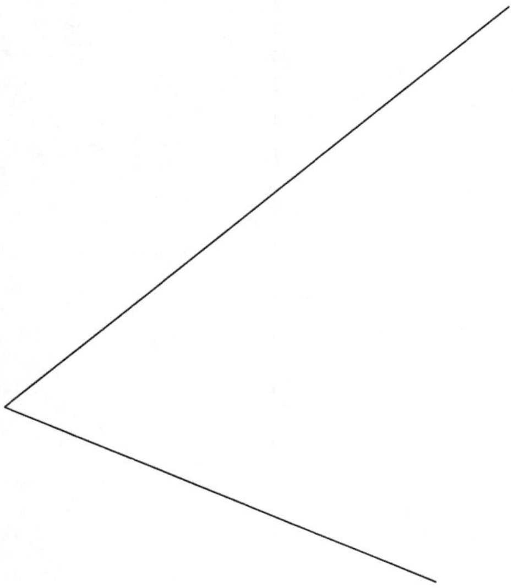

[2]

5 Here are some sketches of four graphs.
Tick (✔) the sketch that could be the graph of $y = {}^{-}2x - 3$.

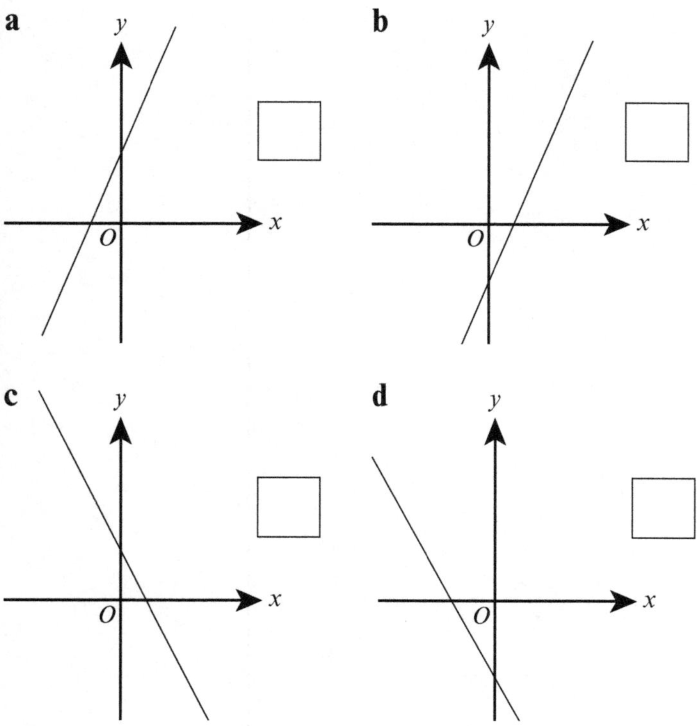

[1]

For Teacher's Use

6 The diagram shows a pentagon with three of its sides extended.

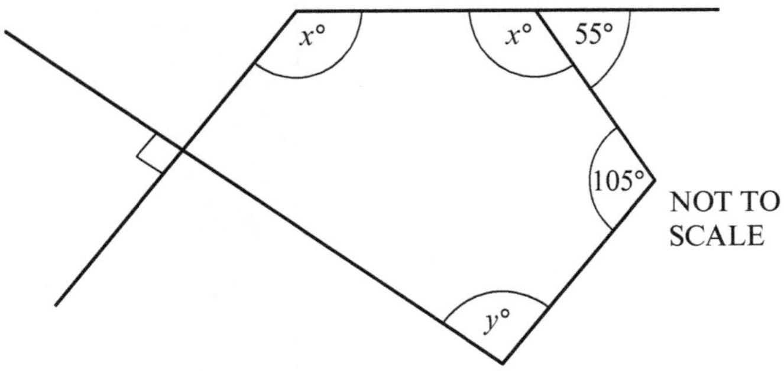

Work out the value of x and the value of y.

x = ...

y = ... [3]

7 Work out the surface area of this square based pyramid.

The top vertex is vertically above the centre of the base.

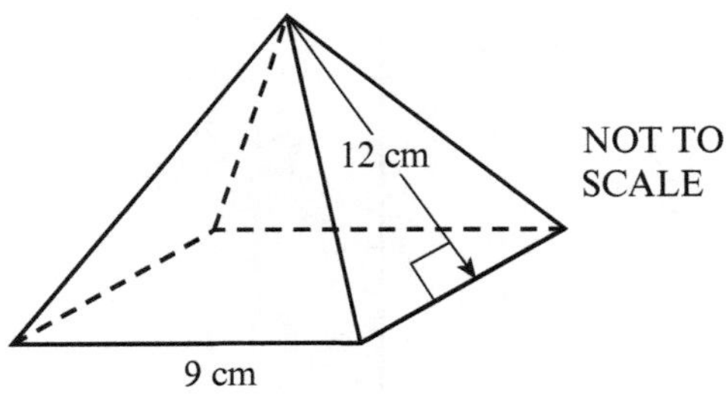

.... cm^2 [3]

For Teacher's Use

8 A number rounded correct to the nearest 10 is 5200.

Sami thinks that the number lies within the bounds $5150 \leq n < 5250$

Tick (✔) to show if Sami is correct or not correct.

Explain your answer.

Correct ☐ Not correct ☐

Because...

.... [1]

9 A ship travels from A to B on a bearing of 210° for 120 km.

(a) Draw a scale drawing of the ship's journey.

Use the scale 1 cm = 20 km

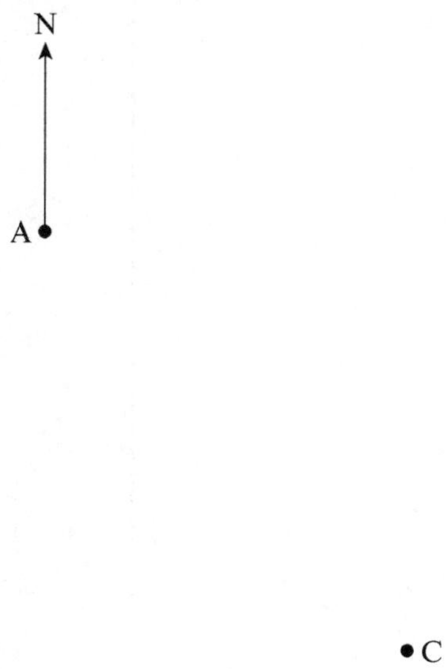

[2]

(b) Use the scale drawing to work out how many kilometres the ship is from the lighthouse at point C.

.... km [1]

For Teacher's Use

10 **(a)** Work out the volume of this cylinder.

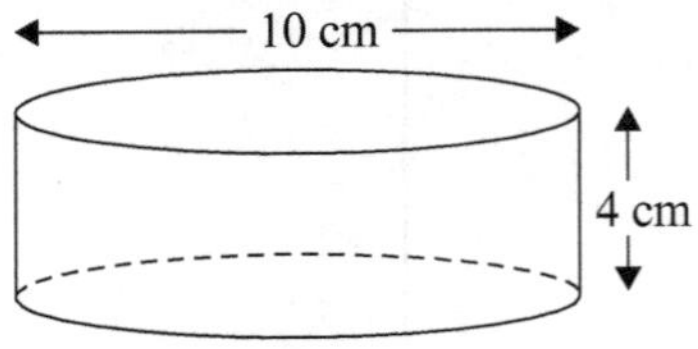

.... cm^3 [2]

(b) A different cylinder has a volume that is double the volume of the cylinder in part **(a)**.

This cylinder also has a height of 4 cm.

What is the radius of the circular cross-section of this cylinder?

.... cm [2]

For Teacher's Use

11 A rectangle has area $145\,cm^2$.

One of the sides of the rectangle is $(x - 8)$ cm long.
The other side is 5 cm.

(a) Draw a ring around the correct equation to represent this.

$\frac{145}{x-8} = 5$ $\quad$ $\frac{5}{x-8} = 145$ $\quad$ $\frac{x-8}{145} = 5$ $\quad$ $\frac{x-8}{5} = 145$

[1]

(b) Work out the value of x.

$x =$ .... [2]

12 Tick (✔) to show if each of these conversions is true or false.

	True	False
50 000 000 ns = 0.05 s	☐	☐
4000 µm = 0.4 m	☐	☐
0.3 t = 300 kg	☐	☐
500 MB = 0.5 GB	☐	☐
1 light year = 9 800 km	☐	☐

[2]

For Teacher's Use

13 Jamil buys a motorcycle for \$8000.

The value of the motorcycle decreases at a rate of 5% each year for the next 3 years.

Calculate the value of the motorcycle after 3 years.

\$.................... [2]

14 Sanjit and Anya have some money.

They share the money between them in the ratio 3 : 7.

Anya gets \$380 more than Sanjit.

What is the total amount of money they share?

\$....................[2]

15 The table shows the heights of 50 plants in centimetres.

For Teacher's Use

Height (cm)	Frequency
$0 \le h < 5$	4
$5 \le h < 10$	7
$10 \le h < 15$	12
$15 \le h < 20$	16
$20 \le h < 25$	8
$25 \le h < 30$	3

(a) Work out an estimate of the mean plant height.

.... cm [3]

(b) Find the modal class interval.

.... [1]

16 Here is an inequality.

$-2 < x + 2 \le 3$

(a) Solve the inequality.

.... [2]

(b) Represent the solution to the inequality on the number line.

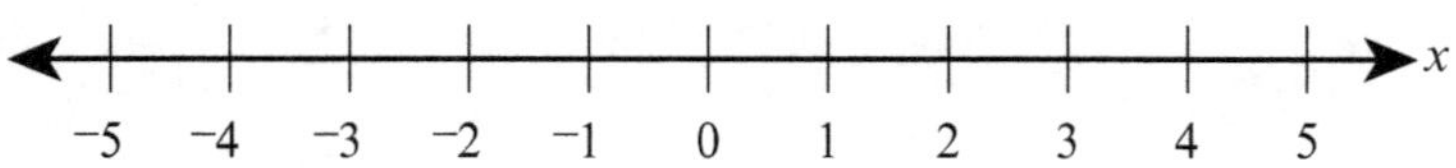

[1]

For Teacher's Use

17 Solve these simultaneous equations.

$4x + 2y = 14$

$3x - y = 18$

$x =$ ....

$y =$ .... [3]

18 Simplify

(a) $\frac{9x^9}{3x^3}$

.... [1]

(b) $10m^6 \div 5m^2$

.... [1]

(c) $\frac{15y^5 + 5y^5}{2y^2}$

.... [1]

19 Calculate the exterior angle of a regular octagon.

.... ° [1]

For Teacher's Use

20 A circle has circumference 37.7 cm.
What is the area of the circle?
Give your answer correct to 3 s.f.

.... cm^2 [3]

21 Write as a single fraction in its simplest form

$\frac{5}{p} + \frac{2x}{3m}$

.... [2]

22 Triangle ABC has been drawn on a square grid.

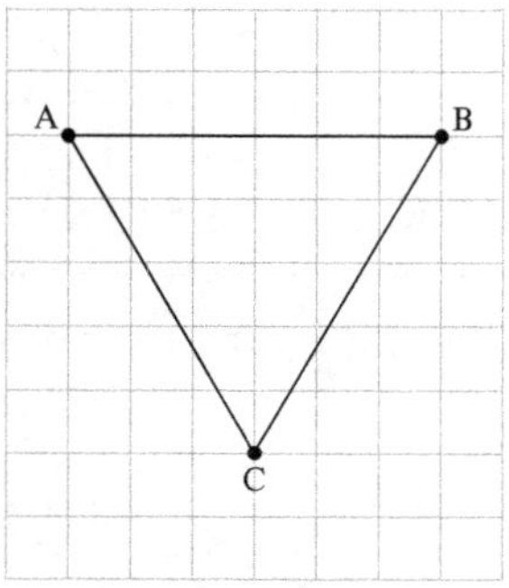

Is the triangle equilateral?

Use Pythagoras' theorem to explain how you know.

....

....

....

....

.... [2]

Mark scheme for paper 2

Cambridge Assessment International Education bears no responsibility for the example answers to test-style questions contained in this publication, which have been developed by the author.

M marks are method marks

A marks are accuracy/answer marks

B marks are independent marks

Question	Marks	Answer	Notes
1	1	59 [B1]	
2	1	18 [B1]	
3	1	22 [B1]	
4	2	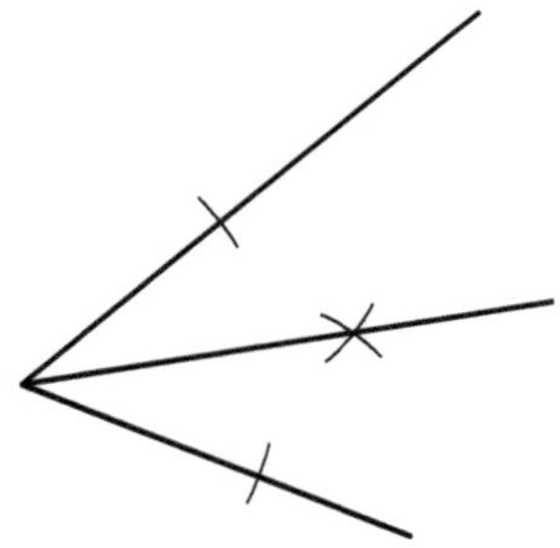 Correct arcs [M1] Correct angle of 30° drawn [A1]	No marks for 30° angle without angle bisector.
5	1	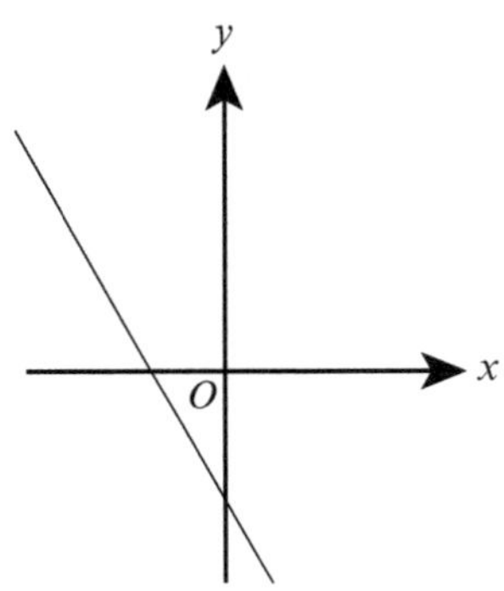 [B1]	Accept any clear indication.
6	3	$x = 125$ [B1] $y = 95$ [B2]	If B2 not scored then B1 for 540 seen.
7	3	$\frac{1}{2}\times 9\times 12$ or better [M1] $9^2 + 4 \times$ *their* triangle area [M1] 297 [A1]	Or better, e.g. 54, $4\times\frac{1}{2}\times 9\times 12$
8	1	Ticks not correct and gives a correct reason e.g. • these are the bounds for rounding to the nearest 100 • the correct bounds are $5195 \le n < 5205$ [B1]	Accept any correct explanation.

<table>
<tr><td>9(a)</td><td>2</td><td>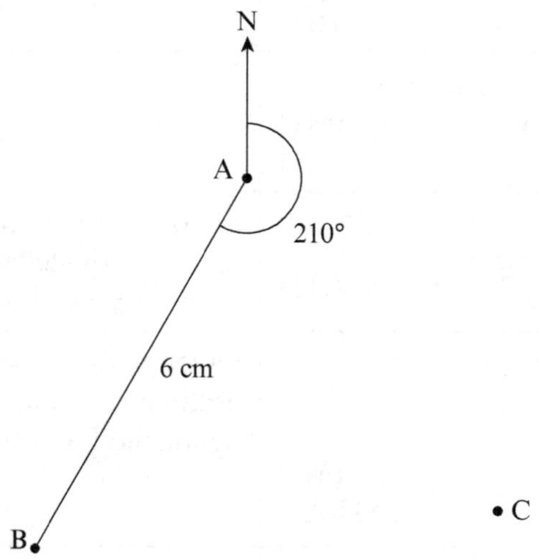

[B2]</td><td>B1 for bearing 210°; allow ± 2°; must be clockwise from north line at A.

B1 for 6 cm line ±2 mm; allow at any angle as the mark is for correct use of the scale.
Condone point B not labelled, assume the end of their drawn line for point B if not labelled.</td></tr>
<tr><td>9(b)</td><td>1</td><td>140 [B1]</td><td>Do not award a mark for 7 cm; it must be converted to km.</td></tr>
<tr><td>10(a)</td><td>2</td><td>$\pi \times 5^2 \times 4$ [M1]
314 [A1]</td><td>Allow anything which rounds to 314.</td></tr>
<tr><td>10(b)</td><td>2</td><td>$314 \times 2 = \pi r^2 \times 4$ [M1]
$r = 7.1$ cm [A1]</td><td>Follow through answer to part (a) for the method mark.
Allow answers rounding to 7 cm, provided there is supporting working.</td></tr>
<tr><td>11(a)</td><td>1</td><td>$\frac{145}{x-8} = 5$ [B1]</td><td></td></tr>
<tr><td>11(b)</td><td>2</td><td>$145 = 5(x-8)$ or better
or correct starting point to solve their chosen equation in (a) [M1]
37 [A1]</td><td>Or better, e.g. $\frac{145}{5} = x - 8$, $29 = x - 8$</td></tr>
<tr><td>12</td><td>2</td><td>50 000 000 ns = 0.05 s True
4000 μm = 0.4 m False
0.3 t = 300 kg True
500 MB = 0.5 GB True
1 light year = 9 800 km False [B2]</td><td>B1 for 3 correct answers.</td></tr>
<tr><td>13</td><td>2</td><td>8000×0.95^3 [M1]
$6859 [A1]</td><td>Accept equivalent methods.</td></tr>
<tr><td>14</td><td>2</td><td>380 ÷ (7 – 3) = 95 [M1]
95 × 10 = $950 [A1]</td><td></td></tr>
<tr><td>15(a)</td><td>3</td><td><table>
<tr><th>x</th><th>f</th><th>xf</th></tr>
<tr><td>2.5</td><td>4</td><td>10</td></tr>
<tr><td>7.5</td><td>7</td><td>52.5</td></tr>
<tr><td>12.5</td><td>12</td><td>150</td></tr>
<tr><td>17.5</td><td>16</td><td>280</td></tr>
<tr><td>22.5</td><td>8</td><td>180</td></tr>
<tr><td>27.5</td><td>3</td><td>82.5</td></tr>
</table>
[M1]
[M1]
$\frac{\Sigma xf}{\Sigma f} = \frac{755}{50} = 15.1$ [A1]</td><td>M1 for correct midpoints (must be at least 5 correct) implied by 5 correct xf values or 755 seen.

M1 for (sum of their midpoints $\times f$) ÷ 50 (their midpoints must be within range including each end of ranges).</td></tr>
</table>

15(b)	1	$15 \le h < 20$ [B1]	
16(a)	2	$-2-2 < x \le 3-2$ or better with at least the left or right side correct [M1] $^-4 < x \le 1$ [A1]	
16(b)	1	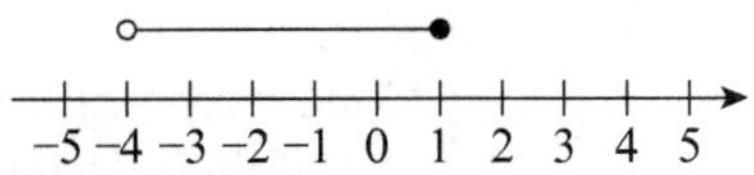 [B1]	Follow through from their inequality in part (a) provided their inequality has both of the signs < and ≤.
17	3	$4x+2y=14$ $+\ 6x-2y=36$ $10x=50$ [M1] $x=5,\ y={}^-3$ [A1][A1]	Other methods are possible; allow one arithmetic slip in the method as long as the method is correct.
18(a)	1	$3x^6$ [B1]	
18(b)	1	$2m^4$ [B1]	
18(c)	1	$10y^3$ [B1]	
19	1	45 [B1]	
20	3	$d=\frac{37.7}{\pi}=12\text{cm}$ $r=6\text{cm}$ [M1] $A=\pi\times 6^2$ [M1] $=113\text{cm}^2$ [A1]	Accept more accurate values for 12 and 6. Must be correct 3 s.f. answer for the final mark. Do not accept other answers.
21	2	$\frac{15m+2px}{3mp}$ [B1][B1]	B1 for the correct numerator and B1 for the correct denominator.
22	2	$5^2+3^2=x^2$ [M1] $\sqrt{5^2+3^2}=5.8$ (1 d.p.) This is not equal to 6. [A1]	For M1 you need only see the 5^2+3^2, students may use any letter or AC^2 or BC^2 or nothing in place of the x^2. Need a decision of 'no' and an explanation for the second mark. The minimum for the explanation is seeing 5.8 (or better accuracy) **and** a statement comparing this to the length of 6, e.g. 5.8 is smaller than side AB.

Worked solutions

The solutions provided here are for some of the more difficult questions from Student Book 9.

Exercise 2C, Question 9 (extension)

(a) $\frac{5(x+3)+4(x-1)}{20}=\frac{5x+15+4x-4}{20}=\frac{9x+11}{20}$

(b) $\frac{7(x-2)-5(2x-1)}{35}=\frac{7x-14-10x+5}{35}=\frac{^{-}3x-9}{35}$

(c) $\frac{4(x+2)+3(x+1)}{(x+1)(x+2)}=\frac{4x+8+3x+3}{(x+1)(x+2)}=\frac{7x+11}{(x+1)(x+2)}$

(d) $\frac{5(x-1)+3(x+3)}{(x-1)(x+3)}=\frac{5x-5+3x+9}{(x-1)(x+3)}=\frac{8x+4}{(x-1)(x+3)}$

(e) $\frac{(x+1)+(x-1)}{(x+1)(x-1)}=\frac{2x}{x^2-1}$

(f) $\frac{6(x-3)-4(x+1)}{(x+1)(x-3)}=\frac{6x-18-4x-4}{(x+1)(x-3)}=\frac{2x-22}{(x+1)(x-3)}$

Exercise 2D, Question 9

$\frac{10x^2+15}{16x^2+24}=\frac{5(2x^2+3)}{8(2x^2+3)}=\frac{5}{8}$ cancelling $2x^2+3$ from the numerator and denominator

$\frac{25x-5x^2}{40x-8x^2}=\frac{5x(5-x)}{8x(5-x)}=\frac{5}{8}$ cancelling $x(5-x)$ from the numerator and denominator

Exercise 2H, Question 1(d)

$(x+4)(x+8)-(x+2)(x+5)$

$=x^2+4x+8x+32-(x^2+2x+5x+10)$

$=x^2+12x+32-(x^2+7x+10)$

$=x^2+12x+32-x^2-7x-10$

$=5x+22$

Exercise 4A, Question 12

(a) $7\times10^5+6\times10^3=700\,000+6000=706\,000$

(b) $8.1\times10^4-25.4\times10^2=81\,000-2540=78\,460$

(c) $12\div10^{-4}-11\times10^3=120\,000-11\,000=109\,000$

(d) $1700\times10^{-2}+0.03\div10^{-3}=17+30=47$

Exercise 4D, Question 11

Upper bound for the length is 9.5 and there are two lengths

Upper bound for the width is 4.5 and there are two widths

$9.5\times2+4.5\times2=28$

Exercise 6C, Question 13

(a) $\sqrt{15^2-12^2}=\sqrt{225-144}=\sqrt{81}=9,\ \frac{1}{2}\times9\times12=54\text{ cm}^2$

(b) $\sqrt{37^2-12^2}=\sqrt{1369-144}=\sqrt{1225}=35,\ \frac{1}{2}\times35\times24=420\text{ mm}^2$

Exercise 6C, Question 15

$\sqrt{(21-5)^2+(14-2)^2}=\sqrt{16^2+12^2}=\sqrt{256+144}=\sqrt{400}=20$

Exercise 6E, Question 9

$8 \times 6 \times 11 = \frac{1}{2} \times x \times (4+7) \times 12$

$528 = 6 \times x \times 11$

$528 = 66x$

$x = 528 \div 66$

$x = 8$

Exercise 7A, Question 7

(a) $\frac{2}{7} \times \frac{5}{3} = \frac{10}{21}$ **(b)** $\frac{2}{5} \times \frac{2}{7} = \frac{4}{35}$

Exercise 7C, Question 1

(a) $5^2 \times 2.65 + 11 = 25 \times \frac{265}{100} + 11 = \frac{265}{4} + 11 = 66\frac{1}{4} + 11 = 77\frac{1}{4}$

(b) $\left(\frac{1}{2} + 5.5\right)^2 - 2\frac{1}{2} = 6^2 - 2\frac{1}{2} = 36 - 2\frac{1}{2} = 33\frac{1}{2}$

(c) $1.5 \times 3.5 \times 2.5 = \frac{3}{2} \times \frac{7}{2} \times \frac{5}{2} = \frac{105}{8} = 13\frac{1}{8}$

(d) $4\frac{2}{9} \times 11 + 4\frac{2}{9} \times 4 = 4\frac{2}{9} \times (11+4) = \frac{38}{9} \times 15 = \frac{38}{3} \times 5 = \frac{190}{3} = 63\frac{1}{3}$

(e) $\left(1\frac{1}{4} + 1.75\right)^3 + \frac{3}{8} = 3^3 + \frac{3}{8} = 27 + \frac{3}{8} = 27\frac{3}{8}$

Exercise 7C, Question 2

(a) $2.63 \times 7 \times \frac{3}{7} \div 3 = 2.63 \times 7 \times \frac{3}{7} \times \frac{1}{3} = 2.63$

(b) $3 \times (7.64 + 1) \times 1\frac{1}{3} \div 4 = 8.64 \times 3 \times \frac{4}{3} \times \frac{1}{4} = 8.64$

Exercise 7C, Question 3

(a) $\left(2\frac{3}{11} + 7\frac{3}{13}\right) - \left(\frac{3}{11} - \frac{1}{13}\right) = 2\frac{3}{11} - \frac{3}{11} + 7\frac{3}{13} + \frac{1}{13} = 2 + 7\frac{4}{13} = 9\frac{4}{13}$

(b) $\left(7.89 + 4\frac{3}{15}\right) - \left(1\frac{1}{15} - 2.11\right)\ 7.89 + 2.11 + 4\frac{3}{15} - 1\frac{1}{15} = 10 + 3\frac{2}{15} = 13\frac{2}{15}$

Exercise 8B, Question 8

(a) $2 = \frac{3x-2}{x-2}$

$2(x-2) = 3x - 2$

$2x - 4 = 3x - 2$

$^-4 = x - 2$

$^-2 = x$

(b) $\frac{4y-5}{2y-1} = 3$

$4y - 5 = 3(2y - 1)$

$4y - 5 = 6y - 3$

$^-5 = 2y - 3$

$^-2 = 2y$

$^-1 = y$

Exercise 8B, Question 9

(a)

$$2 = \frac{2m+10}{m+1} - 1$$

$$3 = \frac{2m+10}{m+1}$$

$$3(m+1) = 2m + 10$$

$$3m + 3 = 2m + 10$$

$$m + 3 = 10$$

$$m = 7$$

(b)

$$\frac{5-2p}{p-1} + 2 = 10$$

$$\frac{5-2p}{p-1} = 8$$

$$5 - 2p = 8(p-1)$$

$$5 - 2p = 8p - 8$$

$$5 = 10p - 8$$

$$13 = 10p$$

$$1.3 = p$$

Exercise 8D, Question 8

(a) $x > 4$ and $x < 8$

5, 6, 7

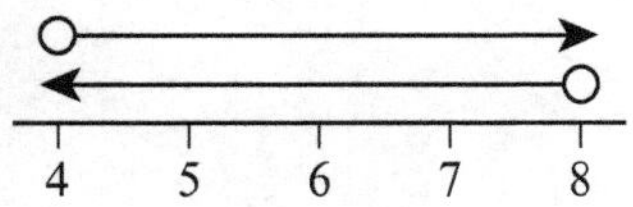

(b) $x > 5$ and $x < 8$

6, 7

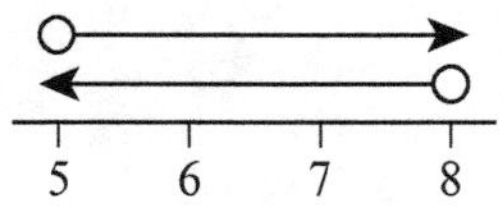

Exercise 8D, Question 9

(a)

$4x + 3 < 55$	$2x - 5 > 13$
$-3]\ \ 4x < 52$	$+5]\ \ 2x > 18$
$\div 4]\ \ x < 13$	$\div 2]\ \ x > 9$

10, 11 and 12

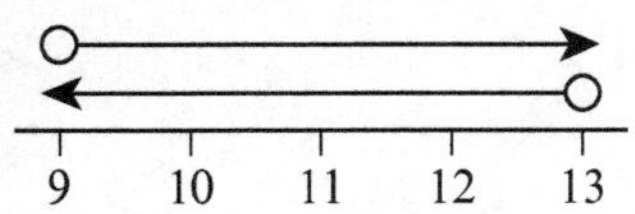

(b)

$3x + 2 > 11$	$5x + 3 < 38$
$-2]\ \ 3x > 9$	$-3]\ \ 5x < 35$
$\div 3]\ \ x > 3$	$\div 5]\ \ x < 7$

4, 5 and 6

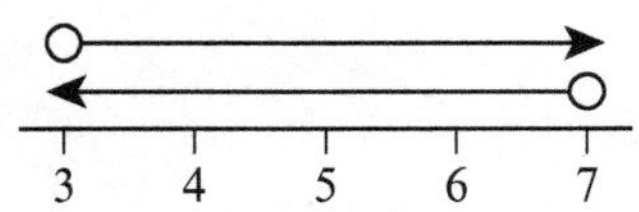

Exercise 8D, Question 10

$3l > 24$

$l > 8$

Length must be greater than 8 cm

Exercise 8D, Question 11

$l < 10$ width $= w$

$2l + 2w = 50$

$l + w = 25$

$l = 25 - w$

$25 - w < 10$

$25 < 10 + w$

$15 < w$

Width must be greater than 15 cm

Exercise 8D, Question 12

$2x + 6 < 24$

$2x < 18$

$x < 9$

Amy is younger than 9

Exercise 8E, Question 11 (extension)

(a) Gradient $\frac{5-2}{1-0}=\frac{3}{1}=3$, intercept at (0,2), so equation $y=3x+2$

(b) Gradient $\frac{8-4}{2-0}=\frac{4}{2}=2$, intercept at (0,4), so equation $y=2x+4$

(c) Gradient $\frac{3-{}^{-}1}{1-0}=\frac{4}{1}=4$, intercept at $(0,{}^{-}1)$, so equation $y=4x-1$

(d) Gradient $\frac{5-{}^{-}3}{2-0}=\frac{8}{2}=4$, intercept at $(0,{}^{-}3)$, so equation $y=4x-3$

Exercise 8G, Question 8

Let x be the width and y be the length of the rectangle

$3x+2y=23$ [1]
$4x+3y=33$ [2]

[1] × 3: $9x+6y=69$ [3]
[2] × 2: $8x+6y=66$ [4]

[3] – [4]: $x=3$
Substitute into [1]:
$9+2y=23$
$2y=14$
$y=7$

Area of rectangle $=3\times 7=21\text{ cm}^2$

Exercise 9B, Question 9

No a regular polygon can't have an interior angle of 110° because the exterior angle would then be 70° and the number of sides would be 360° ÷ 70° = 5.14… which is not possible, as this is not a whole number.

Exercise 9B, Question 10

One possible method is:

Interior angle of equilateral triangle = 60°

Interior angle of regular pentagon = 360° ÷ 5 = 108°

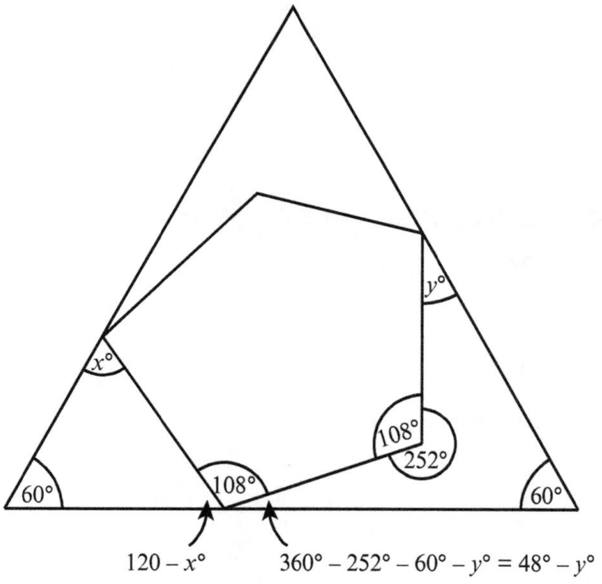

$$120°-x°+108°+48°-y°=180°$$
$$276°-180°=x°+y°$$
$$96°=x°+y°$$

Exercise 11A, Question 5

	flour	:	margarine	:	butter
	4	:	2	:	1
× 200]	800	:	400	:	200

800 g + 400 g + 200 g = 1400 g

Exercise 11A, Question 7

	peach	:	orange	:	mango
	5	:	9	:	4
× 17]	85	:	153	:	68

85 + 153 + 68 = 306 students voted

Exercise 11A, Question 13

If the first number is 1.2

the ratio 5 : 3

× 0.24 × 0.24

becomes 1.2 : 0.72

If the second number is 1.2

the ratio 5 : 3

× 0.4 × 0.4

becomes 2 : 1.2

Exercise 11A, Question 14

Blue : red : yellow

8 : 5

×3 ×3 3 : 2 ×5 ×5

24 : 15 : 10

Blue : yellow is 24 : 10 which simplifies to 12 : 5

Exercise 11D, Question 13

You can feed 80 people for 60 days. You give 60 × 80 = 4800 meals.

For the first 15 days you give 15 × 80 = 1200 meals.

4800 – 1200 = 3600 meals remain

Now there are 80 + 20 = 100 people.

3600 ÷ 100 = 36 days left

36 + 15 = 51 days

Exercise 11D, Question 14

time for 12 goats = time for 16 sheep

time for 1 goat = time for 16 ÷ 12 sheep

18 goats = 18 × (16 ÷ 12) = 24 sheep

18 goats + 4 sheep = 24 + 4 = 28 sheep

42 days = time for 16 sheep

42 × 16 = time for 1 sheep

672 days = time for 1 sheep

672 ÷ 28 = time for 28 sheep

24 days = time for 28 sheep

Consolidation Exercise 11, Question 5

(a) 245 mm : 35 cm = 245 : 350 = 7 : 10

(b) 5.5 ℓ : 8800 ml = 5500 : 8800 = 5 : 8

(c) 6.3 kg : 180 g = 6300 : 180 = 35 : 1

(d) 96 cm² : 4800 mm² = 9600 : 4800 = 2 : 1

Exercise 12G, Question 6 (extension)

gradient of PQ $= \frac{2-1}{7-3} = \frac{1}{4}$

gradient of QR $= \frac{2-6}{7-6} = \frac{^-4}{1} = {^-4}$

gradient of PR $= \frac{6-1}{6-3} = \frac{5}{3}$

$\frac{1}{4} \times {^-4} = {^-1}$ so PQ and QR are perpendicular

$PQ = \sqrt{(2-1)^2 + (7-3)^2} = \sqrt{1^2 + 4^2} = \sqrt{17}$

$PR = \sqrt{(6-3)^2 + (6-1)^2} = \sqrt{3^2 + 5^2} = \sqrt{34}$

$QR = \sqrt{(2-6)^2 + (7-6)^2} = \sqrt{(^-4)^2 + 1^2} = \sqrt{17}$

$\left(\sqrt{17}\right)^2 + \left(\sqrt{17}\right)^2 = \left(\sqrt{34}\right)^2$ so, using Pythagoras' theorem,
PQR is a right-angled triangle and PR is at right angles to PQ.

Exercise 14F, Question 19

5% leaking out gives a multiplier of 0.95 per minute
After n minutes for it to be at least half empty, the multiplier needs to be less than or equal to 0.5.
$0.95n \le 0.5$

Try values of n:
$n = 13$ gives $0.95^{13} = 0.51$ (2 d.p.), too big
$n = 14$ gives $0.95^{14} = 0.49$ (2 d.p.) which is less than 0.5 so the answer is $n = 14$

Consolidation Exercise 14, Question 15

Work out the price after each month and then compare to see which is cheaper:

	Painting A ($)	Painting B ($)
Start	11000	11000
Month 1	9900	10050
Month 2	8910	9100
Month 3	8019	8150
Month 4	7217.1	7200

Exercise 15A, Question 13

		Blue card				
		1	2	3	4	5
Red card	1	1,1	1,2	1,3	1,4	1,5
	2	2,1	2,2	2,3	2,4	2,5
	3	3,1	3,2	3,3	3,4	3,5

Probability of at least one card being even (shown shaded)

$= \frac{9}{15} = \frac{3}{5}$

Exercise 15C, Question 4

P(both plain) $= \frac{5}{8} \times \frac{5}{8}$

P(both striped) $= \frac{3}{8} \times \frac{3}{8}$

If one or the other event happens add the probabilities:

$\frac{5}{8} \times \frac{5}{8} + \frac{3}{8} \times \frac{3}{8} = \frac{25}{64} + \frac{9}{64} = \frac{34}{64} = \frac{17}{32}$

Exercise 15C, Question 8(a)

The probability of a total greater than 4 is the probability of getting 2 on the first spin and 3 on the second, or 3 on the first spin and 2 on the second, or a 3 on both spins.

$P(2,3) = \frac{1}{4} \times \frac{1}{4} = \frac{1}{16}$; $P(3,2) = \frac{1}{4} \times \frac{1}{4} = \frac{1}{16}$; $P(3,3) = \frac{1}{4} \times \frac{1}{4} = \frac{1}{16}$

Add the three probabilities to get: $\frac{1}{16} + \frac{1}{16} + \frac{1}{16} = \frac{3}{16}$

Exercise 15C, Question 8(b)

The probability of a total less than 4 is the probability of getting 1 on the first spin and 2 on the second, or 2 on the first spin and 1 on the second.

$P(1,2) = \frac{1}{2} \times \frac{1}{4} = \frac{1}{8}$; $P(2,1) = \frac{1}{4} \times \frac{1}{2} = \frac{1}{8}$

Add the probabilities to get: $\frac{1}{8} + \frac{1}{8} = \frac{1}{4}$

Exercise 15C, Question 10

(a) P(a vowel then not a vowel) $= \frac{4}{11} \times \frac{4}{6} = \frac{16}{66}$

P(not a vowel then a vowel) $= \frac{7}{11} \times \frac{2}{6} = \frac{14}{66}$

P(exactly one vowel) $= \frac{16}{66} + \frac{14}{66} = \frac{30}{66} = \frac{5}{11}$

(b) Probability of at least one vowel means one or two vowels.

P(exactly one vowel) from part (a) is $\frac{30}{66}$

P(two vowels) $= \frac{4}{11} \times \frac{2}{6} = \frac{8}{66}$

P(at least one vowel) $= \frac{30}{66} + \frac{8}{66} = \frac{38}{66} = \frac{19}{33}$

Answers for Student Book 9

Chapter 1

Check in

1 a 64 **b** 11 **c** 49 **d** 2

2 a 0.6 **b** 0.3 **c** 0.7 **d** 0.1

3 a 102, 15 **b** 102, 15, 0

c 102, -7, 15, 0 **d** 102, -7, 15, $\frac{5}{13}$, 0

Exercise 1A

1 i Rational: -8, 0.7, $\sqrt[3]{27}$, 15.4, 1007, $\sqrt{81}$, $\frac{8}{11}$, $2\frac{3}{5}$, $\sqrt[3]{-8}$, 0.8

ii Irrational: π, $\sqrt{10}$, $\sqrt[3]{5}$, $\sqrt{3}$

iii Not rational or irrational: $\sqrt{-5}$

2 D

3 a -2, 81, $\sqrt{9}$ **b** 0.3, $\frac{2}{9}$ **c** 3π, $\sqrt{5}$

4 a irrational **b** either rational or irrational

c rational

5 a false **b** false, e.g. 0.2 is rational

c true **d** true

e false, e.g. $\sqrt{0.25} = 0.5$ which is rational

6 Students' own explanation e.g. rational numbers can be written as two fractions where *a*, *b*, *c* and *d* are integers and the product of two rational numbers is $\frac{a}{b} \times \frac{c}{d} = \frac{ac}{bd}$

The numerator and denominator are both integers so the answer will be rational as well.

7 Students' own example, e.g. $-\sqrt{2} + \sqrt{2}$, or $-\pi + \pi$ both give 0 as the answer, which is an integer and therefore rational.

Investigation

If the decimal can be written as a fraction with a square number as the numerator and a square number as the denominator, then the square root will be rational. For example, $\sqrt{0.25} = \sqrt{\frac{1}{4}} = \frac{1}{2}$ taking the square root of the numerator and denominator separately.

Exercise 1B

1 Obviously wrong:

a $\sqrt{20} \approx 5.1$: should be less than 5 as 20 is less than 25

c $\sqrt{95} \approx 9.1$: should be bigger than 9.1 as 95 is a lot closer to 100 than it is to 81

d $\sqrt{50} \approx 6.9$: should be bigger than 7 as 50 is bigger than 49

2 a true **b** false **c** true **d** false

e true **f** true

3 a 7 **b** 8 **c** 7.7

4 a 4 **b** 5 **c** -4.6

5 a 12 **b** 13 **c** 12.2

6 a ± 5.7 **b** ± 8.7 **c** ± 2.4 **d** ± 9.5

e ± 10.5 **f** ± 11.6

7 a 2 **b** 3 **c** 2.2

8 a 3.3 **b** 1.6 **c** 4.3 **d** 5.3

e -2.5 **f** -7.4

9 a 5, 6, 7 and 8

b 1, 2, 3, 4, 5, 6 and 7

10 a 0.1 **b** 0.2 **c** 0.17

11 a 0.5 **b** 0.6 **c** -0.55

12 a 41 mm **b** 53 mm

13 a 37 mm **b** 41 mm

14 a ± 0.67 **b** ± 0.96 **c** ± 0.84 **d** ± 0.45

15 a $x = 8.4$ **b** $x = 4.6$

16 Farhad's method gives a rough idea and is correct to 1 d.p. but it is not exactly $\frac{2}{11}$ more.

Investigation

b $\sqrt{16} \div \sqrt{9} = \sqrt{(16 \div 9)}$

and

d $\sqrt{16} \times \sqrt{9} = \sqrt{(16 \times 9)}$

are true and are always true for any pair of numbers.

Exercise 1C

1 a 4 **b** $5 - 6$ **c** 4

2 a $\frac{1}{7}$ **b** $\frac{1}{3}$ **c** $\frac{1}{4}$ **d** $\frac{1}{8}$

e $\frac{1}{5^2} = \frac{1}{25}$ **f** $\frac{1}{3^3} = \frac{1}{27}$

3 a 3 **b** 1 **c** -1 **d** -4 **e** 0

4 a 3^3 **b** 3^{-2} **c** 3^{-1} **d** 3^1

e 3^0 **f** 3^{-4}

5 -1

6 a false **b** false **c** false **d** true

e true **f** false

7 a 6^{-1} **b** 2^{-1} **c** 20^{-1} **d** 100^{-1}

e 4^{-2} **f** 5^{-3}

8 a 5^{-2} **b** 8^{-3}

9 Students' own answers

10 Students' own answers

11 10^{-3}; $\frac{1}{10000} = 10^{-4}$; $\frac{1}{100\,000} = 10^{-5}$

12 a 5^{-2} **b** 4^{-9} **c** 2^{-6} **d** 9^{-3} **e** 6^{-6}

f 12^{-20} **g** 11^{-3} **h** 3^{-8} **i** 7^{-6} **j** 2^4

Consolidation

Exercise 1

1 a 0, 0.6, $\frac{2}{9}$, $\sqrt[3]{8}$ **b** 13π, $\sqrt{55}$

2

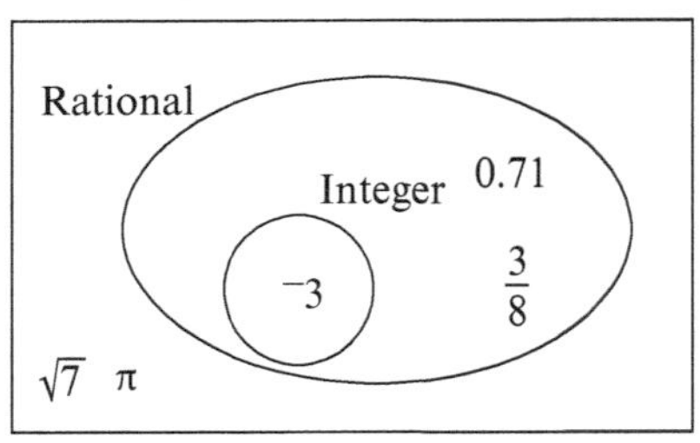

3 a 6.8 **b** ± 3.3 **c** 9.2 **d** 5.4

e ± 12.5 **f** -4.2

4 a 3.7 **b** 2.2 **c** 4.1 **d** 5.2

e -1.9 **f** -4.7

5 5.5 cm

6 3.6 cm

7 a 5^2 **b** 5^{-1} **c** 5^{-3} **d** 5^4 **e** 5^0

8 a 5^7 **b** 8^6 **c** 3^{-5} **d** 4^{13} **e** 4^{-3}

f 6^{-5} **g** 8^{-3} **h** 9^{-12} **i** 1

j 6 **k** 1

Summary

Check out

1 Rational: $^{-}7$, 0.264, $\sqrt[3]{125}$, 10 000, $\sqrt{4}$, $\frac{9}{17}$, $6\frac{5}{11}$, $0.\dot{8}$
Irrational: π, $\sqrt{11}$, $\sqrt[3]{4}$

2 a 4.5 b 2.5 c ± 8.9 d 3.6

3 a 7^{-2} or $\frac{1}{49}$ or $\frac{1}{7^2}$ b 8^8 c 3^{-8} or $\frac{1}{3^8}$ d 9

Chapter 2

Check in

1 a 4 b $^{-}5$ c 3 d $^{-}1$ e $^{-}7$

2 a $^{-}15$ b 12 c $^{-}2$ d 5 e 1

3 a 15 b 13 c 11 d 21

4 a 9^9 b 8^8 c 5 d 3^8

5 a Perimeter $= 6x + 8$, area $= 12x$
b Perimeter $= 12 + 4y$, area $= 12y$

6 a $3(3m + 2)$ b $5(3t - 1)$ c $13(2 - y)$ d $6(4x + 3y)$

Exercise 2A

1 a a^5 b q^8 c r^{10} d s^{15} e p^8
f j^{19} g a^{m+n} h p^{a+b} i m^{a+b+c}
j x^{a+m+c}

2 a q b b^3 c y^4 d $4p^4$ e $3x^4$
f $2y^5$ g $5n^3$ h $\frac{10}{x^2}$

3 a $8a^7$ b $6c^9$ c $6p$ d $2q^4$

4 a x^6 b x^6 c x^9 d $16x^8$ e x^{-8}
f $81x^8$

5 a^{mn}

6 a x^3 b p^9 c y^{-4} d q^3

7 a $\frac{1}{p^5}$ b $\frac{1}{x^3}$ c $\frac{1}{q^9}$ d y^0 e $\frac{1}{k^6}$

8 Students' own examples. For example:
a $3x^{10} \times 4x^6 = 12x^{16}$ b $(2m^3)^4 = 16m^{12}$
c $36y^8 \div 12y^6 = 3y^2$ d $3 \times x^0 = 3$
e $2x^{-3} = \frac{2}{x^3}$

9 They are all wrong.
a $(3p)^3 = 27p^9$ b $8m^8 \div 2m^2 = 4m^6$
c $4t^4 \times 4t^4 = 16t^8$ d $8y^0 = 8$

10 a lm^2 b $\frac{x^5z}{y}$

11 a $\frac{p^2}{q^3}$ b $\frac{x^3}{yz}$ c k^3m^2

12 a $2p^6q^3$ b p c $3p^3$ d $\frac{2pq^5}{3}$

13 $x^2 \times x^3$ ↔ $x^7 \div x^2$
$\frac{x^4}{x^2 \times x} \times x^2$ ↔ $x^{11} \times x^2 \div x^{10}$
$x^4 \times x^2 \times x$ ↔ $\frac{x^{10}}{x^3} \times \frac{x^2}{x \times x}$
$x^5 \div x^8$ ↔ $x^4 \div \frac{(x^4)^2}{x}$

14 $\frac{p^{30}}{(p^2 \times p^8)^2} = p^{10}$ $\frac{(p^5 \times p^3)^2}{(p \times p^3)^2} = p^{12}$ $\frac{(p^4 \times p^2)^3}{p^5} = p^{13}$

15 a p^6 b x c A^6 d m^{10} e x^5
f y g $4T$ h h^{15} i $2g^3$

Exercise 2B

1 a $8x$ b $10a$ c $12b$ d $^{-}3y$ e $9a$
f $3b$ g $^{-}8p$ h $17ab$ i $9a^2$ j $7b^2$
k $11x^2$ l $2y^3$

2 a $4ab^2 + 5ba^2$
b p^2 (the 1 is not needed but not actually incorrect)
c $6x^2$

3 a $2xy - 3z$ b $2ab + 6pq$ c $7a^2 - 2b^2$ d $3a^3 + 4a$

4 a $2p + 6q$ b $3z - z^3$ c $7a^2 - 2b^2$ d $8pq - 3p^2$

5 a $^{-}21y$ b $30p$ c $^{-}12pq$ d $15y^2$
e $6a^3$ f $40x^3$

6 a $3a$ b 2 c 2.5 d $4p$ e 3
f $4p$ g $\frac{5}{x}$ h $\frac{1}{2y}$ i a j a
k $\frac{7b^3}{3a^2}$ l $\frac{4x}{y}$

7 a $22xy$ b $17p^2q + 7lm$ c $pqr + 6abx + mny$
d $4pq - 13ab$ e $4x^2y^2 + 9xy$

8 a $4mn$ b $^{-}3l^2m^2$ c $^{-}16pqr$ d $^{-}15m$
e $8p^3q^4$ f $\frac{7ab^2}{2}$ g $\frac{2n^2}{3l^2}$

Exercise 2C

1 a $\frac{7}{8}$ b $\frac{11}{12}$ c $\frac{1}{2}$ d $\frac{25}{84}$

2 a $\frac{8a}{15}$ b $\frac{19a}{45}$ c $\frac{11a}{3}$ d $\frac{11a}{12}$
e $\frac{-11a}{21}$ f $\frac{5a}{7}$

3 a $\frac{(4x-y)}{6}$ b $\frac{(9y+11x)}{33}$ c $\frac{(5y+4x+5z)}{10}$
d $\frac{(x+2y^2)}{4}$ e $\frac{(6x+10y^2-y)}{4}$ f $\frac{(7x^2+18y)}{63}$

4 a $\frac{1}{2b}$ b $\frac{15}{8y}$ c $\frac{(12+r)}{2pq}$ d $\frac{7x}{5}$
e $\frac{5p}{8}$ f $\frac{(3x+2)}{x}$ g $\frac{(5y-4)}{y}$

5 a $\frac{(p^2+6q^2)}{3pq}$ b $\frac{(15+4l)}{3m}$ c $\frac{(5x+12z)}{4x}$ d $\frac{(r+3pq)}{pqr}$
e $\frac{(xy^2+az^2)}{zy}$ f $\frac{(4a^2b+5bc^2)}{20ac}$

6 $\frac{(48p+51q)}{(p+q)}$

7 $\frac{(10y+14x)}{xy}$

8 a $\$\frac{1.45}{b}$ b $\$\frac{2.35}{a}$ c $\$\frac{(1.45a+2.35b)}{ab}$

9 a $\frac{(9x+11)}{20}$ b $\frac{(^{-}3x-9)}{35}$ c $\frac{(7x+11)}{(x+1)(x+2)}$
d $\frac{(8x+4)}{(x+3)(x-1)}$ e $\frac{2x}{(x^2-1)}$ f $\frac{(2x-22)}{(x+1)(x-3)}$

Investigation

Gap from 1 is $1-\frac{y}{x}$ or $\frac{x}{y}-1$
Write with a single denominator $\frac{x-y}{x}$ and $\frac{x-y}{y}$
The numerators are now the same, $x - y$, so you just need to compare the denominators.

Since $x > y$, the fraction with the denominator of x will be smaller, so $\frac{y}{x}$ will be closer to 1.

Exercise 2D

1 a $\frac{5(6x+3)}{10}=\frac{6x+3}{2}$

b $\frac{4(3y-4)}{40}=\frac{3y-4}{10}$

c $\frac{7}{49p-77}=\frac{7}{7(7p-11)}=\frac{1}{7p-11}$

d $\frac{8(2-9y)}{12}=\frac{2(2-9y)}{3}=\frac{4-18y}{3}$

e $\frac{4(4-18y)}{12}=\frac{4-18y}{3}$

f Choose a factor outside of the brackets that is also a factor of the denominator, rather than the highest factor.

2 a $\frac{9x+1}{3}$ b $6x+8y$ c $\frac{2}{6-y}$ d $\frac{5x+3y-z}{2}$

e $\frac{6p-9}{4}$

3 These are equivalent to $2x+3$: $\frac{2x+15+8x}{5}$, $\frac{1}{4}(8x+12)$, $\frac{2x+3}{1}$, $\frac{x+1.5}{0.5}$

These are not equivalent to $2x+3$: $\frac{2x+9}{3}$, $\frac{1}{2x+3}$, $\frac{20x-(-80x+150)}{50}$, $\frac{4x+12}{2}$

4 It is not fully simplified. Divide through by 5 to give $\frac{5p+2}{4}$

5 a $\frac{x+4}{2}$ b $\frac{2}{2m-5}$

6 She can also take m out as a common factor to give the answer $\frac{4m+3}{3}$

7 a $\frac{6x+5}{16}$ b $\frac{5x-6}{8}$

8 a $\frac{3}{4}$ b 4 c $\frac{3y}{2}$ d $2x$

9 Both fractions simplify to $\frac{5}{8}$

Exercise 2E

1 a $x^2+7x+12$ b x^2+6x+5

2 a $x^2+11x+24$ b $x^2+8x+12$
c $x^2+9x+14$ d x^2+3x+2

3 a $x^2+5x-14$ b x^2+x-56
c $x^2-4x-21$ d x^2-x-6

4 a $x^2-10x+21$ b $x^2-16x+63$
c x^2-4x+4 d x^2-2x+1

5 a x^2+x-42 b $x^2-11x+28$
c $x^2-15x+44$ d $x^2-7x-30$

6 Aisha is correct

7 a $x^2+10x+25$ b $x^2+8x+16$ c x^2-2x+1
d $x^2-14x+49$ e $p^2+18p+81$ f $t^2+8t+16$

8 a $x^2+3x-28$ b $x^2-12x+32$ c $\frac{(x^2+6x+8)}{2}$

10 a x^2-25 b x^2-64 c x^2-49 d $4-x^2$

11 Students' own answers

12 $ac+bc+ad+bd$

Exercise 2F

1 a x^2-4 b x^2-81 c x^2-121 d $36-x^2$

2 a x^2+4x+4 b $x^2+10x+25$
c x^2-6x+9 d $x^2-24x+144$

3 a x^2-100 b x^2-144
c $x^2-22x+121$ d $x^2+26x+169$

4 a $(x+10)^2$ b $(x-8)^2$ c $(x-6)(x+6)$
d $(x-13)(x+13)$ e $(x-y)(x+y)$

5 $(x+1)(x-1)=x^2-1$, so $49\times51=50^2-1=2500-1=2499$

6 $(x+1)^2=x^2+2x+1$ so:

a $91=90^2+2\times90+1=8100+180+1=8281$

b $401^2=400^2+2\times400+1=160\,000+800+1=160\,801$

Exercise 2G

1 a $^-130$ b 2 c 82 d 24
e 0.05 f $^-16$

2 a 21.5 b $^-2.1$ c 0.2 d 0.3
e $^-1.5$ f 0.01053

3 a 5.04 b 1331.1 c 18 d $^-3.2$
e 2.5 f $^-26$ g 50 220

4 $6xy$ and $48x^2-y^2-y$ have the same value

5 a 324 000 b 18 018 c 1000

6 a 285 b 362.5

7 0.1571

8 cylinder

9 isosceles

Investigation

Perpendicular sides are 3 and 4 so $\frac{1}{2}bh=\frac{1}{2}(3\times4)=6$

Using Heron's formula, the perimeter is $3+4+5=12$, half of this is 6 so $s=6$

$A=\sqrt{6(6-3)(6-4)(6-5)}\sqrt{6(3)(2)(1)}=\sqrt{36}=6$

Exercise 2H

1 a x^2+2x-3 b x^2-64
c $x^2+10x+25$ d $5x+22$

2 $n^2+9n+14$

3 $2w^2+7w$

4 $3\times13=39$ and $2\times9+7\times3=39$

5 $11\times11=121$ and $5^2+36=61$

6 $2n^2+12n+18$

7 Both are equal to $2ab+xy-by$. He has found the correct area.

8 As n is even, then $n+1$ will be odd, and so will $n+3$. Their product is: $(n+1)(n+3)=n^2+4n+3$
As n is even, n^2 and $4n$ will also be even. Adding an odd number (here, 3) to an even number always results in an odd number.

9 $\sqrt{x^2+x^2}=\sqrt{2x^2}$ or $\sqrt{2}x$

10 length is $x+8$, $n=5$

Exercise 2I

1 a $r\rightarrow$ square $\rightarrow\times\pi h\rightarrow V$

b $h\rightarrow\times2\rightarrow+r\rightarrow\times\pi r\rightarrow S$

c $a\rightarrow\times t\rightarrow\div2\rightarrow+u\rightarrow\times t\rightarrow s$

2 a $r=\sqrt{\frac{V}{h\pi}}$ b $h=\frac{S-\pi r^2}{2\pi r}$ c $a=\frac{2(s-ut)}{t^2}$

3 a $l\rightarrow\div g\rightarrow$ square root $\rightarrow\times\pi\rightarrow\times2\rightarrow T$

b 0.898

4 a $l\leftarrow\times g\leftarrow$ square $\leftarrow\div\pi\leftarrow\div2\leftarrow T$

b 2

5 a $y=mx\div c$ b $x=cy\div m$

6 a i $r \rightarrow \text{square} \rightarrow \times \pi h \rightarrow \div 3 \rightarrow V$

ii $r \leftarrow \text{square root} \leftarrow \div \pi h \leftarrow \times 3 \leftarrow V$

b 6.74 cm

7 a $A = 14x + 28$ **b** $x = \frac{A-28}{14}$ or $\frac{A}{14} - 2$

c $x = 2.5$ **d** 9

8 a $314.2\,\text{cm}^3$ **b** $R \rightarrow \text{squared} \rightarrow -r^2 \rightarrow \times \pi h \rightarrow V$

c $R = \sqrt{\frac{V}{\pi h} + r^2}$

9 a i $x \rightarrow -s \rightarrow \times r \rightarrow +t \rightarrow y$

ii $x \rightarrow \times n \rightarrow +l \rightarrow \times m \rightarrow y$

b i $x = s + (y - t) \div r$ **ii** $x = \left(\frac{y}{m} - l\right) \div n$

Exercise 2J

1 a $m = \frac{Pt}{(v-u)}$ **b** $r = \sqrt{\frac{V}{\pi h}}$ **c** $u = \sqrt{v^2 - 2as}$

d $x = a + \sqrt{r^2 - y^2}$ **e** $g = \frac{4\pi^2 l}{T^2}$

2 a $\alpha = \frac{P - P_0}{P_0 t}$ **b** $t = \frac{P - P_0}{P_0 \alpha}$

3 a $r = \sqrt{\frac{S}{4\pi}}$ **b** $r = \frac{V}{3\pi h}$

c $r = \frac{2A}{h} - R$ **d** $r = \sqrt{\left(R^2 - \frac{V}{\pi h}\right)}$

4 Both are correct. $\frac{P - 2w}{2} = l$ is the same as $\frac{P}{2} - \frac{2w}{2} = l$ which is $\frac{P}{2} - w = l$

5 a $t = \frac{2s}{(u+v)}$ **b** $t = \frac{(v-u)}{f}$

c $t = \frac{PV}{Rm}$ **d** $t = \frac{(T - k\beta)}{k\alpha}$

6 a $k = d(P - mv)$ **b** $m = \frac{(Pd - k)}{dv}$

c $v = \frac{(Pd - k)}{dm}$ **d** $d = \frac{k}{P - mV}$

7 a $m = {}^-2, c = 9$ **b** $m = \frac{1}{5}, c = {}^-2$

c $m = \frac{2}{5}, c = 3$ **d** $m = 2, c = 0$

e $m = {}^-2, c = 2$

Consolidation

Exercise 2

1 a x^5 **b** y^8 **c** $2x^7$ **d** $3x^9$

e x^9 **f** y^{11}

2 a x **b** 1 **c** x^4 **d** y^5

e x^5 **f** y^2

3 a 1 **b** 5 **c** $4m^6$ **d** $16x^2y^6$

e $\frac{x}{2v^2}$ **f** $\frac{y^3z^2}{x}$

4 a y^{10} **b** a^4 **c** x^{-2} **d** m^8

e p^{-3} **f** $21x^9$ **g** $4m^{-2}$ **h** q^{-8}

i p^2 **j** R^{10}

5 a $x^2 + 8x + 12$ **b** $x^2 - 25$

c $x^2 - 3x - 38$ **d** $x^2 + 6x - 16$

e $x^2 - 7x + 12$ **f** $x^2 - 11x + 10$

g $x^2 + 10x + 25$ **h** $x^2 - 18x + 81$

6 a i $A = x^2 - 2x - 15$ **ii** $A = x^2 - 8x + 12$

b i $P = 4x - 4$ **ii** $P = 4x - 16$

c i $A = 65; P = 36$ **ii** $A = 32; P = 24$

7 a $\frac{d}{4} + \frac{d}{6} = \frac{3d + 2d}{12} = \frac{5d}{12}$ **b** $\frac{3x}{5} + \frac{x}{4} = \frac{12x + 5x}{20} = \frac{17x}{20}$

c $\frac{2m}{3} - \frac{m}{10} = \frac{20m - 3m}{30} = \frac{17m}{30}$

d $\frac{y}{3} + \frac{2y}{5} + \frac{y}{4} = \frac{20y + 24y + 15y}{60} = \frac{59y}{60}$

e $\frac{3}{m} + \frac{1}{2m} = \frac{6+1}{2m} = \frac{7}{2m}$ **f** $1 - \frac{1}{x} = \frac{x-1}{x}$

g $\frac{3}{4b} + \frac{2}{3c} = \frac{9c + 8b}{12bc}$

h $\frac{2}{x+1} + \frac{1}{x+3} = \frac{2(x+3) + (x+1)}{(x+1)(x+3)} = \frac{2x + 6 + x + 1}{(x+1)(x+3)} = \frac{3x+7}{(x+1)(x+3)}$

8 a $v = \frac{3b - p}{2}$ **b** $h = y - 4T$ **c** $m = \frac{3(x-y)}{2}$

d $r = \frac{Y+t}{2} - P$ **e** $n = \frac{ab}{3(j+y)}$ **f** $s = \frac{hy - d}{2}$

9 a $h = 20\left(F - \frac{R}{2}\right)$ **b** $h = 3\left(y^2 - \frac{x}{2}\right)$

c $h = \frac{A}{2\pi r^2}$ **d** $h = \frac{xy}{(x+y)}$

10 a $r = \sqrt[3]{\frac{3V}{4\pi}}$ **b** 3.63 cm

11 a 48 **b** 20.1 **c** $^-4$ **d** 9

e $^-4.38$ **f** $^-12.8$

Summary

Check out

1 a b^8 **b** $3c^2$ **c** $\frac{3a}{2}$ **d** $\frac{5a^3b^2}{c^3}$ **e** $9x^6$

2 a $x = \frac{(y-c)}{m}$ **b** $x = \sqrt{y^2 - c^2}$

c $x = y(p - c) - c$

3 a $x^2 + 6x + 5$ **b** $x^2 + x - 12$ **c** $x^2 - 4x - 12$

d $x^2 - 11x + 24$ **e** $x^2 + 10x + 25$

4 a $\frac{11e}{15}$ **b** $\frac{5y}{12}$ **c** $\frac{11}{2x}$ **d** $\frac{(2p+d)}{8p}$

5 581.25

Chapter 3

Check in

1 a 36° **b** 215°

2 a 2 **b** 4 **c** 5 **d** 1

3

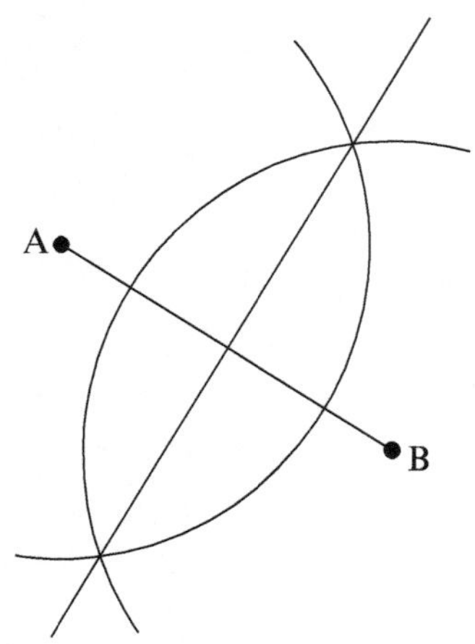

4

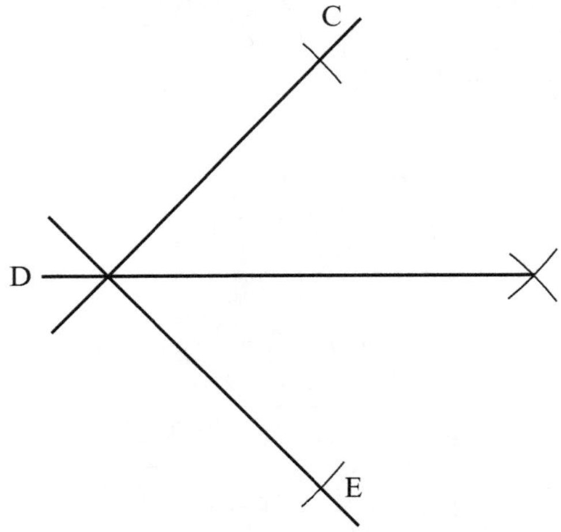

Exercise 3A

1

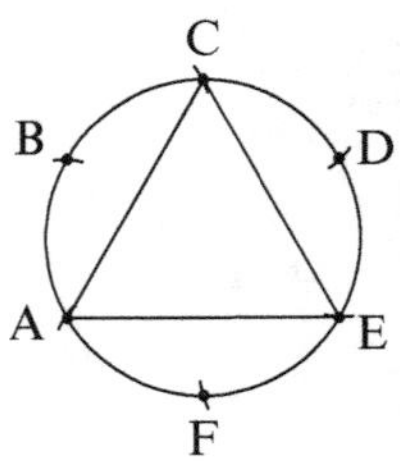

2

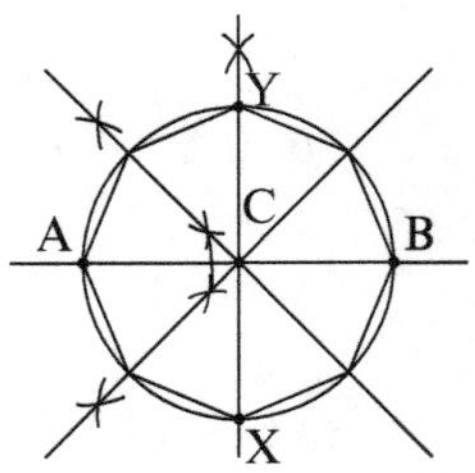

3, 4 Students' own answers

Exercise 3B

1 a

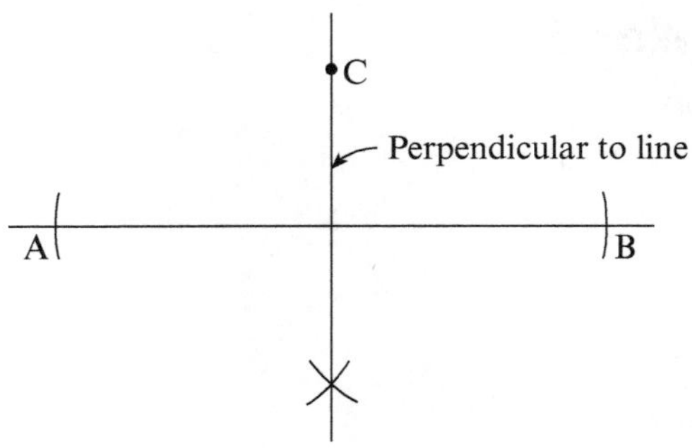

b Any of the four angles can be bisected e.g.

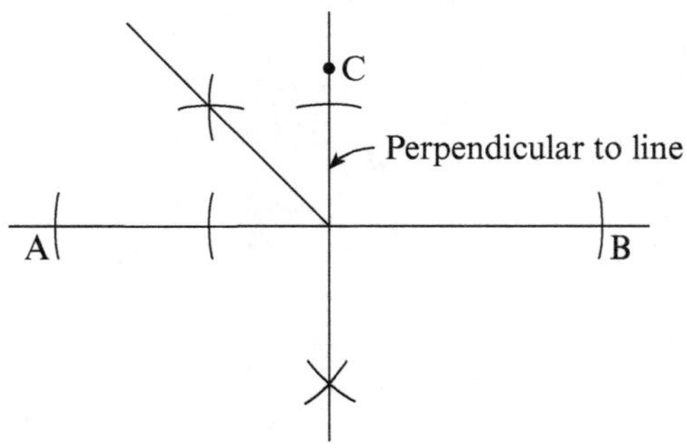

2 a

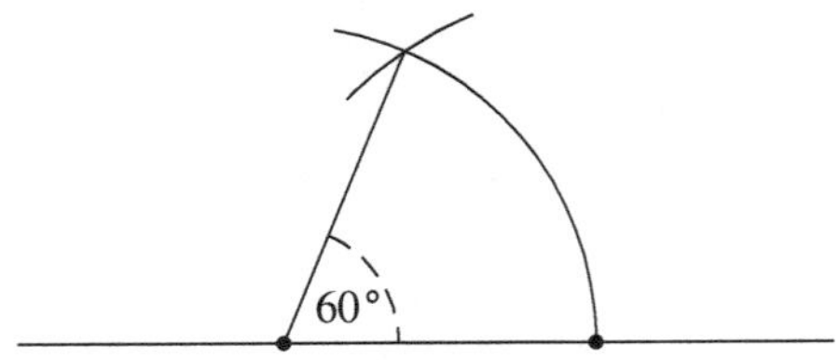

b

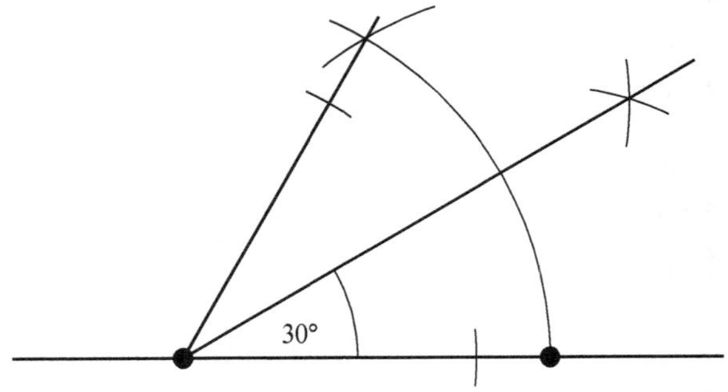

3 a b c

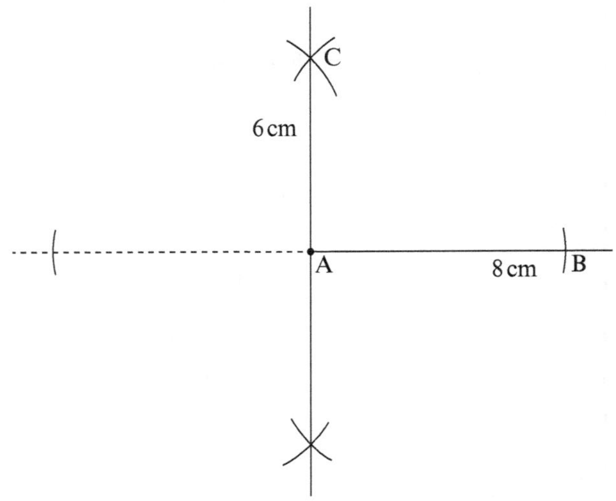

d

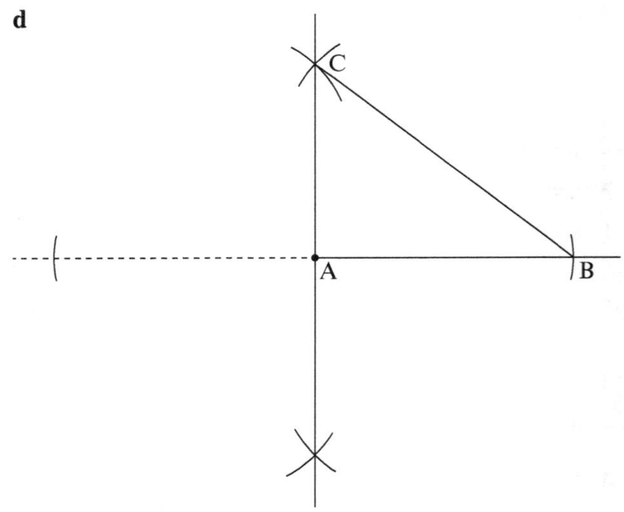

e 10 cm

f

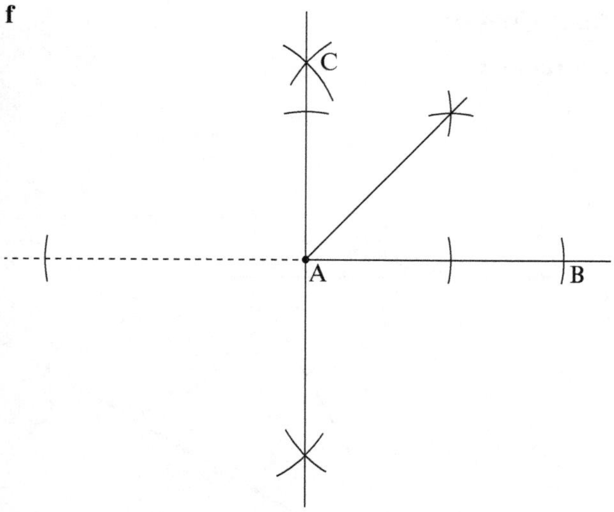

4

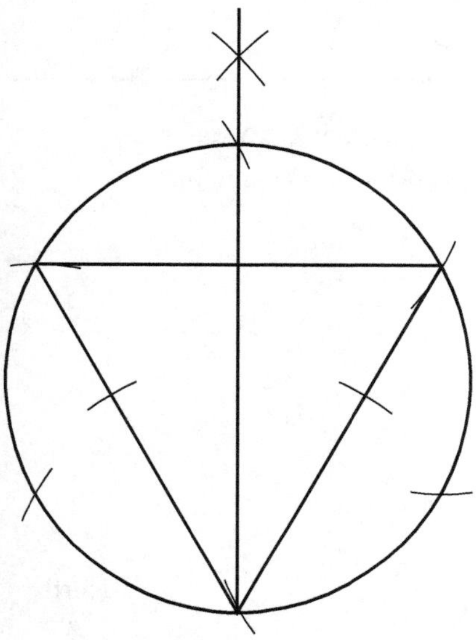

5 **a** Construct an angle of 60°, then bisect, then bisect one of the two angles again.

b Construct an angle of 90°, then bisect, then bisect one of the two angles again.

c Construct a perpendicular bisector of a line then bisect one of the angles of 90°, then one 90° angle plus the 45° angle makes an angle of 135°.

6

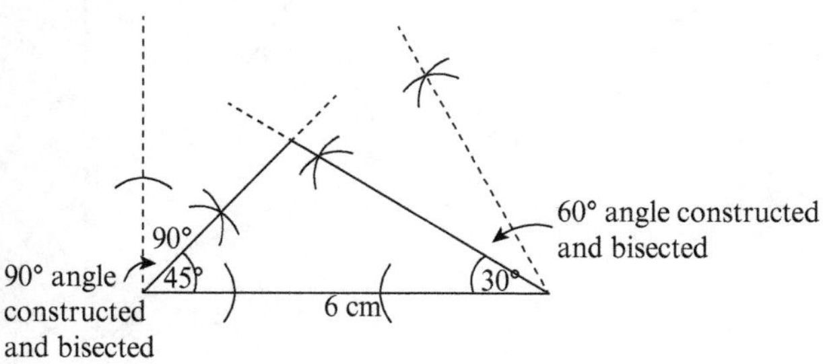

Investigation

When the rectangle base is double the height then the bisectors will meet on the edge of the rectangle. When the base of the rectangle is less than double the height then the bisectors will meet inside the rectangle. When the base of the rectangle is more than double the height then the bisectors will meet outside the rectangle.

When the parallelogram base is double the side then the bisectors will meet on the edge of the parallelogram. When the base of the parallelogram is less than double the side then the bisectors will meet inside the parallelogram. When the base of the parallelogram is more than double the side then the bisectors will meet outside the parallelogram. It makes no difference what the angles are.

Exercise 3C

1 **a** 74 km, 67 km **b** 24 km, 44 km

c 5 km, 19 km

2 **a** 50 km, 87 km **b** 41.5 km, 28 km

c 20 km, 3 km

3

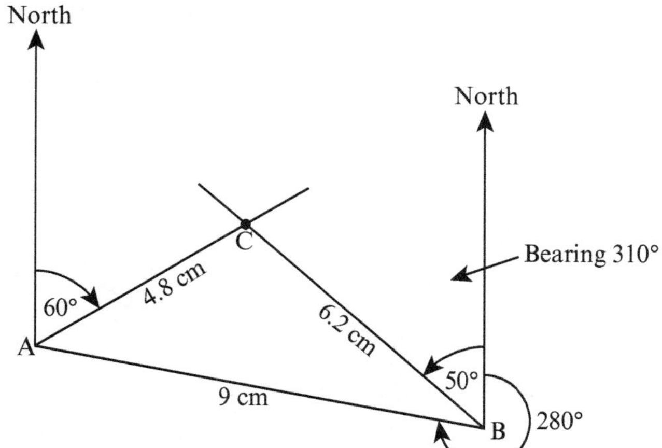

a 19.2 km (allow answer in the range 18.4 to 20 km)

b 24.8 km (allow answer in the range 24 to 25.6 km)

4 **a** **i** 242° **ii** 165° **b** **i** 4.7 km **ii** 6.2 km

5 **a** 153° **b** 22 km

6 **a** 117° **b** 214° **c** 346° **d** 304°

e 117°

7 **a** 9 km East, 4 km North **b** 16 km East, 12 km South

c 39 km West, 10 km South **d** 45 km West, 22 km North

e 16 km West, 19 km North **f** 7 km West, 15 km North

8 **b** **i** 31 km East, 26 km South

ii 35 km West, 20 km South

iii 21 km East, 37 km North

c 17 km East, 9 km South

Exercise 3D

1

2

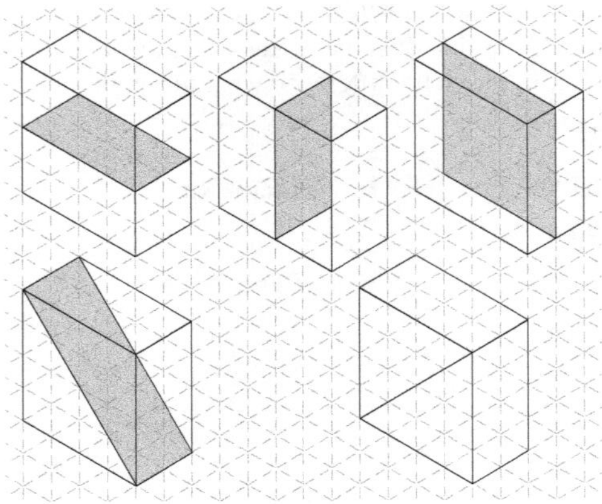

3

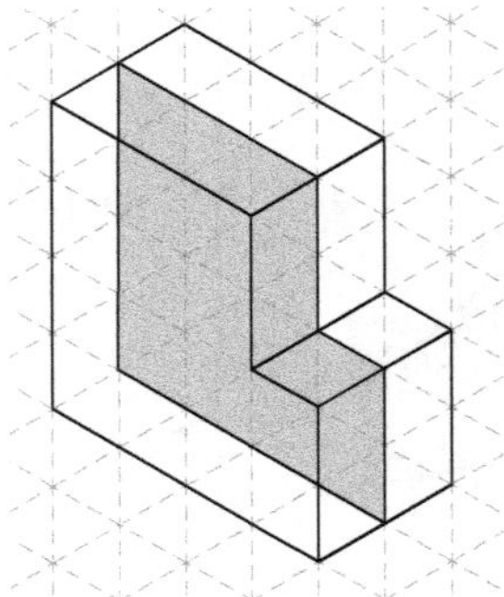

4

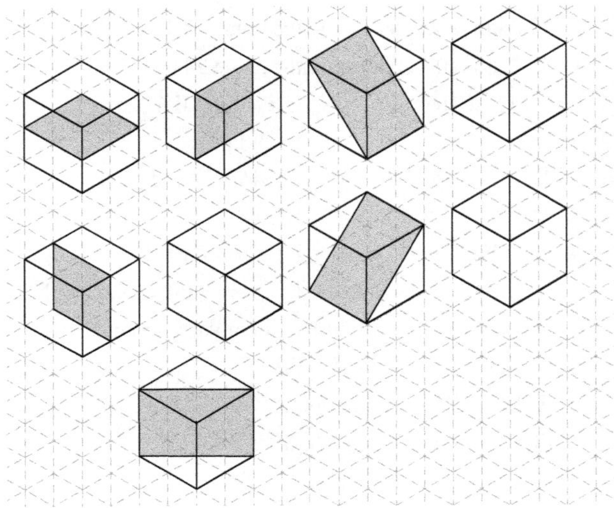

5 **a** 4
b 6
c 7

6 The only correct statement is:

A prism can have exactly two planes of symmetry – e.g. if the cross section is a T shape with one line of symmetry.

Explanations for incorrect statements:

- A prism can have no planes of symmetry – no, it will always have the one that cuts it in half lengthways.
- No prism can have exactly ten planes of symmetry – if the cross section is a regular nonagon (9-sided polygon) it will have ten planes of symmetry.
- A prism with a semicircular cross section has an infinite number of planes of symmetry – no, it just has two.

Consolidation

Exercise 3

1 a

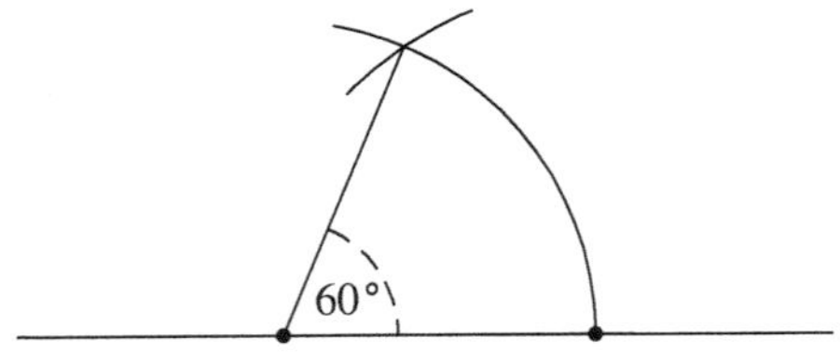

b

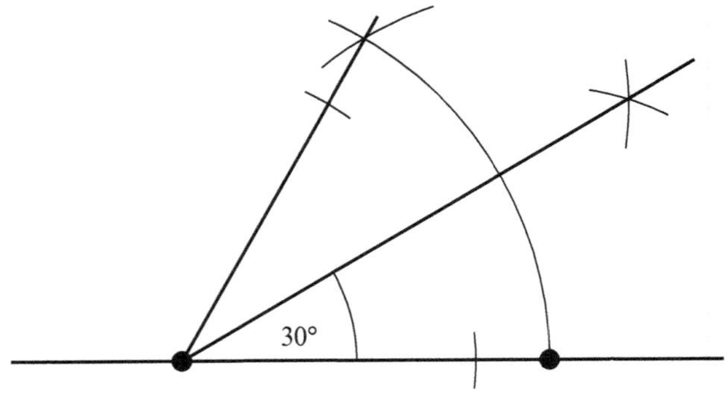

2 b i 7 km East; 12 km West; 20 km West
ii 6 km South;16 km South; 35 km North
c 12 km North, 24 km West

3 a 3 **b** 2 **c** 3

4

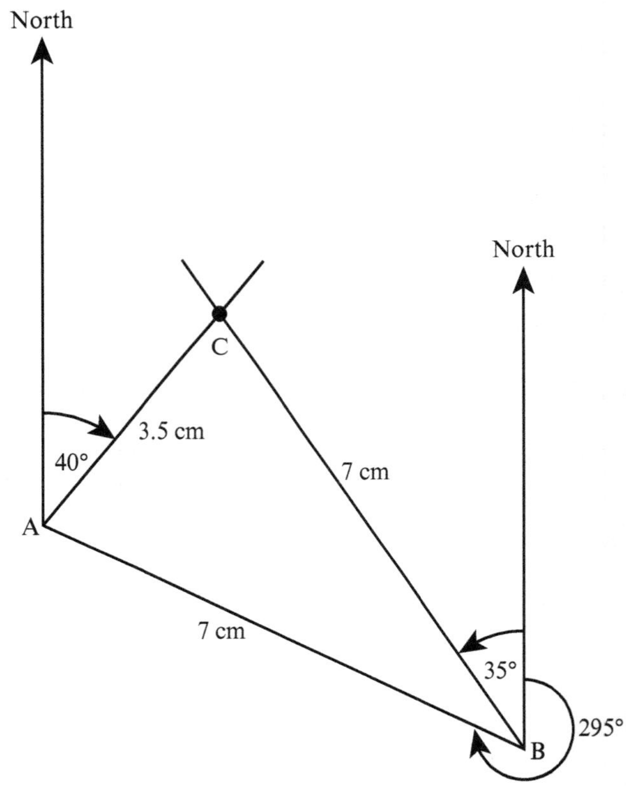

a 21 km (allow answer in the range 19.8 to 22.2 km)
b 42 km (allow answer in the range 40.8 to 43.2 km)

5

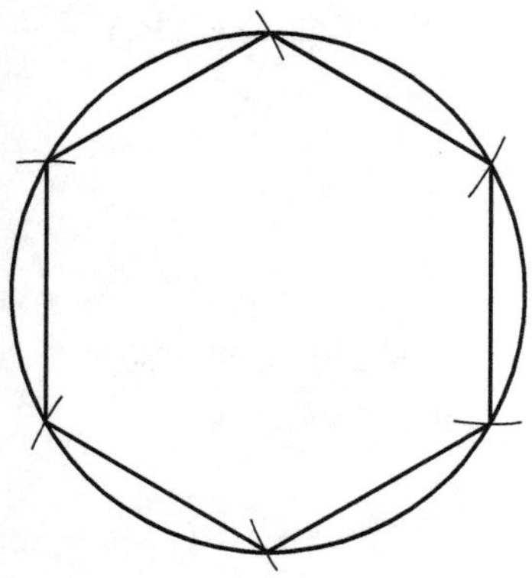

6 a 9 **b** 6 **c** 11

7

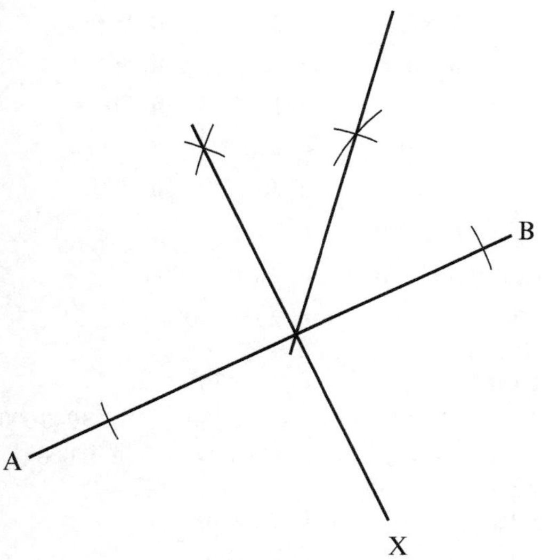

8 114 m

Check out

1 Students' own diagram, e.g.

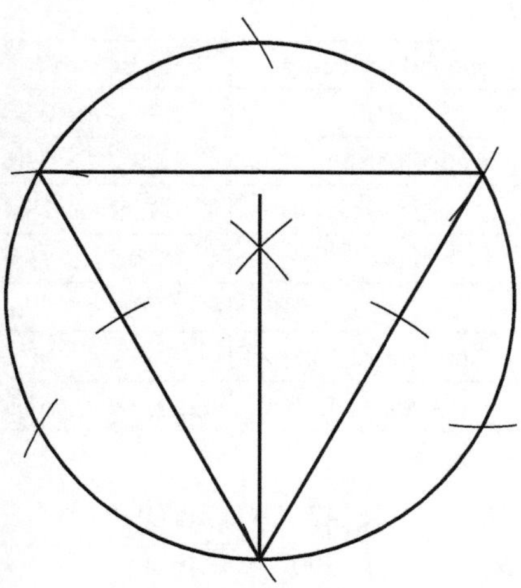

2 a 063° **b** 2 km **c** 315°

3 a i 6
ii 5
iii 8
b 1

Chapter 4

Check in

1 a 10^{11} **b** 0.01 **c** 10^5

2 a 134.56 **b** 38 **c** 0.007 79

Exercise 4A

1

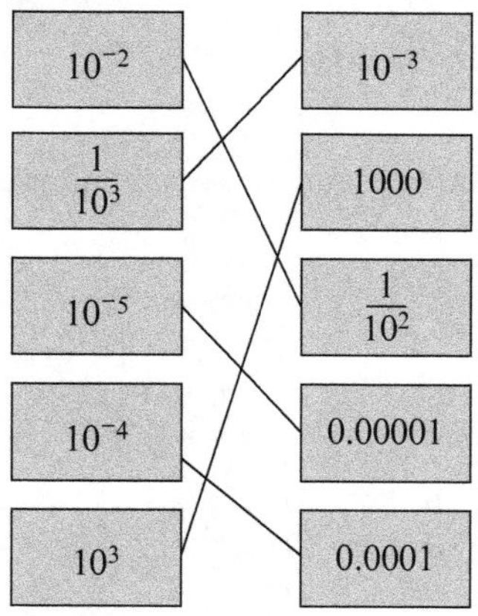

2 a 7000 **b** 28 000 **c** 43 000 **d** 7100

3 a 8.03 **b** 0.42 **c** 5.1 **d** 958

4 a 1000; 1000; 0.8 **b** 2; 100; 100; 35
c 10^5; 100 000; 72 **d** 10^1; 10; 0.2

5 a $\frac{1}{100}$; 100; 7.4
b 5; 100 000; 100 000; 800
c 3; 1000; 1000; 2145
d 1; 10, 10, 3

6 a 100 **b** 1000
c dividing;10 000 **d** multiplying; 100 000

7 a 0.9 **b** 0.287 **c** 8.63 **d** 0.046

8 a 48 **b** 210 **c** 60 **d** 14.3

9 a 3400 **b** 7.13 **c** 8.1 **d** 70

10 One possible solution is
$7 \times \boxed{10^2} = 7 \div \boxed{0.01}$
$5.2 \div \boxed{1000} = 5.2 \times \boxed{10^{-3}}$

11 a $7.1 \times 10^5 = 710\,000$
b $249\,000 \div 10^3 = 249$
c $60\,000 \times 10^{-4} = 6$
d $0.02 \div 10^{-2} = 2$
e $2.84 \times 10^3 = 2840$
f $44\,000 \div 10^4 = 4.4$
g $230 \times 10^{-2} = 2.3$
h $0.007 \div 10^{-6} = 7000$

12 a 706 000 **b** 78 460 **c** 109 000 **d** 47

Exercise 4B

1 a 2×10^3 **b** 4×10^2 **c** 8×10^1
d 9×10^4 **e** 4×10^3 **f** 7×10^5
g 3×10^6 **h** 4×10^7 **i** 1×10^5

2 a 300 **b** 5000 **c** 600 000
d 20 000 000 **e** 3 000 000 **f** 4 000 000 000

3 a 4.2×10^2 **b** 6.3×10^3 **c** 1.7×10^5
d 2.3×10^4 **e** 6.13×10^4 **f** 9.23×10^3
g 4.16×10^2 **h** 9.81×10^4 **i** 6.31×10^6

4 No, as 62 is not between 1 and 10. It should be 6.2×10^6

5 **a** 7×10^3 **b** 6.34×10^4 **c** 8.7×10^4
d 5×10^1

6 **a** 1600 **b** 280 **c** 381
d 475 000 **e** 301 000 000 **f** 1 600 000 000

7 **a** 54.3 is not between 1 and 10
b 0.03 is not between 1 and 10
c Need a single power of 10

8 1.404×10^{10}

9 9.46×10^{17}

10 **a** 5.43×10^8 **b** 3×10^6 **c** 7.1×10^5

Exercise 4C

1 **a** 0.06 **b** 0.4 **c** 0.008 **d** 0.0005
e 0.000 002 **f** 0.000 07

2 **a** 6.3×10^{-1} **b** 7.4×10^{-3}
c 2.8×10^{-2} **d** 1.3×10^{-4}
e 2.356×10^{-2} **f** 8.2×10^{-4}
g 3.91×10^{-6} **h** 1.6×10^{-3}
i 3.83×10^{-3}

3 **a** 3.06×10^{-5} **b** 4.925×10^{-7}
c 4.2831×10^{-3} **d** 9.01×10^{-4}
e 2.5×10^{-10} **f** 8.46×10^{-8}
g 8.6×10^{-8} **h** 3.675×10^{-12}

4 No, standard form should have a multiplication sign not a division sign, it should be 8×10^{-3}

5 **a** 2×10^{-3} **b** 7.2×10^{-2} **c** 8.6×10^{-2} **d** 2×10^{-11}

6 9×10^{-11} g

7 **a** 2.4×10^{12} **b** 4.1×10^9 **c** 9×10^{-3} **d** 3.2×10^{-3}

Exercise 4D

1 **a** 243, 238, 235 **b** 235, 245

2 **a** yes **b** no
c If you round to 1 d.p. this is the same as rounding to the nearest 0.1 so halve this value.
d See what rounding to 2 s.f. is equivalent to; it might be equivalent to rounding to the nearest 100, or nearest whole number. Once you have worked that out you can use the same method.

3

Number, n	Rounded correct to	Bounds
340	nearest 10	$335 \le n < 345$
4500	nearest 100	$4450 \le n < 4550$
7000	nearest 1000	$6500 \le n < 7500$
42 000	nearest 1000	$41\,500 \le n < 42\,500$
83 200	nearest 100	$83\,150 \le n < 83\,250$
690	nearest 10	$685 \le n < 695$

4 **a** These are the bounds if the number had been rounded to the nearest 100.
b $4795 \le n < 4805$

5

Number, n	Rounded correct to	Bounds
0.7	1 d.p.	$0.65 \le n < 0.75$
3.78	2 d.p.	$3.775 \le n < 3.785$
3400	2 s.f.	$3350 \le n < 3450$
4000	1 s.f.	$3500 \le n < 4500$

6 Any number in the interval $524 < x < 525$, e.g. 524.5

7 $827.5 \le h < 828.5$

8 **a** could be true
b definitely false
c could be true

9 $325 \le n < 375$

10 **a** $22.5 \le n < 27.5$ **b** $67\,250 \le n < 67\,750$
c $7.25 \le n < 7.75$ **d** $270 \le n < 290$

11 28 cm

Consolidation

Exercise 4

1 **a** false **b** false **c** false **d** false

2 **a** 42 000 **b** 30 **c** 5.3 **d** 49
e 610 **f** 2.1 **g** 1 **h** 70

3 **a** 2 **b** 6.03 **c** 21 **d** 4.35
e 0.05 **f** 200 **g** 360 **h** 560

4 **a** 4×10^3 **b** 2×10^4 **c** 7×10^7

5 **a** 30 000 000 **b** 4 000 000 **c** 7 000 000 000

6 **a** 4×10^4 **b** 2.17×10^3
c 3.401×10^3 **d** 9.08×10^4
e 1.12×10^4 **f** 2.5×10^4

7 **a** 1500 **b** 632 **c** 16 280 **d** 3 940 000

8 **a** 0.7 **b** 0.03 **c** 0.004 **d** 0.0009
e 0.000 06 **f** 0.000 005

9 **a** 1.6×10^{-1} **b** 3.7×10^{-1}
c 2.2×10^{-2} **d** 9.4×10^{-2}
e 6.4×10^{-3} **f** 3.7×10^{-3}

10 **a** > **b** = **c** <

11 **a** 67.5 m
b $106.5 \le l < 107.5$

12

Number, n	Rounded correct to	Bounds
0.3	1 d.p.	$0.25 \le n < 0.35$
23	the nearest whole number	$22.5 \le n < 23.5$
18.45	2 d.p.	$18.445 \le n < 18.455$
730	2 s.f.	$725 \le n < 735$
900	1 s.f.	$850 \le n < 950$
900	the nearest 10	$895 \le n < 905$
55	the nearest 5	$52.5 \le n < 57.5$

Check out

1 **a** 2300 **b** 21 **c** 47 **d** 19.2
e 34 **f** 36 **g** 290 **h** 4.1

2 **a** 8×10^3 **b** 7.236×10^3 **c** 3.8×10^{-2}
d 1.83×10^5 **e** 6.2×10^{-5} **f** 3×10^{-2}

3 **a** $21.5 \le h < 22.5$ **b** $72.25 \le m < 72.35$

Chapter 5

Check in

1 **a** continuous **b** discrete **c** discrete
d continuous **e** continuous **f** continuous

2 mean = 6, median = 5, mode = 4

3 range = 9 – 4 = 5

4 $46.40

Exercise 5A

1 a

Height (h, cm)	Frequency, f	Midpoint, x	fx
$10 \le h < 20$	4	15	60
$20 \le h < 30$	8	25	200
$30 \le h < 40$	11	35	385
$40 \le h < 50$	7	45	315
	Total = 30		Total = 960

b mean = $\frac{\text{total } fx}{\text{total } f} = \frac{960}{30} = 32$

c 40

d They are estimates because you do not know exact values; you have to use midpoints/class boundaries.

e $30 \le h < 40$

f The median lies in position $\frac{30+1}{2} = 15.5$ th position.

There are 12 values smaller than 30.

The 13th value up to the 24th value lie in the class $30 \le h < 40$ so the median is in this class.

2 a $125 \le m < 130$

b Sarah's mean is lower than the lowest value in the table so this is not a possible value of the mean. There are only 2 oranges in the class $140 \le m < 145$. The others are all under 140 so a mean above 140 is not possible.

c 130.25

d $125 \le m < 130$

3 a 2.5 b 1.6 km c $1 \le d < 2$ d 5

e $1 \le d < 2$

4 a 25 b $0 \le y < 10$ c 80 d 31.4

e $20 \le y < 30$ f $0 \le y < 20$

5 a $20 \le y < 30$ (any letter can be used in place of y)

b 23.5

6 a 27.48 seconds

b The modal class is now $27 \le t < 39$, and 24 is not in that class.

Time (t, seconds)	Frequency
$15 \le t < 27$	23
$27 \le t < 39$	27

7

Length (l, cm)	Frequency
$150 \le l < 160$	3
$160 \le l < 170$	7
$170 \le l < 180$	3
$180 \le l < 190$	2

The modal class is $160 \le l < 170$, but the mode is 152 which does not lie in this class.

Exercise 5B

1 Questions **1**, **2**, **3**, **5** and **6**: students' own answers, discussion or teacher marked suggested

4 a primary b secondary c secondary d primary

Exercise 5C

In many of these questions, students' answers may vary from these and still be valid, these are examples.

1 a quantitative b qualitative

c quantitative d qualitative

e quantitative

2 a question **i**

b *Advantage* – covers all different types of problems with the phone – cracked case, cracked screen, scratches.

Disadvantage – it can be subjective e.g. one person's idea of 'poor' is different to another person's.

c *Advantage* – easy to answer, you just have to count.

Disadvantage – it only covers one type of problem, there may be other problems other than scratches.

3 a They didn't want to give their age.

b They didn't understand the question (due to its ambiguity).

c and **d** Students' own answers, discussion or teacher marked suggested.

4 a He could go to a different place rather than outside the pool.

b He could ask more people.

5 a First question is leading/biased, and who is Dr Raphael? The second question is too subjective and it is not specified what it is compared to. The third question is too subjective and no comparison to times gone by.

It might be worth making it clearer what 4, 3 and 2 mean as only 5 and 1 are defined.

b Students' own answers taking on board most answers from part **a**, discussion or teacher marked suggested.

6 i This is a broad question. You may only eat one type of vegetable but you could still answer 'yes'. Someone may eat vegetables, but not salad, so what would they answer?

ii This is a leading question and students are going to feel under pressure to agree.

iii There is no time frame to this question e.g. students will not know if this is per day, per week, per month, etc.

iv This is a subjective question. One person's idea of a balanced diet may not be the same as another. Also, Beth gives her survey to only 10 people, which is a very small sample.

She gives the survey only to friends, who may all have similar views on healthy eating to her.

7 He could include yes/no answer boxes to tick to avoid open answers which are difficult to analyse. Asking 'what do you think?' in the third question is too open and allows for too many possible answers. He should give response boxes again.

8 a 30% of his sample are managers, but only 10% of the staff in the office are managers.

b Ask 5 managers and 45 ordinary office staff.

c Managers might work longer hours and be less satisfied about this. He might want to bias his survey to this point of view so the boss reduces everyone's hours.

9 Students' own answers

Consolidation

Exercise 5

1 a $12 \le h < 15$ b 15 c 10.86

d $9 \le h < 12$

2 a No, because the modal class is $64 \le h < 66$

b The mean, 63.35, is in the same interval containing the median: $62 \le h < 64$

c 10

3 a i secondary ii primary iii secondary

b i The last question is likely to return the same answer as the first question.

ii Gender is possibly not relevant.

iii 'How old are you?' They *could* answer with the qualitative answers, e.g. 'old' or 'young'.

c i There are twice as many people in her town as the neighbouring town so the sample is not in proportion to the population.

ii It is fair if she wants to ask each person 'which town is better at sport?' to have the same number from each town in the sample, as people are likely to be biased towards their town.

4 Students' own answers, discussion or teacher marked suggested

5 **i** This is a subjective question – one person's idea of a lot of shopping may not be the same as another.

ii and **iv** are leading questions – people are going to feel under pressure to agree/answer positively.

iii There is no time frame to this question, e.g. people will not know if this is per week or per month.

Sunil gives his survey to just 20 people, which is a very small sample for an entire town.

He doesn't have an equal number of men and women and the town is likely to be more equally balanced than this.

He gives the survey only to people leaving one shop so it is likely to be biased to people who do like shopping in shops (specifically sports shops) and not towards people who might prefer shopping online.

Check out

1 47.5, 52.5, 57.5, 62.5

2 **a** 49.2 **b** $40 \le m < 60$

3 **a** 25 **b** $25 \le t < 30$

4 **a** quantitative **b** qualitative
c quantitative **d** quantitative

5 **i** This is a leading question and students are going to feel under pressure to agree.

ii This is a subjective question. One person's idea of a lot of exercise may not be the same as another.

iii There is no time frame to this question.

Henry gives his survey to only 10 students, which is a very small sample.

He gives the survey only to people in the sports hall, so it is likely to be biased.

Review A

1 Rational $^-9, 0.2, \sqrt[3]{8}, 12, \sqrt{25}, \frac{5}{11}, 7\frac{1}{5}, 0.\dot{4}$
Irrational $\pi, \sqrt{7}$

2 **a** m^6 **b** p^3 **c** $20t^5$ **d** y^{-2} **e** x^{12}

3 **a** 7 **b** 2

4 $^-2$

5 **a** quantitative **b** qualitative
c quantitative **d** quantitative
e qualitative

6 **a** $x^2 + 7x + 12$ **b** $x^2 + 5x - 6$
c $x^2 - 6x + 5$ **d** $x^2 + 14x + 49$

7 **a** 3 **b** 8.04 **c** 2.7 **d** 79

8 **a** true **b** true **c** false

9 **a** 9×10^4 **b** 2.8×10^5 **c** 3×10^{-5} **d** 5.21×10^{-3}

10 **a** $40x^9$ **b** $3x^4$ **c** $4t^4y$

11 **a** true **b** false **c** false **d** false
e true **f** false

12 **a** 3 **b** 8

13 **a** 058° **b** 1 cm = 2 km
c 4 km **d** 310°

14 **a** $x^2 + 2x - 15$ **b** $p^2 - 12p + 35$
c $m^2 - 8m + 16$ **d** $y^2 - 49$

15 **a** $90 \le m < 100$
b No, the mean is 100.5
c 50
d $90 \le m < 100$

16 **a** 4^{-3} **b** 3^{-6} **c** 5^{-6} **d** 120^{-2}
e 5^{-6} **f** 3^{-5}

17 **a** $x = \frac{4F+3}{y}$ **b** $x = \frac{5m}{4} - b$ **c** $\frac{Vm}{(d-p)} = x$ **d** $x = \frac{5}{(R-d)}$

18 2, 3

19 20 is a small sample size for a club with 500 members.
50% of his sample are children but 80% of the members are adults.
Wednesday may be a school day and so older children may be in school and just very young children may be asked, and they might not understand the questions.
Question **1** is biased as it leads the person to agree.
Question **2** is too open; it would also be easier to analyse answers if they had response boxes to choose from. Also it is not really relevant to what he is trying to find out.

20 **a** 1 **b** $^-5$ **c** 7^2 or 49
d $^-3$ **e** $\frac{1}{1000}$ or 0.001 **f** $^-1$

21

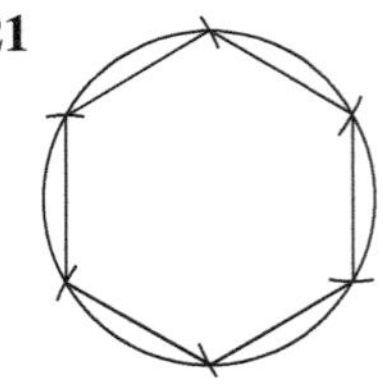

22 **a** $(x + 8)^2$ **b** $(x - 11)^2$
c $(x + 5)(x - 5)$ **d** $(x + 10)(x - 10)$

23 The question "How much pocket money do you get?" doesn't state a time period. It isn't really relevant what class they are in. "Do you earn any money?" is a closed question with a yes or no answer that may require a follow up question "how much?" This may or may not be relevant to the amount of pocket money someone gets.

24 **a** 4 **b** 3180 **c** $^-4$ **d** $^-4$
e 4.4 **f** 3 **g** 18 000 **h** 0.009

25 A regular octagonal prism

26 **a** 3×10^5 **b** 2.24×10^4 **c** 3.1×10^6
d 4×10^2 **e** 5.3×10^{-5} **f** 2.1×10^{-4}

27 The mean is 26.84 which lies in the class $23 \le t < 27$ and the modal class is $27 \le t < 31$

28 **a**

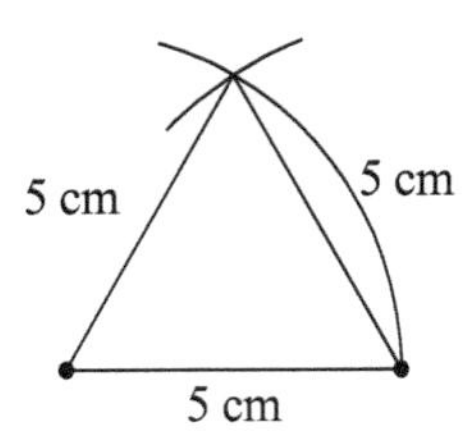

b

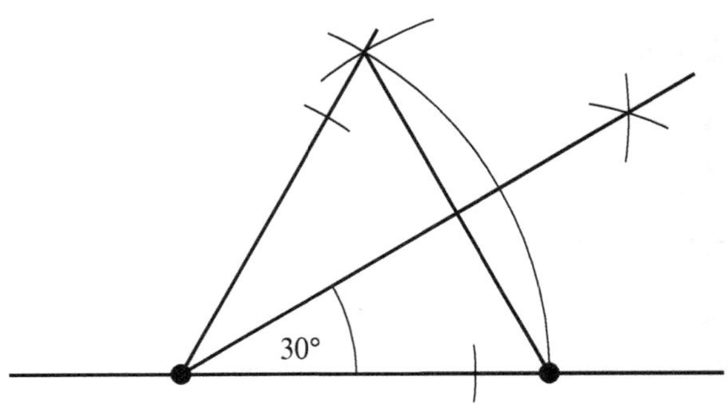

29

Number, n.	Rounded correct to...	Bounds
440	nearest 10	$435 \le n < 445$
6200	nearest 100	$6150 \le n < 6250$
42 000	nearest 1000	$41\,500 \le n < 42\,500$
756 200	nearest 100	$756\,150 \le n < 756\,250$

30 a $\frac{7w}{12}$ **b** $\frac{29e}{35}$ **c** $\frac{2}{3c}$ **d** $\frac{7}{4d}$

e $\frac{7x}{4}$ **f** $\frac{5h+4}{h}$

31 a ± 5.5 **b** 6.7 **c** ± 2.2 **d** 10.7

e $^-7.9$ **f** 13.2

32 a $\frac{4x+1}{3}$ **b** $\frac{x+5y-3z}{2}$ **c** $\frac{2x+7y}{2}$ **d** $\frac{1}{9-3m}$

e $\frac{12-4t}{9}$

33 a 4.1 **b** 1.7 **c** 3.1 **d** 5.4

e $^-2.3$ **f** $^-7.9$

34 a For example, he could go to different places in the school rather than just outside the sports hall and he could ask more people. He could change the time he asks people. His sample is biased because of where he is, people coming out of the sports hall at lunchtime might be likely to enjoy sports as they could be coming out of a lunchtime sports club.

b For example, Mike might want the teachers to put on a tennis club in his school so he might want them to think there are more students than there are who enjoy tennis.

35 a $(x+5)(x+8) - x(x+6)$ **b** $7x + 40$

36

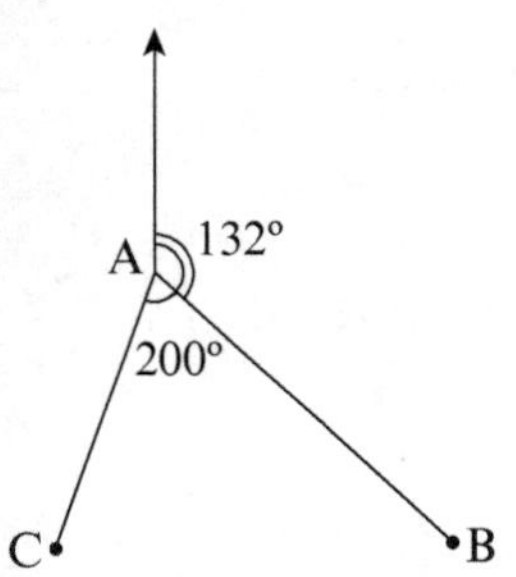

37 a $\frac{3k+dgh}{ghk}$ **b** $\frac{4a^2b+15c^2}{12a}$

38 a 1.84×10^{12} **b** 3.2×10^{-5}

39 Ask the first 50 customers who come out of the cinema looking happy. Ask leading/biased questions such as, 'Do you agree we are the friendliest local cinema?', 'Why would you recommend us to a friend?'

Chapter 6

Check in

1 a 225 cm² **b** 216 m² **c** 135 m²
d 35 mm² **e** 56 cm²

2 a circumference
b double it **c** halve it

Exercise 6A

1 a i 19.6 cm² **ii** 15.7 cm
b i 28.3 cm² **ii** 18.8 cm
c i 0.126 m² **ii** 1.26 m
d i 38.5 cm² **ii** 22.0 cm

2 a 23.6 cm, 88.4 cm² **b** 9.42 cm, 28.3 cm²
c 37.7 cm, 150.8 cm²

3 622 cm²

4 3.78 m

5 a $A = 89.3$ cm², $P = 35.7$ cm
b $A = 25.6$ cm², $P = 18.8$ cm
c $A = 64.3$ cm², $P = 33.4$ cm
d $A = 9.14$ cm², $P = 12.3$ cm

6 17.8 mm

7 2001 (to the nearest whole number of revolutions)

8 38.5 cm²

9 a 10.7 cm² **b** 186 cm² **c** 28.3 cm² **d** 56.5 cm²

10 Area = 9426 m², therefore 2356.5 kg of seed

11 a 3.46 m² **b i** 2 **ii** \$18.33

12 a 1.26 cm, 2.51 cm² **b** 15.3 cm, 53.5 cm²
c 49.5 cm, 223 cm²

Exercise 6B

1 a 9 cm², 16 cm²
b 25 cm²
c $9 + 16 = 25$

2 a 1 square, 9 squares
b 10 squares
c $1 + 9 = 10$

3 d The area of the larger square is equal to the sum of the areas of the two smaller squares.

5 a 20 mm **b** 50 mm **c** 13 mm **d** 29 mm
e 37 mm

6

	a	b	c	a^2	b^2	$a^2 + b^2$	c^2
a	12	16	20	144	256	400	400
b	30	40	50	900	1600	2500	2500
c	5	12	13	25	144	169	169
d	20	21	29	400	441	841	841
e	12	35	37	144	1225	1369	1369

The numbers in the last two columns are the same.

Exercise 6C

1 $h^2 = 7^2 + 24^2$, $h^2 = 49 + 576$, $h^2 = 625$, $h = \sqrt{625} = 25$

2 a 13 cm **b** 17 cm **c** 26 cm **d** 20 cm

3 a 3.61 cm **b** 9.43 cm **c** 5.66 cm **d** 6.08 cm

4 $a^2 + 4^2 = 5^2$, $a^2 + 16 = 25$, $a^2 = 9$, $a = \sqrt{9}$, so $a = 3$

5 a 8 cm **b** 6 cm **c** 9 cm **d** 16 cm

6 a 7.81 cm **b** 8.94 cm **c** 9.54 cm **d** 10.30 cm

7 a i 7.07 cm **ii** 4.24 cm **b i** 8.06 cm **ii** 9.22 cm

8 24.5 cm

9 12 cm

10 6 cm and 11.4 cm

11 26 and 16.6 cm

12 39.4 cm

13 a 54 cm² **b** 420 mm²

14 $29^2 = 20^2 + 21^2$ and $41^2 = 40^2 + 9^2$, so A and C are right-angled.
$18^2 \neq 15^2 + 9^2$ so B is not right-angled.

15 20 units

Exercise 6D

1 a 92 cm² **b** 870 mm²

2 a 480 cm² **b** 246 m²

3 a i 314 cm² **ii** 113 cm² **iii** 565 cm²
b i 471 cm² **ii** 170 cm² **iii** 919 cm²

4 The cylinder with the smallest surface area is A (641 cm²)
The cylinder with the largest surface area is B (858 cm²)
The two cylinders with the same surface area are C and D (855 cm²)

5 Yes, the surface area is 2070 cm^2

6 a 240 cm^2 b 22.56 m^2

7 Students' answers will vary.

8 a 12.3 m^2 b 1.1 litres

9 a Doubling the length only doubles the area of the rectangles, the areas of the triangular faces will be unchanged.
b 1560 cm^2

10 Area of triangular faces is $\frac{1}{2} \times 3 \times 4 \times 2 = 12$ cm^2
Hypotenuse in triangle is 5 cm from Pythagoras ($3^2 + 4^2 = 5^2$)
Area of rectangular faces is $3x + 4x + 5x = 12x$
Total surface area $= 12x + 12 = 12(x + 1)$ and since x is an integer then $x + 1$ is an integer so it is a multiple of 12.

Exercise 6E

1 a 1130 cm^3 b 176 cm^3

2 a 26 cm^2 b 78 cm^3

3 a 416 cm^3 b 432 cm^3 c 259 cm^3

4 a i 181.0 cm^3 ii 128.7 cm^3
b i 12 cm ii 6 cm

5 a 172.8 cm^3 b 202.5 cm^3 c 154 m^3

6 a 311 018 cm^3 b 311 litres c 426

7 a 200 cm^3 b 202.5 cm^3

8 12 cm

9 8 cm

10 a 88 cm^2 b 17 600 cm^3

11 a 628 cm^3 b 72 cm^3 c 556 cm^3

12 a 28.3 cm^2 b 226 cm^3

Exercise 6F

1 a light years b tonnes
c micrometres d micrograms
e megametres f nanograms
g terabytes h microlitres
i gigabytes

2 a true b false, 500 000 μm = 0.5 m
c true d true
e false, 0.000 04 t = 40 000 000 μg
f false, 2 GB = 2000 MB

3 a i 2 MB ii 2 TB b i 0.7 ng ii 0.7 t
c i 3 μm ii 3 ly

4 a 38 b 5000
c heavyweight paper and thick card d 3

5 250

6 a Yes, he eats 7.2 + 19.5 + 64 = 90.7 mg of vitamin C.
b No, she eats 64 + 3.5 + 2.9 = 70.4 mg of vitamin C.

7 a 6×10^{12} g b 6×10^9 kg
c 6×10^6 t d 6 Mt

8 a 6 g b 300 μg

9 a 500 t b 2 Mt

Consolidation

Exercise 6

1 a i 25.1 cm ii 50.3 cm^2
b i 18.8 cm ii 28.3 cm^2

2 a 20 cm b 39 cm c 8.06 cm (2 d.p.)
d 20 cm e 13.86 cm (2 d.p.) f 8.66 cm (2 d.p.)

3 a 56.5 cm^2 b 11.8 cm^2

4 6.18 cm

5 3.61 m

6 6.71 m

7 7.75 m, i.e. 1.04 m higher

8 6.93 m

9 a i 136 m ii 1204 m^2
b i 190 cm ii 438 cm^2
c i 26.5 cm ii 39.4 cm^2

10 a 252 cm^3 b 308.4 cm^2

11 a 700 b 0.3 c 0.07 d 0.4

12 134 mm^2

13 a 6.28 cm^3 b 1539 cm^3 c 942 mm^3 d 0.0707 m^3

14 35.75 m^3

15 a 63.6 m^2 b 113 m^2 c 49.5 m^2

16 a i 0.3 ng ii 0.3 t
b i 0.05 GB ii 5000 MB
c i 0.000 05 mm ii 0.005 ly

17 a 176.4 cm^3 b 3695 cm^3

18 a 132 cm b 1500

19 17.5 cm

20 a 517.5 m^3 b 432.5 m^2

21 125

22 a $r = 1, h = 0.2$ b 62.8 cm^3

23 20

24 a 68.2 cm^2 b 2047 cm^3

25 a 2 m b 4 s

26 a 86 394 cm^3 b 0.0864 m^3

Check out

1 a 63.6 cm^2 b 26.4 cm^2

2 a 5 b 11.3 (1 d.p.) c 28.9 (1 d.p.)

3 b is a prism

4 972 m^3

5 1176 cm^2

6 603 cm^2

7 a 34 000 ns = 34 μs
b 4 GB = 4000 MB
c 70 mg = 70 000 μg
d 800000000 km = 800 Gm

Chapter 7

Check in

1 a $\frac{7}{20}$ b $\frac{3}{8}$ c $\frac{13}{63}$ d $\frac{11}{56}$
e $\frac{1}{4}$ f $\frac{1}{4}$

2 a $5\frac{7}{12}$ b $2\frac{1}{4}$ c $9\frac{5}{12}$ d $2\frac{11}{35}$

3 a 16 b 27 c 16

Exercise 7A

1 a 1, 3, 5, 15 b 1, 2, 3, 4, 6, 8, 12, 24
c 1, 2, 3, 5, 6, 10, 15, 30 d 1, 5, 25

2 Factors of 60 are 1, 2, 3, 4, 5, 6, 10, 12, 15, 20, 30, 60
Factors of 84 are 1, 2, 3, 4, 6, 7, 12, 14, 21, 28, 42, 84
HCF of 60 and 84 is therefore 12

3 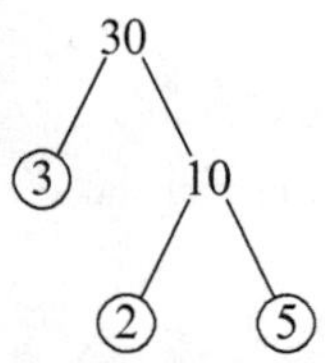

Rani's working is correct.

4 Students' own answers

5 **a** 10 **b** 4 **c** 6 **d** 18

6 **a** $\frac{2}{3}$ **b** $\frac{7}{15}$ **c** $\frac{3}{7}$

7 **a** $\frac{10}{21}$ **b** $\frac{4}{35}$

8 28 m

9 My number could be odd or even.

10 60 cm

11 You can tell a number is prime because it has only two factors.
8 is not prime because it has 1, 2, 4 and 8 as its factors, 8 is a composite number.
16 is a square number because it has an odd number of factors 1, 2, 4, 8, 16.
7 is a prime number because it has only 1 and 7 as its factors.
12 is not a square number because it has an even number of factors.
1 is not prime because its only factor is 1.

Exercise 7B

1 **a i** smaller **ii** $\frac{1}{10}$ **b i** bigger **ii** $\frac{14}{15}$

c i smaller **ii** $1\frac{1}{3}$ **d i** bigger **ii** $2\frac{2}{3}$

2 **a** $4\frac{1}{2}$ **b** $2\frac{1}{4}$ **c** $2\frac{13}{16}$ **d** $9\frac{1}{6}$ **e** $6\frac{1}{4}$ **f** $8\frac{29}{32}$

g $15\frac{5}{6}$ **h** $20\frac{20}{33}$

3 **a i** bigger **ii** $\frac{9}{11}$

b i bigger **ii** $5\frac{2}{5}$

c i bigger **ii** $6\frac{1}{4}$

d i smaller **ii** $\frac{2}{5}$

4 The correct working is:

$$3\frac{2}{3}\div 1\frac{1}{11}$$
$$=\frac{11}{3}\div\frac{12}{11}$$
$$=\frac{11}{3}\times\frac{11}{12}$$
$$=\frac{121}{36}$$
$$=3\frac{13}{36}$$

5 **a** $\frac{1}{2}$ **b** 2 **c** $2\frac{5}{8}$ **d** $1\frac{11}{21}$

e $2\frac{16}{25}$ **f** $2\frac{5}{13}$ **g** $2\frac{23}{90}$ **h** $\frac{9}{16}$

6 **a** $5\frac{5}{6}$ m **b** $24\frac{3}{10}$ kg **c** $24\frac{8}{15}$ ml **d** $2\frac{1}{6}$ t

e $\frac{1}{4}$ **f** $\frac{2}{15}$ **g** $\frac{49}{64}$

7 $\frac{4}{5}$ kg

8 $11\frac{9}{10}$ cm^2

9 $12\frac{4}{9}$ m

10 **a** 219 **b** $\frac{433}{658}$

11 **a i** $11\frac{1}{5}$ and $5\frac{3}{5}$ **ii** $2\frac{3}{5}$ and $1\frac{1}{15}$ **iii** $2\frac{3}{5}$ and $1\frac{2}{13}$

b i $11\frac{1}{5}$ and $1\frac{1}{15}$ **ii** $2\frac{3}{5}$ and $5\frac{3}{5}$ **iii** $11\frac{1}{5}$ and $5\frac{3}{5}$

12 **a** $-3\frac{7}{16}$ **b** $^-1$ **c** $\frac{-7}{32}$ **d** $3\frac{6}{25}$

13 **a** $2\frac{3}{5}$ **b** $1\frac{1}{8}$ **c** 3 **d** $5\frac{2}{3}$

e $\frac{9}{25}$ **f** $1\frac{3}{5}$

14 **a** 20 **b** 2

d the second number is 10 times bigger than the first

Exercise 7C

1 **a** $77\frac{1}{4}$ **b** $33\frac{1}{2}$ **c** $13\frac{1}{8}$ **d** $63\frac{1}{3}$

e $27\frac{3}{8}$

2 **a** 2.63 **b** 8.64

3 **a** $9\frac{4}{13}$ **b** $13\frac{2}{15}$

4 **a** $2\frac{1}{4}$ **b** $1\frac{1}{8}$ **c** 19 **d** $1\frac{25}{48}$

e $2\frac{3}{8}$ **f** $\frac{13}{72}$ **g** $5\frac{7}{15}$ **h** 0

i $4\frac{2}{3}$ **j** 1 **k** $\frac{8}{15}$ **l** $1\frac{3}{5}$

5 **a** $2\frac{1}{3}$ **b** $4\frac{2}{5}$ **c** $\times\frac{1}{2}$ or $\div 2$

6 \$31.50

7 **a** $\frac{15}{44}$ **b** $\frac{27}{32}$ **c** $1\frac{2}{3}$ **d** $1\frac{2}{3}$

e $\frac{14}{15}$ **f** $\frac{5}{12}$ **g** $1\frac{23}{26}$ **h** $\frac{10}{27}$

i 0 **j** $\frac{7}{20}$ **k** $\frac{77}{150}$

Exercise 7D

1 **a** $0.\dot{6}$ **b** 7.125 **c** $0.8\dot{3}$ **d** 5.68

e $0.\dot{1}\dot{9}$ **f** $0.58\dot{3}$

2 **a** $0.\dot{5}$ **b** $0.\dot{2}$ **c** $0.\dot{1}\dot{2}$ **d** $0.\dot{0}\dot{8}$

e $0.\dot{2}\dot{5}$ **f** $0.\dot{6}\dot{8}$

3 Recurring decimals: $\frac{1}{3}, \frac{11}{9}, \frac{4}{7}, \frac{11}{14}, \frac{11}{12}, \frac{6}{17}, \frac{5}{18}, 12\frac{2}{3}, \frac{155}{99}$

Terminating decimals: $\frac{9}{12}, \frac{8}{250}, \frac{4}{25}, \frac{3}{6}, \frac{423}{1000}, \frac{13}{26}, \frac{9}{36}, \frac{15}{8}, 6\frac{103}{125}$

4 **a** $0.\dot{4}$ **b** $0.\dot{2}\dot{3}$ **c** $0.\dot{8}7\dot{2}$ **d** $0.\dot{5}46\dot{7}$

Consolidation

Exercise 7

1 **a** $\frac{2}{5}$ **b** $\frac{33}{43}$ **c** $1\frac{2}{7}$ **d** $\frac{12}{25}$ **e** $1\frac{47}{48}$

2 **a** $5\frac{1}{12}$ **b** $\frac{33}{46}$ **c** $\frac{1}{3}$ **d** $\frac{1}{2}$ **e** $2\frac{7}{19}$

3 $1\frac{4}{5}$ m

4 $2\frac{1}{2}$ hours

5 $\frac{5}{26}$

6 6

7 8

8 3

9 **a** $\frac{13}{34}$ **b** $\frac{7}{32}$

10 \$280

11 **a** 0 **b** 81 **c** 29 **d** 1

12 **a** $0.\dot{8}$ **b** $0.\dot{4}$ **c** $0.\dot{3}\dot{1}$ **d** $0.\dot{0}\dot{4}$

13 Recurring decimals: $\frac{2}{3}, \frac{14}{9}, \frac{8}{14}, \frac{7}{12}, \frac{6}{17}, 167\frac{1}{3}$

Terminating decimals: $\frac{3}{12}, \frac{17}{250}, \frac{467}{1000}, \frac{14}{28}, \frac{27}{36}, 10\frac{13}{125}$

Check out

1 a 6 b $6\frac{1}{4}$ c $2\frac{1}{5}$ d $1\frac{40}{81}$
e $1\frac{3}{32}$

2 a $41\frac{1}{2}$ b $5\frac{3}{4}$ c 45 d 78
e $7\frac{2}{5}$

3 a $0.\dot{3}$ b $0.\dot{7}\dot{8}$ c $0.\dot{0}\dot{2}$

4 $\frac{8}{18}, \frac{5}{12}, \frac{25}{3}, \frac{31}{56}$

Chapter 8

Check in

1 a < b > c < d >
e < then ≤

2 a $m+2t$ b $5x-5y$
c $^{-}2p+11+4v$ d $^{-}2f-4g-8$

3 a i $15p-6$ ii $^{-}24+18x$
iii x^2+5x iv $15p^2-12p$
b i $52p-36m$ ii $30+30T$

4 a Equation b Expression
c Expression d Formula/function

5 a $x=\frac{y-1}{2}$ b $x=\frac{2y-5}{3}$

Exercise 8A

1 a $x=8$ b $x=^{-}1$

2 a $x=3$ b $x=3$

3 a $x=6$ b $x=24$ c $x=^{-}27$

4 a $p=5$ b $x=9$

5 a $d=2$ b $x=72$

6 a $x=4$ b $x=4$

7 a $x=3$ b $x=2$ c $n=12$

8 a $x=5$ b $y=2$

9 a $y=1$ b $x=3$

10 a $4(x+3)=32$ b $x=5$

11 a $x+x+2+x+4=6x$ b $2=x$, sides are 2, 4 and 6 cm

12 a $2y+8=32$ b 12 cm

13 a $x+20$ b $2x+20$ c 55

14 a $2x+19$ cm b 5.5

15 5 cm and 9 cm

16 9 years old and 12 years old

17 23, 24 and 25

18 21, 23 and 25

19 112 and 114

20 a Let x = cost of a child's ticket. Then $2x$ = the cost of an adult ticket.
Mr and Mrs Brown + 5 children:
$2x+2x+5x=315$, or $9x=315$
b Child's ticket = 35 cents; adult's ticket = 70 cents

21 $2000-5t=75$, $t=385$ (Thus, the cost of each ticket is \$3.85)

22 $2(3n+7)=80$, $n=11$

23 $50x-900=900$, therefore $x=36$

Exercise 8B

1 a 2 b 10 c 0.5 d 2.5

2 Enri forgot the brackets. This is the correct working and solution:
$2=\frac{22}{x+4}$ [multiply by $x+4$]
$2(x+4)=22$ [divide by 2]
$x+4=11$ [subtract 4]
$x=7$

3 a 4 b 7 c 5 d 13
e 4 f 17

4 $\frac{570}{x+7}=15$, so $x=31$

5 18

6 a $\frac{145}{x-8}=5$
b 37

7 a 14 b 25

8 a $^{-}2$ b $^{-}1$

9 a 7 b 1.3

Exercise 8C

1 a $x>3$ b $x>5$ c $x\le 3$ d $x<3$
e $x>2$ f $x\ge 1.5$ g $x>1.5$ h $x<3$

2 a $x<5$ b $x\ge 9$ c $x>3$ d $x>7$
e $x\ge 3$ f $x\le 2$ g $x>8$

3 $3\times 10-5=25$
$2\times 10+7=27$
$25 \not> 27$

4 a She has forgotten to reverse the inequality sign when dividing by a negative number.
b $5-2\times{}^{-}4=5+8=13$ which is greater than 11

5 a $x\le 6$ b $x>^{-}6$

6 $20+0.5d\ge 55$

7 a yes b yes c yes d no

8 a $x<^{-}2$ b $x\ge 5$ c $x\le ^{-}4$ d $x>^{-}6$ e $x\le ^{-}3$

9 a $x\le 3$ b $x>5$ c $x\ge 2$ d $x\le 6.5$ e $x<10$

10 $x<7$

Exercise 8D

1 For Question **1** of Exercise 8C:

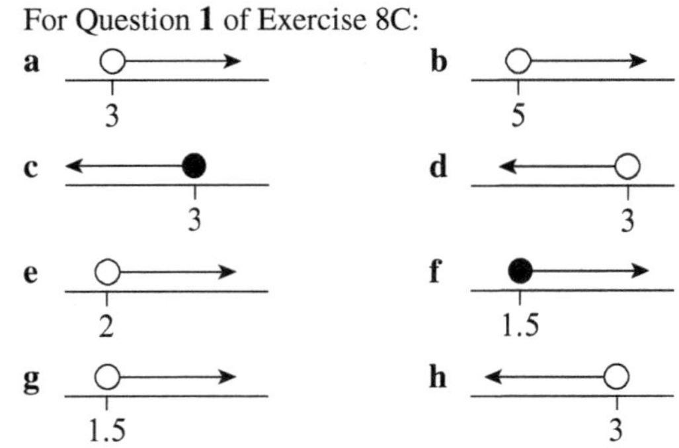

For Question **2** of Exercise 8C:

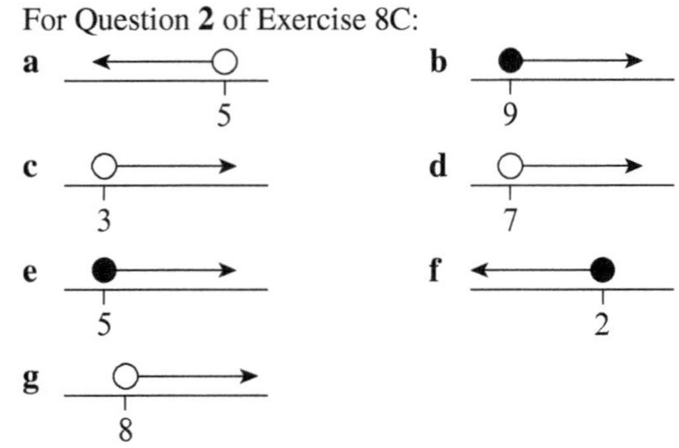

2 **a** 5, 6, 7, 8, 9 **b** 1, 2 **c** 6, 7, 8, 9
d 1, 2, 3 **e** 7, 8, 9 **f** 1, 2, 3, 4

3 **a** 16, 17, 18, 19, 20 **b** 18, 19, 20 **c** 1, 2, 3
d 1, 2, 3, 4, 5 **e** 1, 2, 3 **f** 15, 16, 17, 18, 19, 20

4 **a** $x < 2$ **b** $x \geq {}^{-}2$ **c** ${}^{-}1 < x < 2$
d ${}^{-}5 \leq x < 10$ **e** $0 \leq x \leq 30$

5 **a** ${}^{-}2 < x < 6$ **b** $9 < x \leq 16$ **c** $1 \leq x \leq 7$
d $4 \leq x < 10$ **e** $1 < x \leq 4$

6 **a** **i**
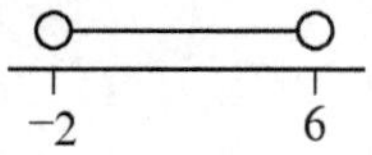

ii ${}^{-}1$, 0, 1, 2, 3, 4, 5

b **i**
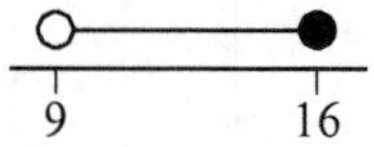

ii 10, 11, 12, 13, 14, 15, 16

c **i**
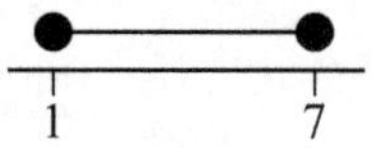

ii 1, 2, 3, 4, 5, 6, 7

d **i**
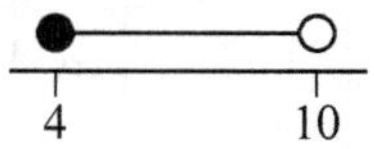

ii 4, 5, 6, 7, 8, 9

e **i**
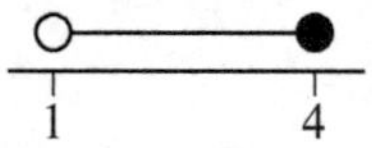

ii 2, 3, 4

7 **a** 24, 25 **b** 21, 22, 23 **c** 22, 23, 24, 25

8 **a** 5, 6, 7 **b** 6, 7

9 **a** 10, 11, 12 **b** 4, 5, 6

10 length must be greater than 8 cm

11 width must be greater than 15 cm

12 Amy is younger than 9

13 **a**
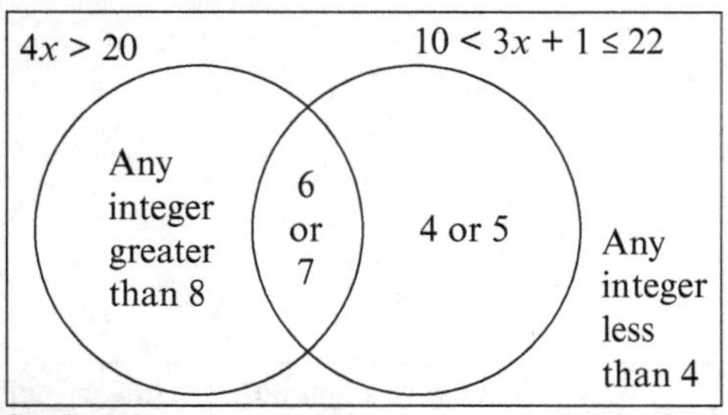

b Students' own answers

14 Students' own answers

15 **a** $2 \leq x < 6$ **b** ${}^{-}1 \leq x < 3$ **c** ${}^{-}3 < x \leq 9$

Exercise 8E

1 Only graph B could be that of $y = 3x - 4$

2 **a** $y = 3x + 5$ **b** $y = 2x - 3$
c $y = 7 - x$ **d** $y = {}^{-}1 - 2x$

3 **a** y-intercept = 3, gradient = 2
b y-intercept = ${}^{-}2$, gradient = 3
c y-intercept = $\frac{1}{3}$, gradient = $\frac{4}{3}$
d y-intercept = 4, gradient = ${}^{-}2$
e y-intercept = $\frac{1}{3}$, gradient = $\frac{{}^{-}1}{2}$
f y-intercept = $\frac{{}^{-}5}{2}$, gradient = 2
g y-intercept = 14, gradient = −6
h y-intercept = $\frac{8}{3}$, gradient = $\frac{{}^{-}1}{6}$

4 **b**, **c** and **e**

5 **a** and **c**, **b** and **f**

6 **a** $y = 2x$ **b** $y = 2x + 1$ **c** $y = 2x + 5$
d $y = 2x - 3$ **e** $y = 2x + \frac{3}{2}$

7 **a** $y = 3x + 2$ **b** $y = 5x + 2$ **c** $y = \frac{x}{2} + 2$
d $y = {}^{-}2x + 2$ **e** $y = 2$

8 **a** $y = 3x + 5$ **b** $y = 4x - 2$ **c** $y = {}^{-}x + 1$

9 **a** 2 **b** $\frac{1}{2}$ **c** ${}^{-}2$ **d** 2
e $\frac{1}{2}$ **f** ${}^{-}2$

10 **a** a and d, b and e, c and f
b c and f

11 **a** $y = 3x + 2$ **b** $y = 2x + 4$ **c** $y = 4x - 1$
d $y = 4x - 3$

Exercise 8F

1 **a** $x = 1, y = 1$ **b** $x = 2, y = 1$ **c** $c = 1, d = 1$
d $x = 3, y = 1$ **e** $x = 3, y = 5$ **f** $a = 8, b = 2$

2 **a** $x = 1, y = 1$ **b** $x = 1, y = 1$ **c** $x = 1, y = 3$
d $a = 1, b = 1$ **e** $x = 4, y = 1$ **f** $m = 5, g = {}^{-}2$

3 $x = 7, y = 8$

4 **a** $6x + 3y = 60$ (result of [1] – [2]) or ${}^{-}6x - 3y = {}^{-}60$ (result of [2] – [1])
b $x = 5, y = 10$

5 When the coefficients of the variable you are eliminating are the same sign, you subtract; when the coefficients of the variable you are eliminating are the opposite sign, you add.

6 Small 250 ml, large 400 ml

7 Both are about the same level of difficulty, with adding being just slightly easier.

8 In Greg's method, all the values end up positive for the resulting equation.

9 12 and 7

Exercise 8G

1 **a** $x = 4, y = 1$ **b** $x = 1, y = 1$
c $x = 3.8, y = 1.6$ **d** $x = 2, y = 2$

2 **a** $x = 1, y = 1$ **b** $x = 1, y = 1$ **c** $x = 5, y = 2$
d $x = 1, y = 2$ **e** $x = 1, y = {}^{-}1$ **f** $x = 2, y = 2$
g $x = 0.5, y = 1.5$

3, 4 Students' own answers

5 12 children and 4 sheep

6 15 bottles of still and 8 bottles of sparkling

7

8 21 cm²

Investigation

✸ = 5 ○ = 3 ◆ = 7

So the second row totals 20

Exercise 8H

1 **a** $x=4, y=1$ **b** $x=1, y=1$ **c** $x=1, y={}^{-}1$
d $x=1, y=3$

2 **a** $x=2, y=1$ **b** $x=3, y=3$

3 **a-d** Students' own answers

Consolidation

Exercise 8

1 **a** $x=15$ **b** $x=12$ **c** $x=13$ **d** $x=4$
e $x=5$ **f** $x=40$ **g** $x=12$

2 **a** $\frac{266}{x+10}=7$ is correct **b** $x=28$

3 **a** $x<2$ **b** $x>4$ **c** $x<12$ **d** $x<7$
e $x\geq 3$ **f** $x<3$ **g** $x>0.5$ **h** $x\leq 6$
i $x<2$

4 **a** $x=7, y=3$ **b** $x=2.5, y={}^{-}3.5$
c $x=8, y=3$ **d** $x=1, y=0.5$
e $x=1, y={}^{-}1$ **f** $x=1, y=1$

5 **a** $x\geq 11$ **b** $x<12$ **c** $x<4$ **d** $x>4$
e $f<2.5$ **f** $y\leq 2$ **g** $t>4$

6 **a** $80w+150\leq 500$ **b** 4 **c** \$30

7 **a** $x=9, y=2$ **b** $r=6, t={}^{-}1$

8 $e=23.5, n=18$

9 **a** \$30 **b** \$20

10 **a** 1 **b** $\frac{3}{2}$ **c** $\frac{4}{5}$ **d** $-\frac{3}{5}$ **e** $-\frac{1}{2}$

11

Not a linear graph	A linear graph with a positive gradient	A linear graph with a negative gradient
7y × 3x = ⁻2	7y - 3x - 2 = 0	7y + 3x = 2

12 $b=17, a=15$, so the cost of 7 bananas and 8 apples = \$2.39

13 **a** $y=2x+3$ **b** $y={}^{-}x+5$ **c** $y=3x-6$

Check out

1 **a** 2 **b** 3 **c** 11

2 **a** $x<3$ **b** $x>4$ **c** $x\leq 0.5$ **d** $x\geq 3$

3 **a** $\frac{4}{3}$ **b** $\frac{{}^{-}3}{2}$ **c** 3

4 **a** $y=\frac{3}{2}x-1$ **b** $y=x+1$

5 **a** $x=2, y=1$ **b** $x=3, y=1$
c $x={}^{-}1, y=2$ **d** $x={}^{-}18, y={}^{-}15$

6 **a** $x=5, y=1$ **b** $m=9, p={}^{-}3$

Chapter 9

Check in

1 **a** 52° **b** 45° **c** 131°

2 **a** 24° **b** 40° **c** 68°

3 **a** 60° **b** 50°

Exercise 9A

1 Students' own answers

2 a 1080° b 900° c 1440° d 1260°

3 a

Shape	Number of sides, n	Sum of interior angles	$n - 2$	$(n - 2) \times 180°$
Triangle	3	180°	1	180°
Quadrilateral	4	360°	2	360°
Pentagon	5	540°	3	540°
Hexagon	6	720°	4	720°
Heptagon	7	900°	5	900°
Octagon	8	1080°	6	1080°
Nonagon	9	1260°	7	1260°
Decagon	10	1440°	8	1440°

b They are the same

c i 1620° ii 3240° d 17

4 a 120° b 126° c 107° d 90°

5 Any two angles, each less than 180° with a sum of 240°

6 $180(n-2)°$

7 No, with a correct explanation e.g.

- a 12-sided shape has interior angles summing to $(12 - 2) \times 180 = 1800$ and $3 \times 360 = 1080$
- 3 lots of 360 is the same as 6 lots of 180 which gives the sum of interior angles of an octagon

8 $x = 140°, y = 125°, z = 65°$

9 It is not a multiple of 180°

10 160°

11 $x = 20$

Exercise 9B

1 interior = 108°, exterior = 72°

2 a i 60° ii 45° iii 40° iv 51.4°
b i 120° ii 135° iii 140° iv 128.6°

3 Equilateral triangle 3, 120°, 60°; Square: 4, 90°, 90°; Hendecagon: 11, 32.7°, 147.3°; Dodecagon: 12, 30°, 150°

4 exterior = 10°, interior = 170°

5 a $a = 72°, b = 108°$ b $x = 56°$ c $m = 95°$
d $d = 115°, e = 120°$

6 a 16 b 30

7 a 20° b 18

8 42

9 No (see Teacher Book for explanation)

10 96°

Exercise 9C

1-3 Students' own answers

4 There are gaps between the circles.

5 b i, ii and iv tessellate

7 a $a = e = 60°, b = 90°, c = d = 30°$
b $a = d = e = 70°, b = c = 110°$
c $a = c = d = e = 70°, b = 40°$

8 rotation and translation

9 a yes b 360° c 60°

Exercise 9D

The quadrilateral will be a square, because the exterior angles of two octagons add to 270°, thus requiring a further 90° to make 360°.

1 Interior angle of a pentagon is 108°. $3 \times 108 = 324$ and $4 \times 108 = 432$. Neither of these answers are 360°, to tessellate the angles must make 360°.

2 Interior angle of an octagon is 135°. $3 \times 135 = 405$ and $2 \times 135 = 270$. Neither of these answers are 360°, to tessellate the angles must make 360°.

3 Interior angle of a heptagon is 128.6°. $3 \times 128.6 = 385.7$ and $2 \times 128.6 = 257.1$. Neither of these answers are 360°, to tessellate the angles must make 360°.

4 a rhombus
b 36°, 36°, 144°, 144°

5 square because two octagons make 270° need 90° to make that up to 360°

6 a rhombus and an equilateral triangle.

Exercise 9E

1 a $x = 120°, y = 60°, z = 80°$
b $p = 103°, q = 77°, r = 103°, s = 77°$
c $l = m = n = 78°$
d $r = 123°, u = 135°, v = 57°, w = 135°$

2 $n = 60°, o = 78°$

3 $a = 123°, b = 83°, c = 140°, d = 97°$

4 $w = 80°, x = 35°, y = 65°, z = 94°$

5 a 67 b 120 c 66 d 72

6 $x = 65, y = 58$

7 There are many possible explanations, here is one:
DEC = 60°, angles on a line add up to 180°
EDC = 30°, angles in a triangle add up to 180°
BCD = 30°, alternate angle to EDC
ADC = 125°, angles on a line add up to 180°
ABC = 125°, angles in a quadrilateral add up to 360°
Since ADC = ABC and AD = AB then ABCD is a kite.

8 a $a = 113°, b = 67°, c = 126°$
b $d = 66°, e = 48°, f = 73°$

9 72°, 108°, 72°, 108°

10 $a = 36°$ $b = 72°$

11 $x = 30°$ $y = 120°$ $z = 60°$

Exercise 9F

1 a (4, 6)
b i (−1, 13) ii (5, 9)
c i (0, 4) ii (−2, 1)
d i (−2, 11) ii (2, 5)

2 a (8, −4) b (3, 3) c (58, 68)

3 The sketch is wrong. It should look like this:

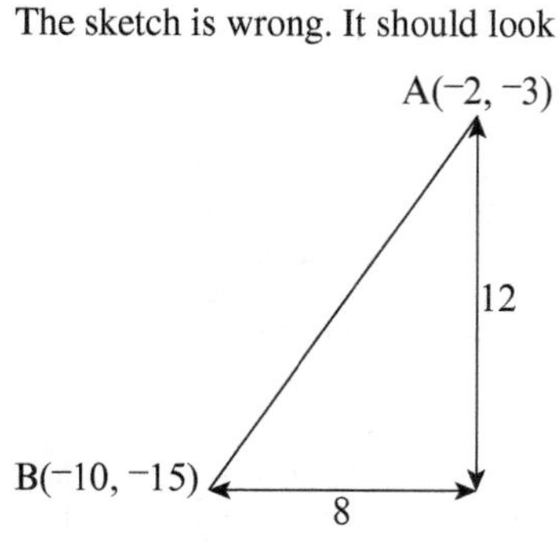

He should be subtracting:
x-coordinate is $^{-}2 - 8 \times \frac{1}{4} = -4$
y-coordinate is $^{-}3 - 12 \times \frac{1}{4} = {}^{-}6$
The correct point is ($^{-}$4, $^{-}$6)

4 **a** (5, $^{-}$11) **b** (9, $^{-}$15) **c** (31, $^{-}$5)

5 (6, $^{-}$4)

6 **a i** (6, 10) **ii** (24, 40) **iii** (42, 70)
b Q
c 72 is not a multiple of 5
d A

7 No, and correct explanation, e.g.
- She subtracted the y-coordinates the wrong way round.
- She should have subtracted 15 from the y-coordinate not added it.

The working/answer should be
$^{-}12 - 8 = -20$
$^{-}20 \times \frac{3}{4} = {}^{-}15$
$(21 + 9, 8 - 15) = (30, {}^{-}7)$

Consolidation

Exercise 9

1 **a** 63 **b** 115

2 144°

3 **a** 15 **b** 18

4 **a** (1, 4) **b** ($^{-}$3, $^{-}$2)

5 1960 is not a multiple of 180

6 **a** 58 **b** 10

7 **a** 156° **b** 24°

8 No, as angles do not add up to 360° when placed together.

9 **a** (8, $^{-}$8) **b** (12, $^{-}$12) **c** (42, 6)

10 1260°

11 **a** (17, $^{-}$6) **b** (1, $^{-}$11)

12 Four quadrilaterals can be fitted together to make an angle of 360°.

Check out

1 **a i** 720° **ii** 1080°
b i 360° **ii** 360°

2 exterior = 30°, interior = 150°

3 Students' own constructions

4 **a** ($^{-}$10, 10) **b** ($^{-}$9, 8)

Chapter 10

Check in

1 mean 18, median 18, mode 22, range 10

2 **a**

1	7
2	6 6 7
3	0 1 8
4	0 2 5 6
5	1 5 7
6	0 1

Key: 2|7 = 27 years

b **i** 41 **ii** 26 **iii** 44

3 **a** **i** discrete data takes only certain values in a given range (e.g. things you count), continuous data takes all values in a given range (e.g. things you measure)
ii primary data is data you collect yourself, secondary data has been collected by someone else
b data collection sheets are like a simplified version of a questionnaire, often used in interviews to quickly collect a lot of data all on a small number of sheets of paper

4 **a** **Results of a hockey team**

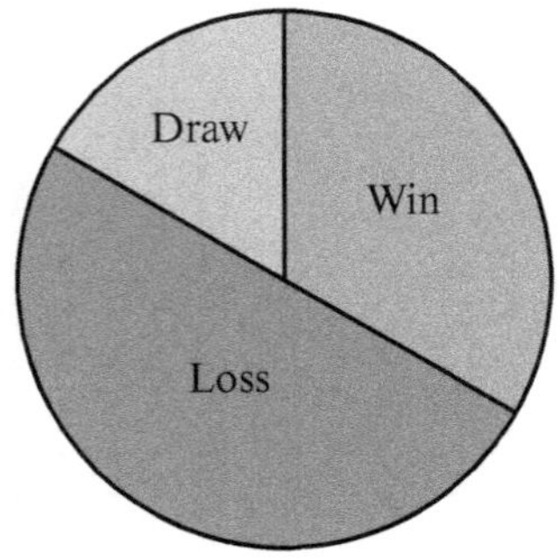

b **Results of a hockey team**

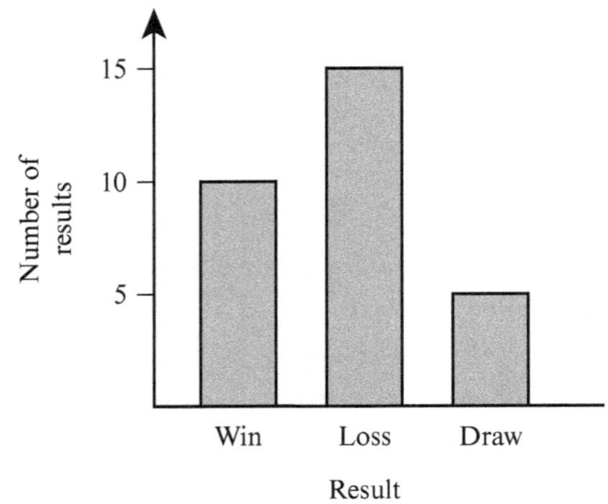

c **Results of a hockey team**

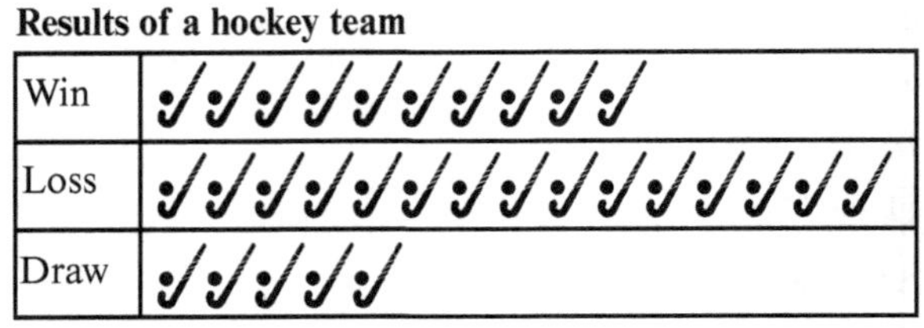

Scale: ✓ represents five results

d The bar chart was easiest to read.

Exercise 10A

1 Emma should use the midpoints of the interval and join with straight lines.

2

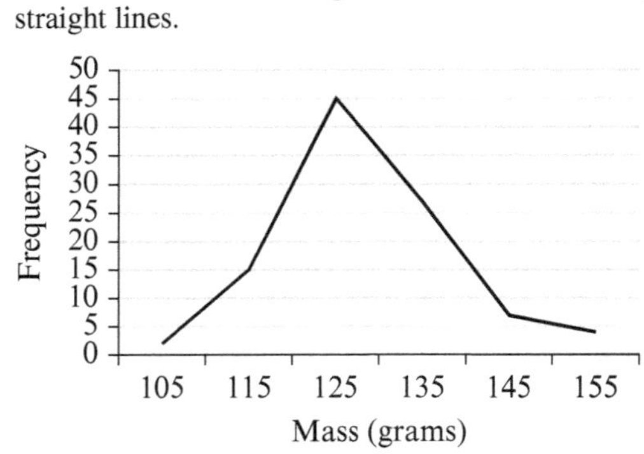

3 **a** 160 **b** 22.5 mm

c

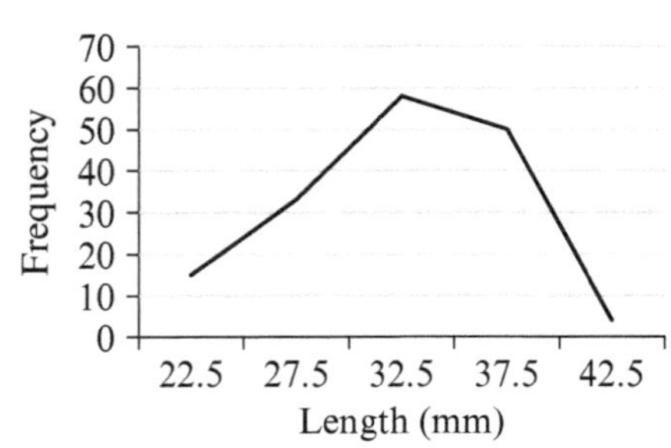

4 a 525 kg

b

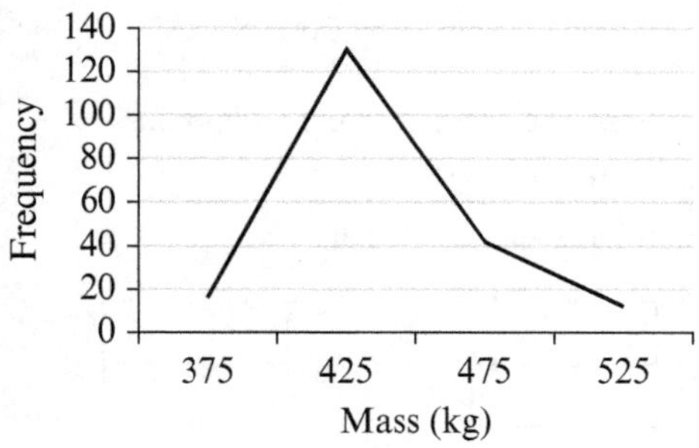

5 a

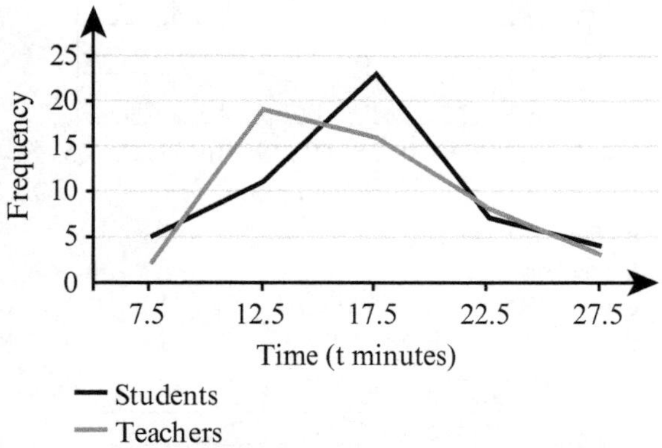

b Teachers generally take less time to travel to school.

c Any reason why it may not be fair to compare these distributions, for example: Modes of transport may not be similar (more teachers may drive and more students may walk.)

6 a

Cost ($d)	Frequency
$1500 \le d < 2000$	4
$2000 \le d < 2500$	14
$2500 \le d < 3000$	13
$3000 \le d < 3500$	4
$3500 \le d < 4000$	5

b We only know there were 4 cars sold in the interval $1500 \le d < 2000$, we do not know the exact value of the cars.

c Could possibly be true: **i**, **iii**
Definitely not true: **ii**, **iv**

Exercise 10B

1 a machine A as it has more washers over 3 mm

b i 4.8 mm **ii** 4.8 mm

c i 1.6 mm **ii** 1.7 mm

d i 3.5 mm **ii** 2.8 mm

e i 2.5 mm **ii** 2.8 mm

f **d** is the only one that really supports the conjecture that machine A produces thicker washers

g the median is the most reliable and meaningful

h The range for A is 32, and for B is 31. They are fairly consistent with each other in the washer thickness they make.

2 a

Maths		Geography
8 5 3	4	2
7 4 2 1 0	5	1 7
9 6 3	6	0 2 5
1 0	7	2 3 5 9
2	8	0 8 8
0	9	1 7

Key
4 | 2 = 42 and 3 | 4 | = 43

b Geography looks easier as more students got higher marks

c Maths 60.7 Geography 72, yes

3 a i 14 **ii** 1

b i 0 **ii** 9

c club B, it has more older people

e i 25 **ii** 47

iii Ages in club A are more consistent with ages in club B

4 a

KMA School		Highfield Academy
9 8 8	0	6 7 8
8 8 7 6 5 5 0	1	3 3 5 6 8 9
5 5 2 1	2	0 1 4 4
3 0	3	2 3 8
6 1	4	1 4

Key
0 | 6 = 0.6 km and 8 | 0 | = 0.8km

b There isn't much difference between the two schools

Exercise 10C

1 Always true **d**, **f**, **i**; Sometimes true **b**, **c**, **e**; False **a**, **g**, **h**

2 a i positive correlation **ii** the taller you are the longer your arm length

b i positive correlation **ii** the longer your leg length the faster your maximum speed

c i no correlation **ii** mass and history score are not related

d i negative correlation **ii** the bigger the car the lower the fuel economy

3 a \$32 000 **b** positive **c** 27

4 a negative correlation

b, d

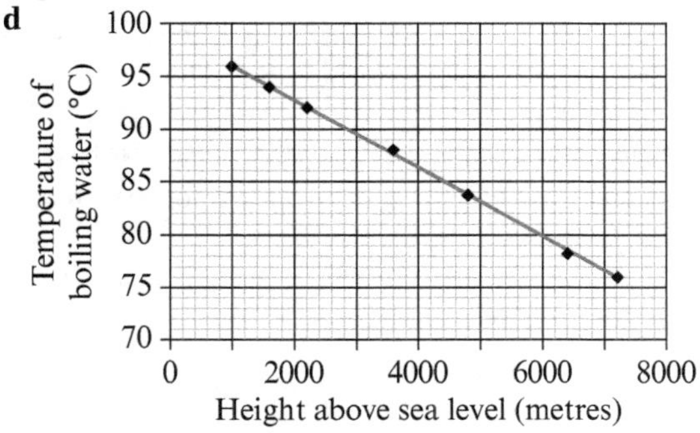

c supports **e** 90°C **f** 6000 m

5 a positive correlation

b, d

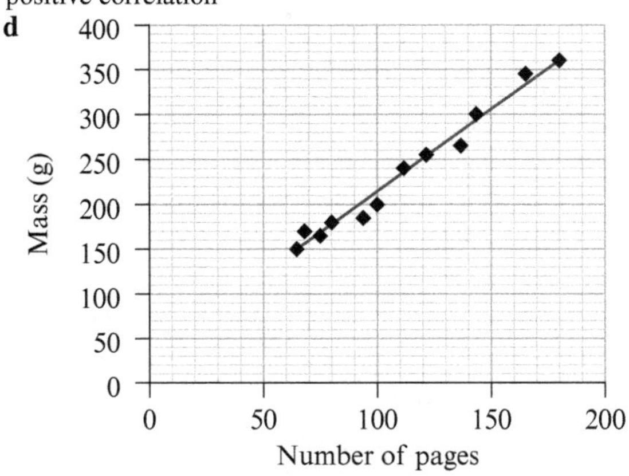

c support

e 160 pages

f 75 g

g Part **e** is more reliable because you used interpolation; extrapolation was used in **f** so it is less reliable.

6 a no correlation

b

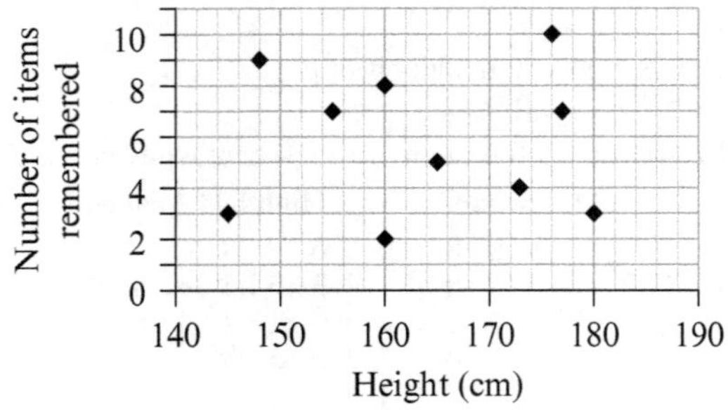

c support

7 a

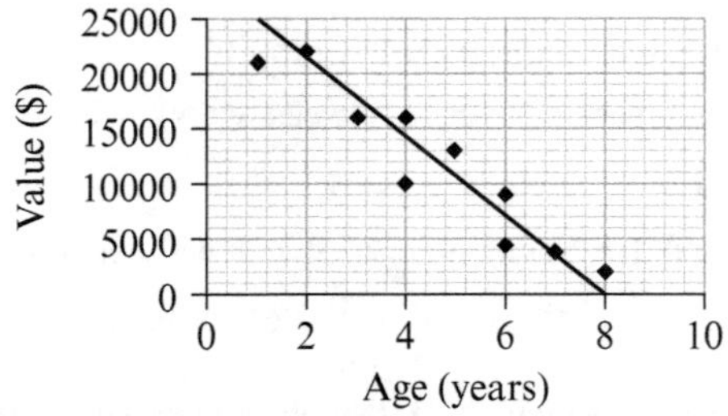

b it does decrease as the car ages. The correlation is negative, but not strong (or weak)

c make and model of the car and the condition it is in and the number of miles it has travelled

8 a the line does not follow the trend of the points since the student has tried to make sure the line goes through the origin when it shouldn't. There is clearly negative correlation but this line shows positive

b this is fine

c the line does follow the trend but it is too high, there is only one point above the line and eight below it

d this is fine

e there is no clear correlation so a line should not be drawn at all

f the line does follow the trend but it is too low, there is only one point below the line and several above it

9 a might be positive correlation as the more letters in the name the longer it may take

b

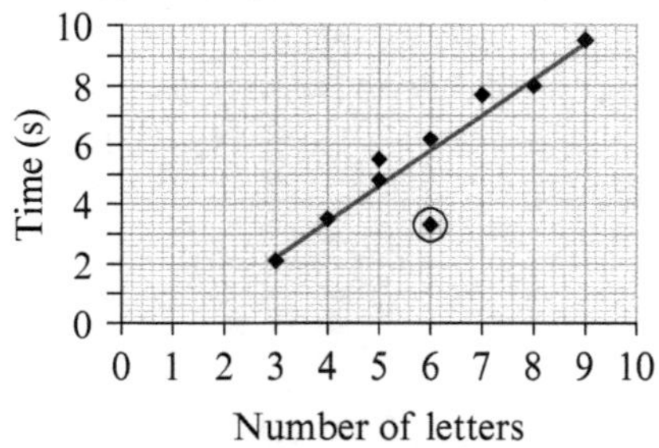

c not particularly better or worse as he is on the line of best fit

d Hannah's name is the same forwards as backwards

e outlier

10 Layla is right, the sample size is very small and this can cause unexpected results showing false correlation

11 Students' own answers

12 Students' own example of two variables that have positive correlation but one does not cause the other, e.g. sales in scarves, sales in hot drinks.

Exercise 10D

1 a The sale of 20 soft drinks

b 80 **c** 50 **d** Saturday **e** 380

f

Day	Mon	Tue	Wed	Thur	Fri	Sat
Number of soft drinks sold	60	80	50	40	50	100

g Thursday, it had the lowest sales

2 a 36 **b** 113 **c** 24 **d** infer

3 a Sector angles: 162°, 90°, 36°, 36°, 18°, 18°

b $60 000

c Pie charts are useful for showing proportions.

4 a 15.3% **b** 54°, 131°

5 a Data is grouped and continuous.

b

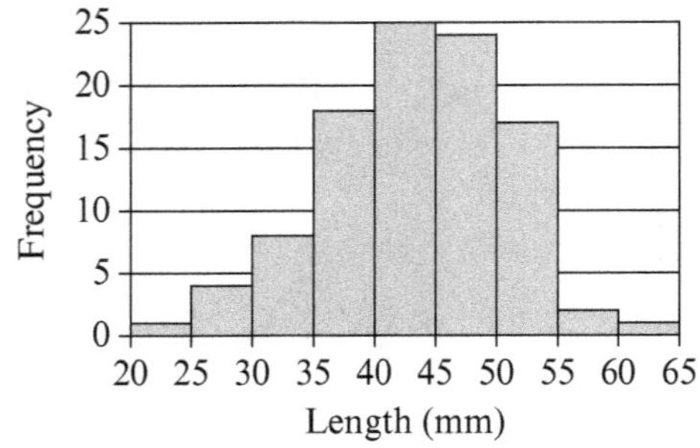

6 a first year, steepest drop

b decreasing

c i $12500 **ii** $8500

d i $5000 **ii** $11200

7

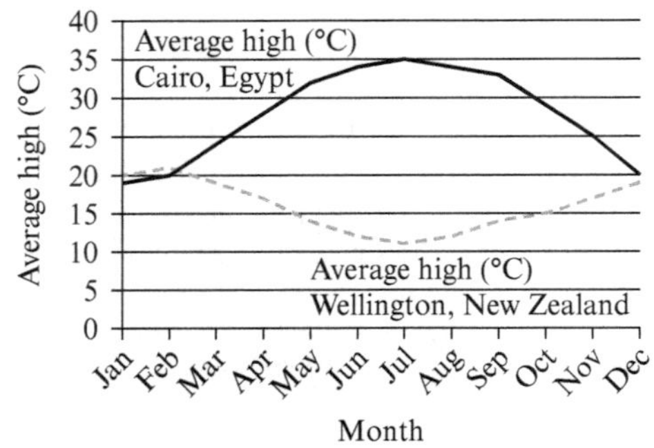

Cairo and Wellington have their warm seasons at opposite times of the year.

8 a Students' own answers

b To look for a correlation/relationship between two variables this is easiest to see on a scatter graph.

c

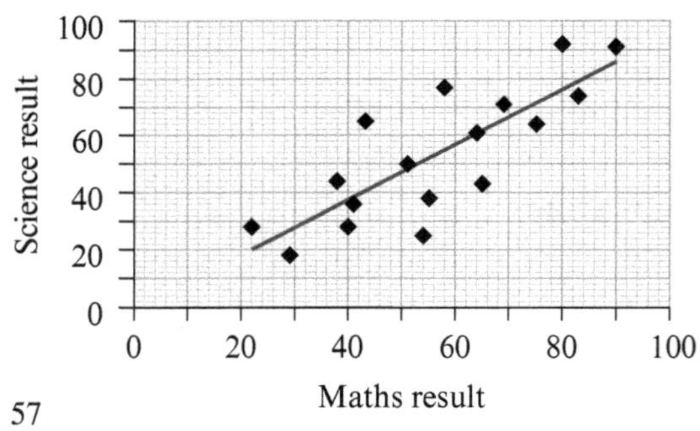

d 57

9 pie chart, bar chart, pictogram as data is discrete

10 a scatter graph

b expect no correlation (or possibly weak positive)

c

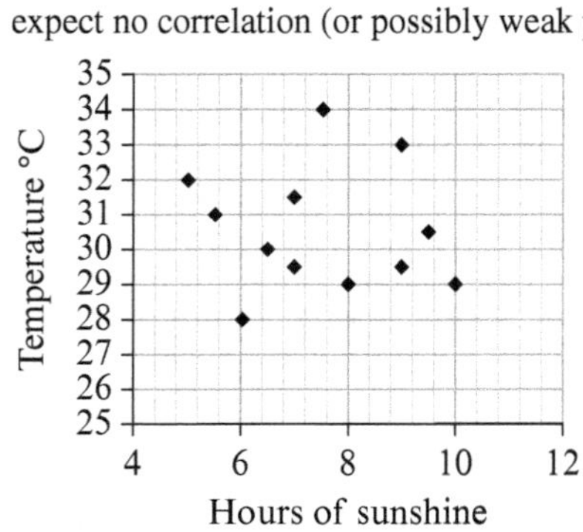

d a day can be very sunny but cold, also the range of temperatures is only from 28 to 34 which is not a big range

e, f mean or median better

11 a histogram, frequency polygon
b bar chart, pictogram, pie chart
c scatter graph

12 a This is only the temperature for 7 months. To see if global warming is a factor, you need to look over a period of years for the same month, or average the whole year.

b Area rather than height has been used to represent the price in this infographic, which exaggerates the cost.

c The improvement in times is only about 0.4 seconds, which is not that significant, but the axis does not start at 0, so it gives a false impression of the difference in sprinting times.

13 a spider monkey
b spider monkey
c i is likely to be a spider monkey
ii could be either a spider monkey or a mandrill
iii is likely to be a mandrill
d i mandrill
ii cannot say as don't know the mean mass of all Old World or all New World monkeys

Consolidation

Exercise 10

1 a, c

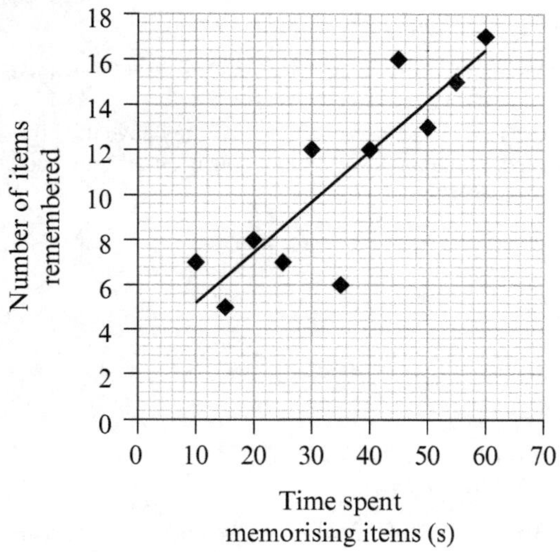

b positive
d 11
e it is within the data set so yes, but the correlation isn't that strong so be wary as there are obviously other factors involved (e.g. age)

2 a 38.5 cm
b

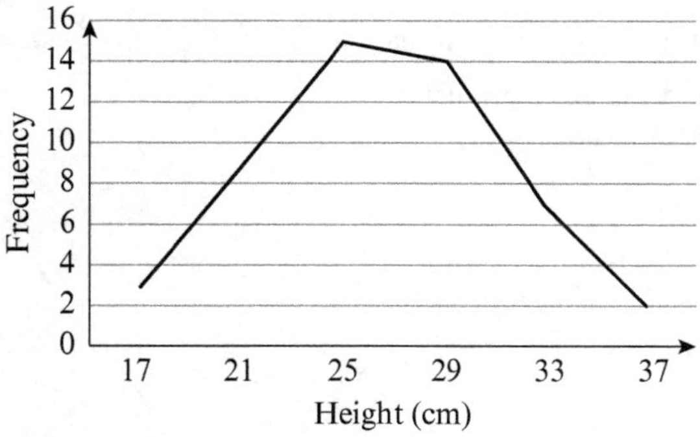

3 a i

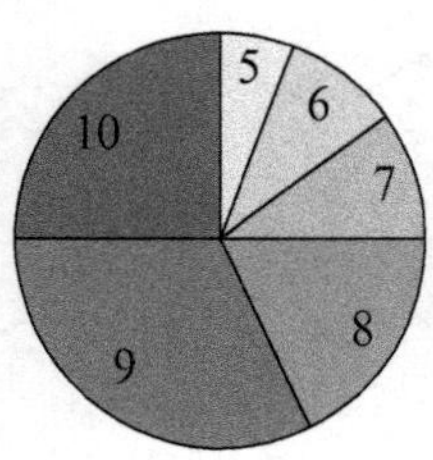

ii

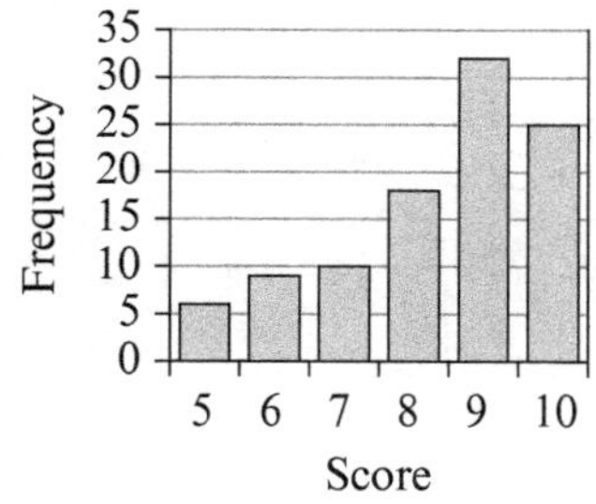

b the bar chart

4 a histogram or frequency polygon possible.
Frequency polygon:

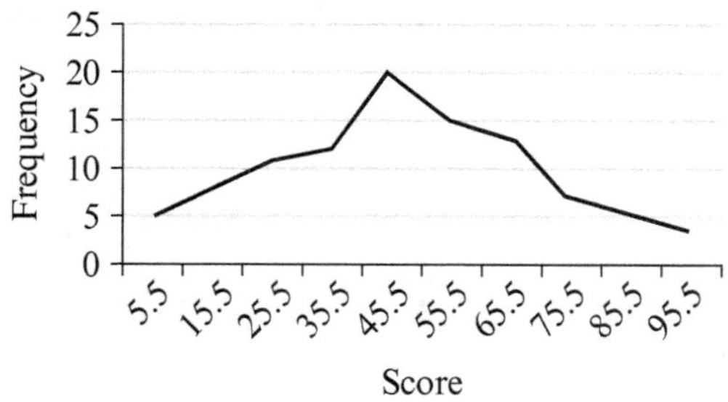

b 72% **c** 36%

d You cannot plot grouped data on a stem-and-leaf diagram.

5 a

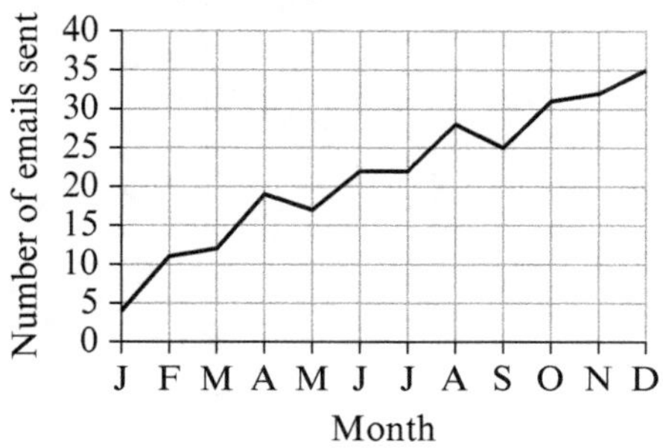

b increasing

6

7

Recovering well		Those still unwell
8	36	2
9 8 7 6 6 6 5 5 5 5 2 1	37	5 7
2 1	38	3 4 4 6 7 8 8 9
	39	2 2 4 5

Key
36 | 2 = 36.2 °C and 8 | 36 | = 36.8 °C

b i 1.4° C **ii** 3.3° C

c i 37.6° C **ii** 38.7° C

d generally those who are still unwell have higher temperatures as there are more patients with temperatures 38° or above in the unwell section

8

Class	f
75-125	30
125-175	50
175-225	40
225-275	10

9 **a** 20 – 25

b no as they could have any result between 0 and 5

c 32%

d 4%

e approximately normal distribution

10

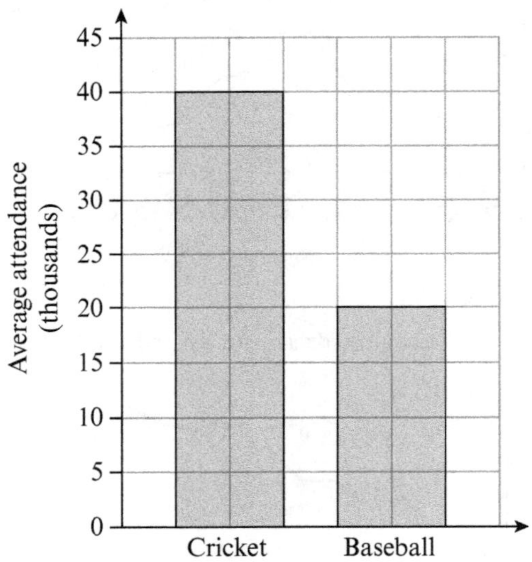

Check out

1

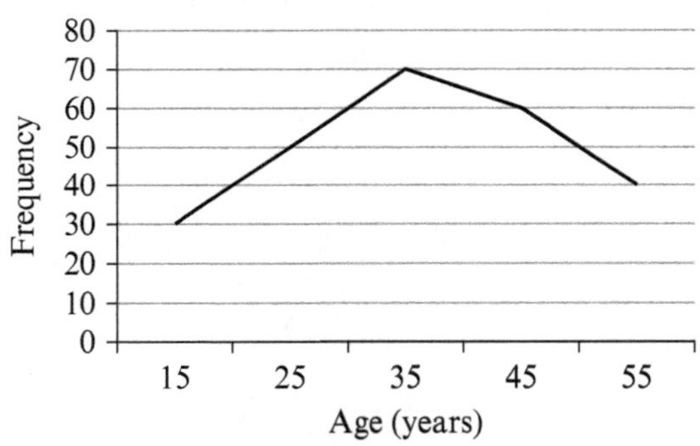

2 **a** **i** 2.6 km **ii** 3.7 km

b **i** 3.2 km **ii** 3.7 km

c **i** 1 km **ii** 4.5 km

3 and **4 a**

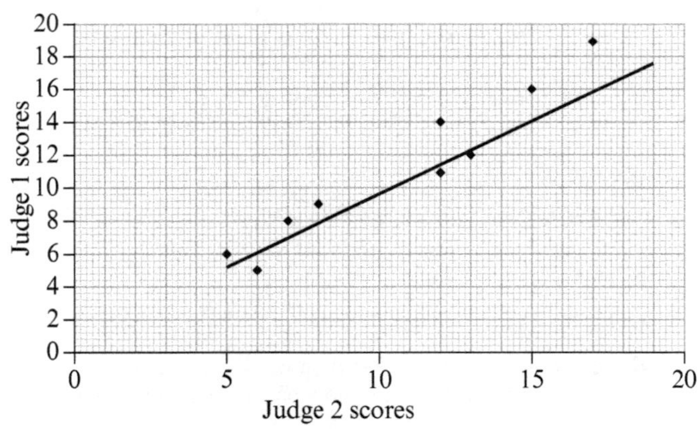

4 **b** 9.5

Review B

1 **a** 78.5 cm² **b** 38.5 cm²

2 **a** $\frac{3}{4}$ **b** $\frac{6}{35}$ **c** $12\frac{3}{4}$ **d** 24

e $5\frac{5}{7}$ **f** 6

3 **a** 5 **b** 20

4 **a** 5 cm **b** 10.4 cm **c** 12 cm **d** 8.5 cm

5 **a** Recurring $\frac{2}{3}, \frac{17}{9}, \frac{4}{7}, \frac{11}{12}, 5\frac{4}{9}$

b Terminating $\frac{3}{12}, \frac{19}{250}, \frac{3}{6}, \frac{137}{1000}, \frac{13}{26}, 1\frac{7}{8}$

6 **a** 84.9 m² **b** 138.9 cm²

7 **a** 8 **b** 4

8 **a** 10 cm **b** 26 cm **c** 25 cm

9 **a**

Class A		Class B
	0	7
9	1	4 6 6 8 9 9
5 4 4	2	0 2 2 7 7 8 9 9
7 6 6 6 6 5 3 2 1 1	3	1 2 3 3
9 3 2 1 0 0	4	3 5
0	5	
	6	

Key
0 | 7 – 7 marks and 9 | 1 | – 19 marks

b Median Class A = 36, Median Class B = 27, Range Class A = 31, Range Class B = 38

Class A did better on average as they have the higher median. Class A were more consistent as they had the lower range.

10 **a** 49 **b** 2 **c** $\frac{15}{32}$ **d** 4

e $2\frac{1}{2}$ **f** $\frac{3}{10}$

11 **a** $x = 1, y = 1$ **b** $x = 1, y = 0.5$ **c** $x = 1, y = 2$

12 814 cm²

13 **a** $d = 18°, e = 18°, f = 21°$

b $a = 110°, b = 135°$

14 **a** 3 m **b** 28.3 m² **c** 94.2 m² **d** 56.5 m³

15 **a** 2 **b** 7 **c** 2 **d** 6

16 **a** $a = 65°, b = 115°$ **b** $c = 34°, d = 63°$

c $f = 92°$ **d** $g = 72°, h = 108°$

e $i = 51.4°, j = 128.6°$

17 **a** **i** $y = \frac{3}{2}x - \frac{7}{2}$ **ii** $y = \frac{-2}{3}x - \frac{5}{3}$ **iii** $y = \frac{3}{2}x - 4$

b **i** $m = \frac{3}{2}, c = -\frac{7}{2}$ **ii** $m = -\frac{2}{3}, c = \frac{5}{3}$ **iii** $m = \frac{3}{3}, c = {}^{-}4$

c **i** and **iii** are parallel

d **ii** is perpendicular to **i** and **iii**

e

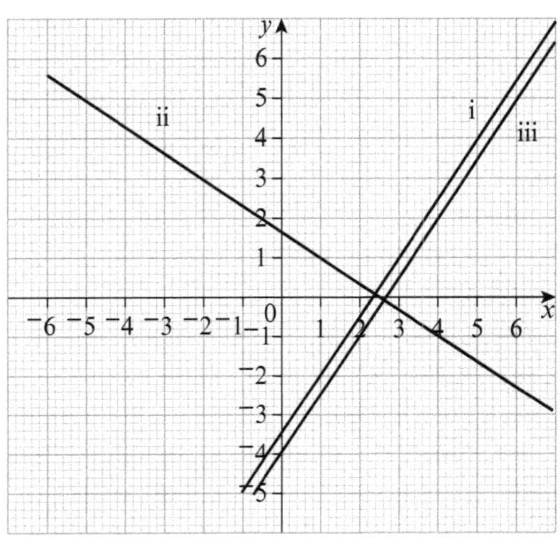

18 a Perimeter = 14.28 cm, Area = 12.56 cm^2
b Perimeter = 25.7 cm, Area = 39.25 cm^2
c Perimeter = 20.13 cm, Area = 21.2 cm^2
d Perimeter = 17.85 cm, Area = 9.81 cm^2

19 164 kg

20 $\frac{5}{9} = 0.\dot{5}$, $\frac{73}{99} = 0.\dot{7}\dot{3}$

21 a 0, 1, 2, 3, 4 **b** 4, 5, 6, 7, 8, 9, 10
c 0, 1, 2, 3 **d** 7, 8, 9, 10

22 a 144° **b** 36°

23 a

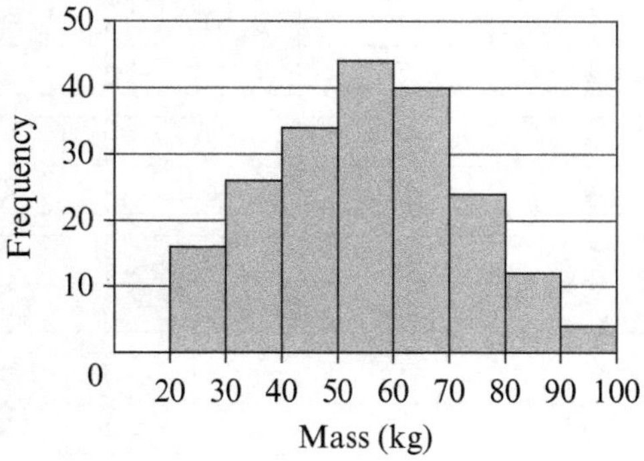

b

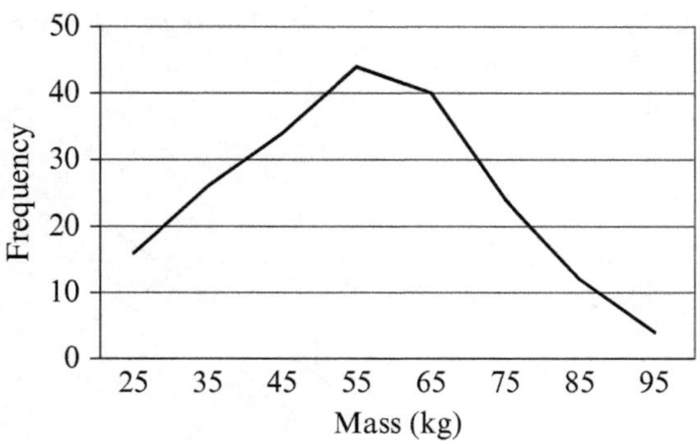

24 22

25 3.6 cm

26 a (9, $^-$5) **b** (3, 4)

27 504 cm^3

28 a $x = 6, y = 4$ **b** $x = {}^-2, y = 4$ **c** $x = 5, y = {}^-3$

29 3142 cm^3

30 a 11 cm **b** 0.9 cm

31 c

32 triangle, square, hexagon

33 a $43\frac{1}{2}$ **b** 5.869

34

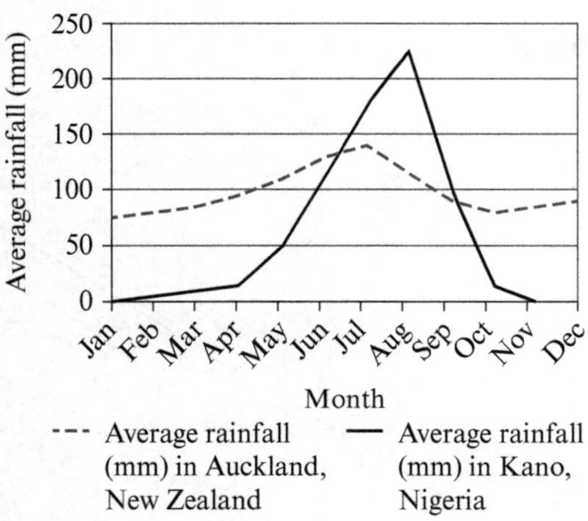

Rainfall more steady in Auckland with a small peak in June/July.

Very clear rainy season in Kano between June and September and little rain at other times.

35 a i 5 MB **ii** 5 TB
b i 1.2 ng **ii** 1.2 t
c i 0.4 μm **ii** 0.4 ly

36 10.6 cm

37 a 15 **b** 18

38 40 adults and 15 children

39 $13\frac{3}{4}$ cm^3

40 $180(n - 2)°$

41 **b**, **c**, **d**, **e** and **f**

42 180 cm^3

43 a 3 **b** $2\frac{14}{25}$

44 a

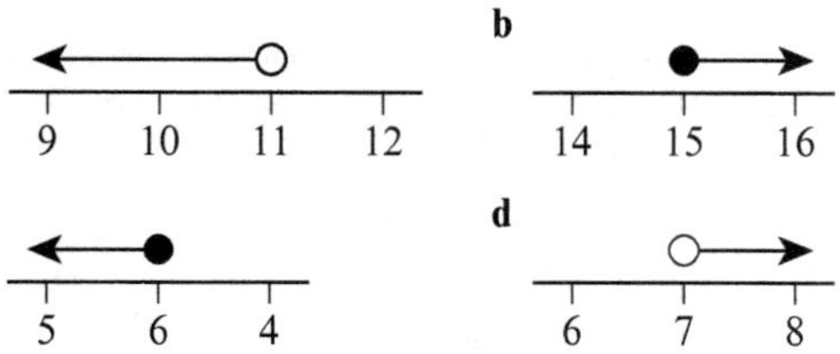

45 a Team A = 94, Team B = 65
b Team A = 76, Team B = 91
c Team A has the higher modal score, team B has the higher median score
d Team B scored more baskets than team A, the median is the more meaningful measure.

46 a

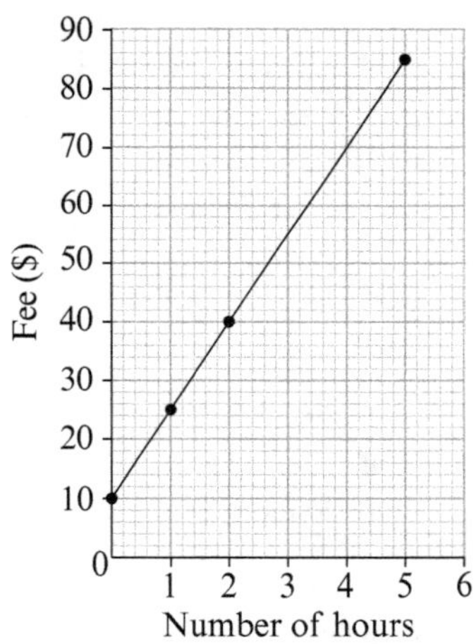

b 15, the charge is $15 per hour
c 10, the call out fee is $10
d No, the line does not pass through the origin

47 a 160 m **b** 25.48 m

48 a $x = 4, y = 3$ **b** $m = 3, n = -1$

49 24 cm^2

50 a 1 **b** 2
c Negative correlation; the farther the weekly distance ran, the lower the resting pulse rate
d Someone with a resting pulse rate of 55 who ran 28 km per week. This could be because they normally run more, but are recovering from an injury and are running less until they are fully recovered.
e 63 beats per minute
f 14 km per week

51 a i $^-2 < x \le 5$ **ii** $0 \le x < 4$

b

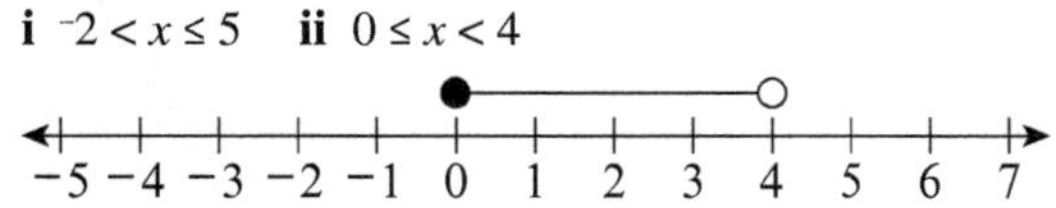

52 a 9000 **b** 20 **c** 600 000 **d** 0.7

53 a

Distance from home in km (d)	Time of journey in hours (t)
17	1
19	2
21	3
23	4
25	5
37	11

b gradient 2, travel at a speed of 2 km/h
c intercept is 15, start at a distance 15 km from home
d $d = 2t + 15$

54 a $1\frac{1}{9}$ **b** 18

55 You can't fit the angles of a pentagon together to make 360°, since the interior angle of a regular pentagon is 108° ($3 \times 108° = 324°$ and $4 \times 108° = 432°$).

56 a 1.41 cm **b** 14.1 cm **c** 141 cm

57 a P = 37.7 cm, A = 102.5 cm^2
b P = 27.9 cm, A = 44.6 cm^2

58 a $0.8\dot{4}$ **b** $0.2\dot{1}$

59 5.83 m

60 a 5 **b** 3.2

Chapter 11

Check in

1 a mm, cm, m, km **b** g, kg, t **c** ml, ℓ

2 a Unlimited number of answers, for example 10 : 6, 15 : 9, 500 : 300
b Unlimited number of answers, for example 8 : 14, 16 : 28, 40 : 70

3 a 5 : 8 **b** 2 : 3 **c** 4 : 5 **d** 4 : 1 : 3

Exercise 11A

1 a 189 cm, 216 cm **b** $95, $19, $114
c 90 g, 150 g **d** 32 cm, 48 cm
e 390 ml, 520 ml **f** 24 mins, 56 mins, 40 mins

2 A = 45°, B = 75°, C = 60°

3 a she has divided by the total number of shares not just the shares there are more of
b $56 \div 2 = 28$
Kamil gets $3 \times 28 = \$84$
Sean gets $5 \times 28 = \$140$
(Check $140 - 84 = 56$)

4 3

5 1400 g

6 117, 52

7 306

8 Isosceles

9 a i 0.5 **ii** 0.5
b 15 red, 40 blue, 25 black

10 $120 , $84

11 30°, 165°, 105°, 60°

12 165

13 2 and 0.72

14 12 : 5

15 a False. The proportion of women is $\frac{15}{39}$
b False. The correct ratio is 14 : 15
c True
d Not possible to say, because we don't know the actual numbers of people in the group. The 3 : 2 : 3 ratio would apply only if there were 40 people in total, including the additional child.

Investigation

Amira $480, Beth $400, Cassie $320

Exercise 11B

1 a

x	6	30	1
y	24	120	4

b

x	2	9	8
y	14	63	56

c

x	5	8	11
y	1.5	2.4	3.3

d

x	12	3.6	⁻2
y	18	5.4	⁻3

2 a $122.50 **b** $9.03 **c** 3.375 kg **d** 105 m^2

3 a €21.60 **b** $400
c, d

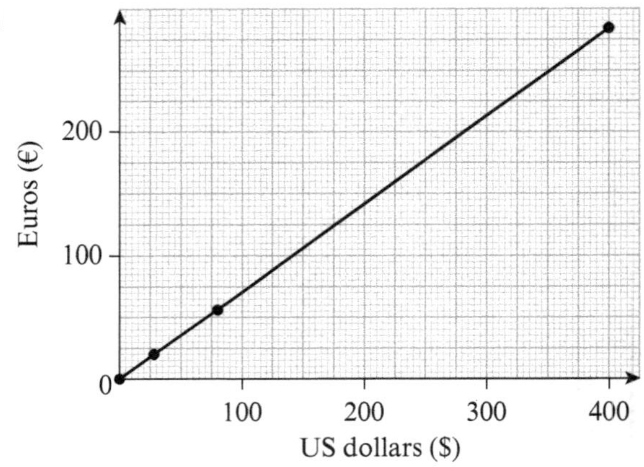

e yes **f** €53

4 $4.22

5 a

US dollars ($)	70	82	104
British pounds (£)	46.20	54.12	68.64

b i $350 **ii** R4149

6 83.2 litres

7 2.4 cm

Exercise 11C

1 a 3
b

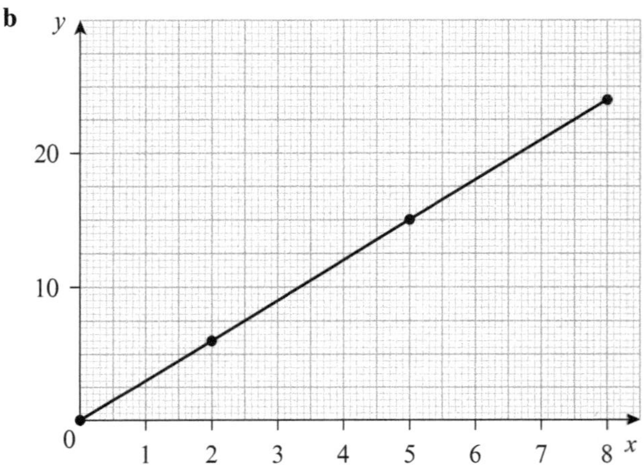

c yes **d** 3 **e** yes **f** $y = 3x$ **g** 21 **h** 11

2 a

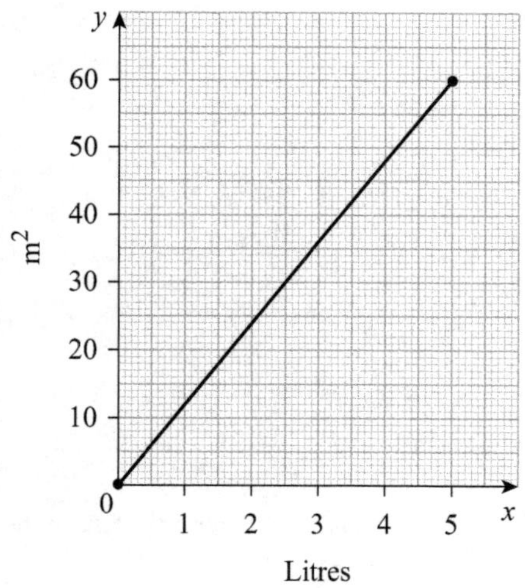

b $y = 12x$ **c** $96\,m^2$ **d** 4ℓ
e $12\,m^2$ **f** same

3 a

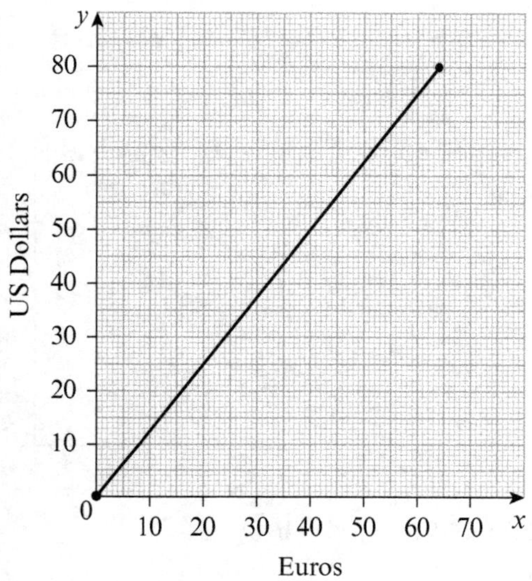

b $y = 1.25x$
c i €58 **ii** $39
d i €58.40 **ii** $38.75; equation more accurate

4 a $y = 2.5x$
b i 12.5 **ii** 9.2

5 Barrel B was not full. Line 1 is barrel B, line 2 is barrel A. It takes 35 seconds to fill barrel A and 18 seconds for barrel B to have 80 cm of water in it. Barrel A is 80 cm tall. Water fills barrel B at a rate of 4.4 cm per second and barrel A at a rate of 2.3 cm per second.

Exercise 11D

1 a 45 hours **b** 9 hours

2 a direct proportion, variables are in the same ratio
b inverse proportion, variables have the same product
c inverse proportion, variables have the same product

3 15 days

4 a inversely **b** directly **c** directly **d** inversely

5 4 minutes

6 10 buckets

7 5 chapters

8 20 minutes

9 length × width = area and this is constant

10 a

m	20	5	12
b	36	144	60

b $m = \frac{720}{b}$

11 50 minutes

12 $hk = 8$

13 51 days

14 24 days

15 18

Investigation

The graph would look like this:

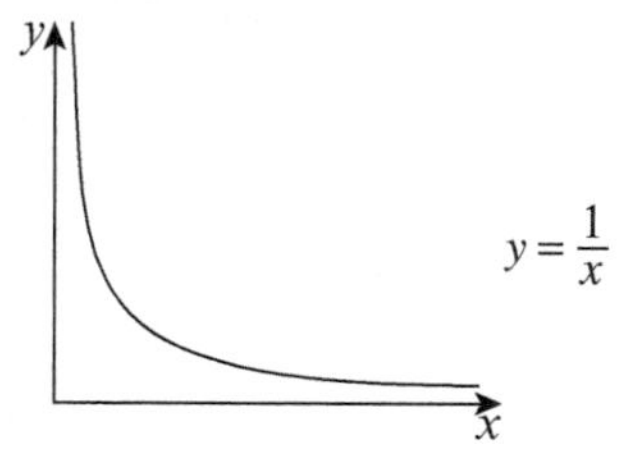

Consolidation

Exercise 11

1 a 4 : 7 **b** 7 : 6 **c** 2 : 5 : 7

2 a $90, $150 **b** 35 g, 40 g
c 14 cm, 56 cm **d** 700 m, 3500 m, 2100 m
e 48 mins, 36 mins, 96 mins

3 a 2 : 5 **b** 5 : 6 **c** 4 : 7 : 1 **d** 2 : 3 : 6

4 208

5 a 7 : 10 **b** 5 : 8 **c** 35 : 1 **d** 2 : 1

6 255

7 a 6 kg
b i 2.34 kg **ii** 3.9 kg

8 no

9 Amir 39, Rita 91

10 a $17.82 **b** 9

11 90 ml acid, 495 ml water

12 a

x	3	4	20
y	81	108	540

b

x	8	10	22
y	28	35	77

13 12 : 20 : 21

14 0.5 and 0.08

15 10 days

16 3 slices

17 9 minutes (6 taps would take 36 minutes, so 36 – 27 = 9)

18 a $y = 3.5x$ **b** gradient 3.5, intercept 0

19 a 19.50, 26.00, 32.50, 39.00

b

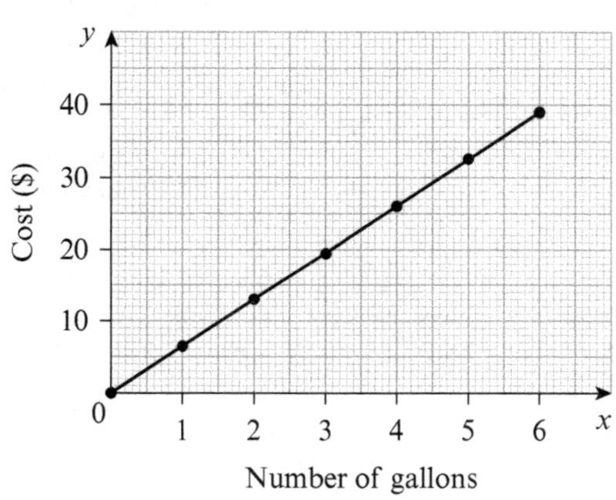

c **i** \$29.25 **ii** a little more than 3 gallons
d $c = 6.5g$
e yes, as it is a straight line passing through the origin

Check out

1 **a** **i** 40° **ii** 80° **b** Gill \$96, Nat \$36

2 **a** ₹11 946 **b** \$900

3 $y = 4.5x$ is a straight line of gradient 4.5 passing through the origin

4 **a** 30 **b** 10 **c** 4

Chapter 12

Check in

1 X = 1.2, Y = 2.8, Z = 4.2

2 **a** X(1.4, 2), Y(2.2, −7.5), Z(−2.2, 7)
b Points correctly plotted

3 **a** $x = 2$ **b** $x = 5$ **c** $x = 5$ **d** $x = 13$

4 **a** parallel **b** perpendicular
c perpendicular **d** neither

Exercise 12A

1 **i** linear: **a, b, d** **ii** geometric: **c, e, f** **iii** quadratic: **g**

2 **a** 22, 25 **b** 3.5, 4 **c** 243, 729 **d** −12, −16
e 320, 640 **f** $\frac{1}{8}, \frac{1}{16}$ **g** 25, 36

3 **a** Add 3 **b** Add 0.5
c Multiply by 3 **d** Subtract 4
e Multiply by 2 **f** Divide by 2
g Square

4 **a** 34 **b** 5.5 **c** 19 683 **d** −28
e 5120 **f** $\frac{1}{128}$ **g** 100

5 **a** 3, 8, 23, 68 **b** 2, 4, 8, 16
c −4, 8, 20, 32 **d** 2, 3, 8, 63
e 0, 2, 6, 38 **f** 2, 4, 16, 256

6 **a** 0.5, 1, 1.5, 2, 2.5 **b** −2, 1, 6, 13, 22
c 5, 8, 13, 20, 29 **d** 1, 8, 27, 64, 125

7

Position-to-term rule	Position	Term
Square then add 3	10	103
Square then subtract 5	20	395
Cube	6	216
Cube	10	1000
Square then add 2	8	66
Square then subtract 1	12	143

8 The second differences are the same each time.

9 The first term is 3, the next two terms are 13 and 173 and the numbers are getting bigger so 163 can't be in the sequence.

10 **a** Yes. If you add 2 to 79 you get 81, taking the square root of this number is the integer 9.
b No. If you add 2 to 53 you get 55, taking the square root of this number is not an integer so it is not in the sequence.

11 17

12 **a** 27 **b** 10 002 **c** square and add 2
d The second difference is constant. (Other correct explanations are possible, check with your teacher.)

Exercise 12B

1 **a** $3n + 1$ **b** $5n - 2$ **c** $6n - 4$ **d** $25 - 5n$
e $0.5n + 6.5$ **f** $0.25n + 4$

2 **i** **a** 31 **b** 48 **c** 56 **d** −25 **e** 11.5 **f** 6.5
ii **a** 61 **b** 98 **c** 116 **d** −75 **e** 16.5 **f** 9

3 **a** 6, 9, 14, 21 **b** $\frac{1}{2}$, 1, $1\frac{1}{2}$, 2
c 0, 3, 8, 15

4

$n^2 + 2$	3, 6, 11, 18, 27, …
$n^2 - 2$	−1, 2, 7, 14, 23, …
n^3	1, 8, 27, 64, 125, …
$2n + 3$	5, 7, 9, 11, 13, ….
$8 - 2n$	6, 4, 2, 0, −2, …

5 Yes. The 7th term is 49.5 and the 5th term is 25.5, so the difference is 49.5 − 25.5 = 24

6 **a** $3n + 5$ **b** 155 **c** 31st

7 **a** $10n - 9$ **b** 991 **c** No, it doesn't end in 1
d 17th

8 **a** n^2 **b** $n^2 + 6$ **c** $n^2 - 5$

9 $2\frac{1}{2}$

10
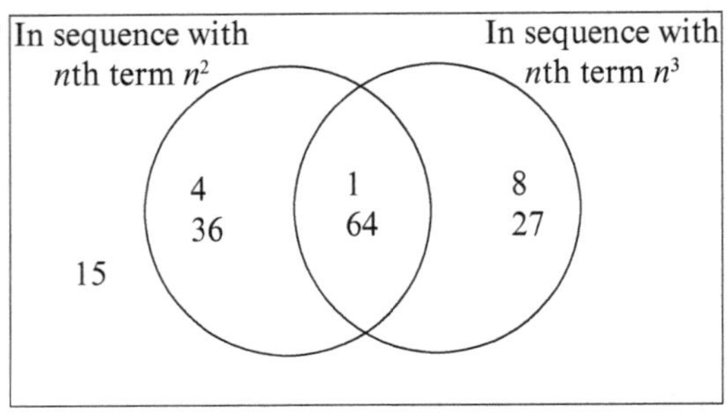

11 **a** $\frac{n}{7}$ **b** $\frac{n}{11}$ **c** $\frac{n}{6}$

12 **a** 3, 11, 19, 27, 35 **b** $8n - 5$

13 **a** 7, 11, 15, 19, 23 **b** $4n + 3$

14 All of them

Exercise 12C

1 **a** 45
b
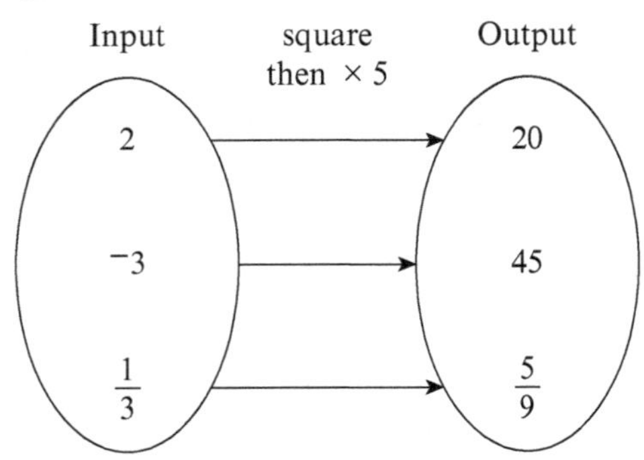

c
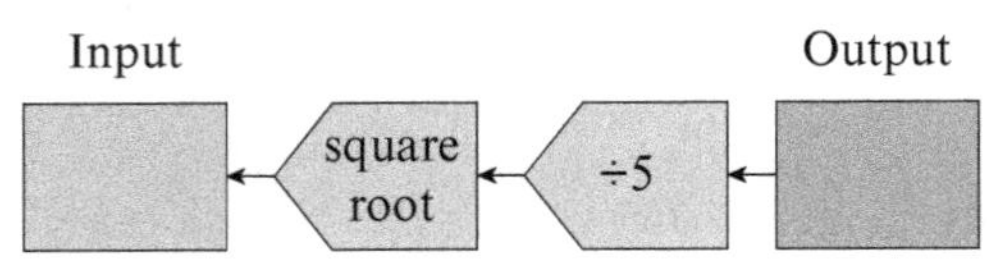

d 4 and −4

2

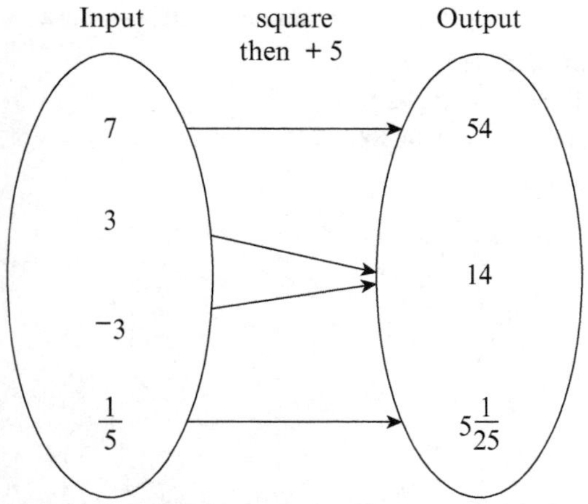

3 **a**

x	-3	-2	-1	0	1	2	3
y	8	3	0	-1	0	3	8

b

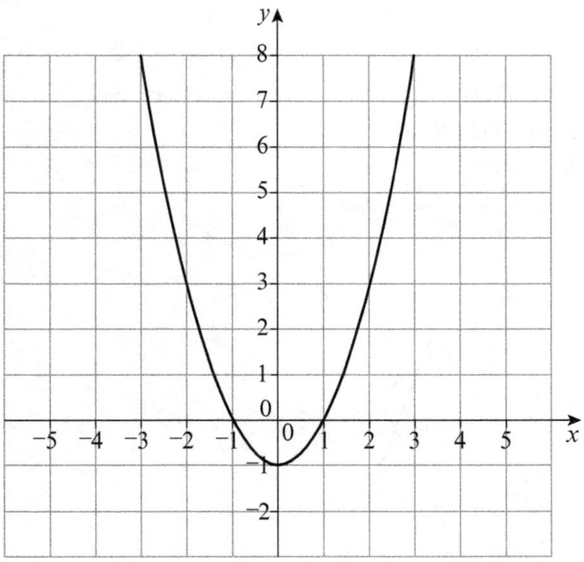

4 **a**

x	-3	-2	-1	0	1	2	3
y	10	5	2	1	2	5	10

b

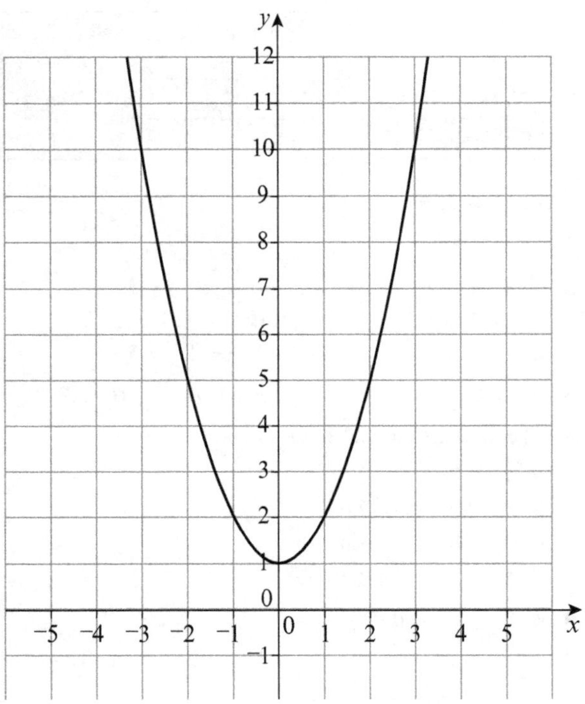

5 $a = 20$

6

Coordinates below the graph of $y = x^2 + 2$	Coordinates on the graph of $y = x^2 + 2$	Coordinates above the graph of $y = x^2 + 2$
(0, -2) (-2, -2) (-1, 1)	(2, 6) given (-1, 3) (0, 2) (4, 18)	(0, 5) (1, 7)

Exercise 12D

1 **a** points accurately plotted on Cartesian grid

b set **iii** does not represent a linear function

2 **a** $3x$ row: -3, 3, 6, 12,
+2 row should all be +2,
y row: -1, 2, 5, 8, 11

b

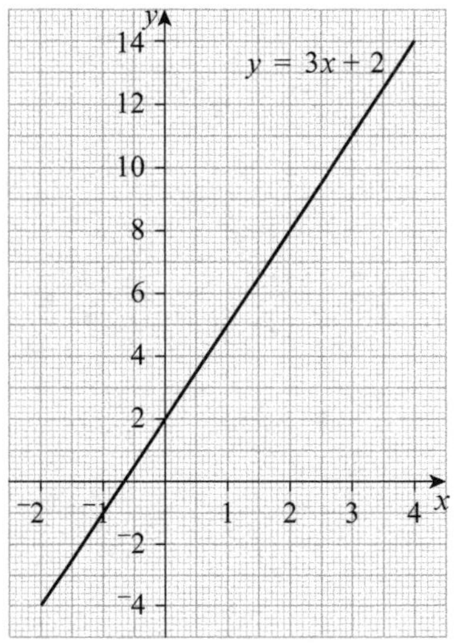

3 **a** **i**

x	-2	-1	0	1	2	3	4
y	2	3	4	5	6	7	8

ii

x	-2	-1	0	1	2	3	4
y	0	2	4	6	8	10	12

b

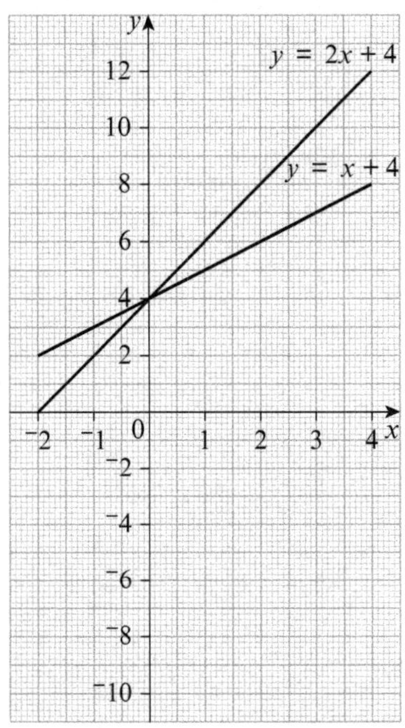

4

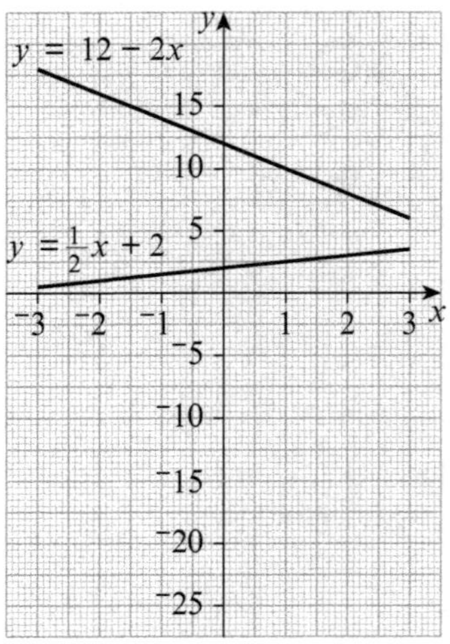

5 **a** $y = 8 - x$

b

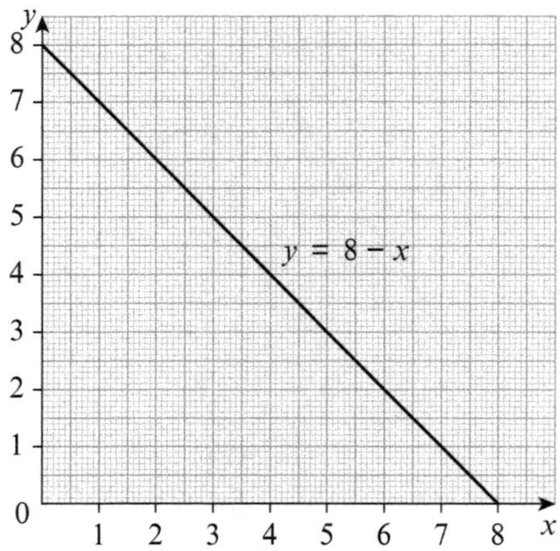

6 **a** $2y = 8 - x, y = 4 - \frac{x}{2}$

b 4 row should all be 4, $\frac{^-1}{2}x$ row: $\frac{1}{2}, 0, ^-1, ^-1\frac{1}{2}$, y row: $4, 3\frac{1}{2}, 2\frac{1}{2}, 2$

c

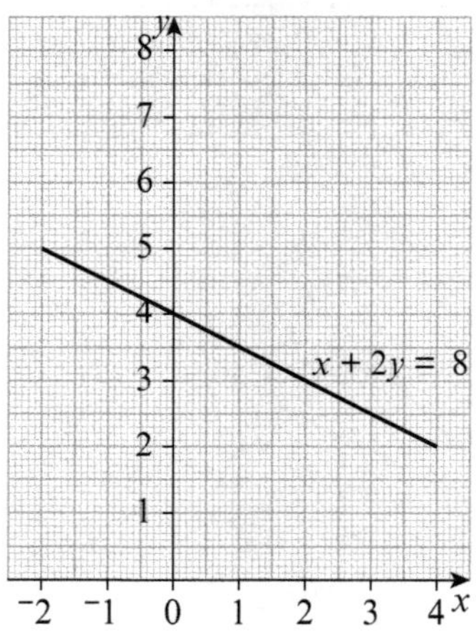

7

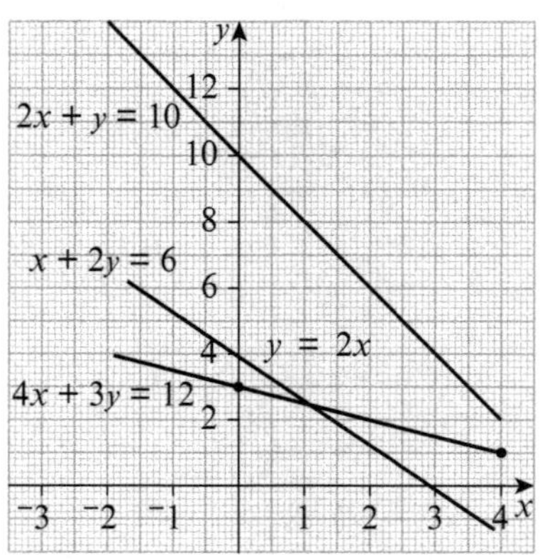

8 **a** linear **b** linear **c** non-linear **d** non-linear

9 **a** (3, 249), (5, 415), (7, 581)

b (2, 40), (4, 75), (5, 92.5), (7, 127.5)

c (0, 200), (2, 144), (4, 88), (7, 4)

Exercise 12E

1

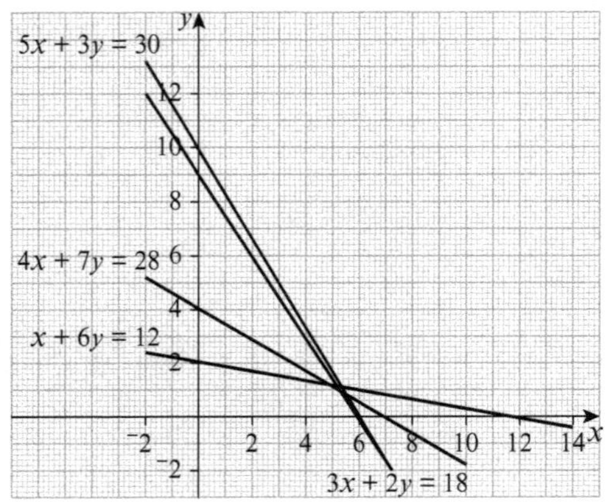

2 **a** **i** $m = 3, c = 4$ **ii** $m = 5, c = ^-2$

iii $m = \frac{1}{2}, c = 4$ **iv** $m = ^-1, c = 3$

b Students' drawings

5

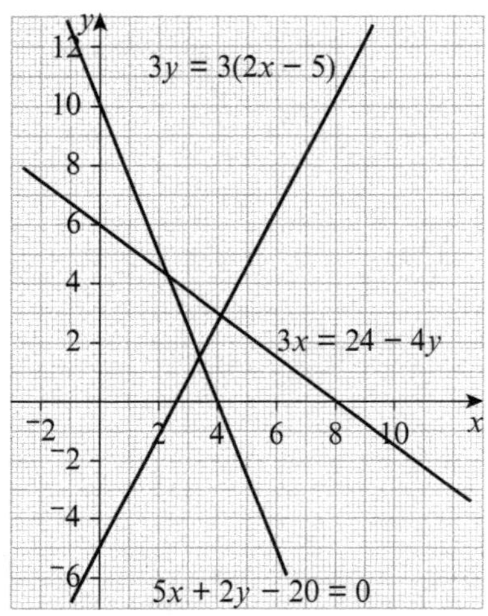

6 **a** $(^-1, 5)$ **b** (6, 7) **c** (0, 4) **d** $(^-4, 0)$

7

	On the line $2y - x = 8$	**Not** on the line $2y - x = 8$
Above the x-axis	$(^-2, 3)$ (2, 5)	$(^-5, 1)$
Below the x-axis	$(^-10, ^-1)$	$(3, ^-1)$

Exercise 12F

1 $x = 2, y = 3$

2 **a** $x = 1, y = 5$ **b** $x = 0, y = 3$

c $x = 1, y = 2$ **d** $x = ^-6, y = ^-11$

e $x = 2, y = 4$ **f** $x = 2, y = 5$

3 **a** (3, 6) **b** (5, 6) **c** (3, 7) **d** (2.5, 7)

4 The lines are parallel so they do not intersect.

5 **a** $x = 3.7$ $y = 2.5$

b $x = 4.6$ $y = 2.6$

c $x = 5.8$ $y = 1.4$

d $x = 8.7$ $y = 4.5$

6 **a** $2\frac{3}{4}$ **b** $1\frac{3}{4}$

7 **a** $y = 2x, y = 6 - x$ **b** $y = 2x + 3, y = 3x - 2$

8 **a** (1.5, 5.5); $y = 7 - x, y = 3x + 1$

b (2.5, 6); $y = 2x + 1, y = 4x - 4$

Exercise 12G

1 a 3 b 3 c $\frac{-1}{3}$

2 a i 2 ii $\frac{-1}{2}$

b i $^-4$ ii $\frac{1}{4}$

c i $^-5$ ii $\frac{1}{5}$

3 **a**, **b** and **f** are parallel
c, **d** and **e** are parallel

4 **a** and **c**, **b** and **f**, **d** and **e**

5 a

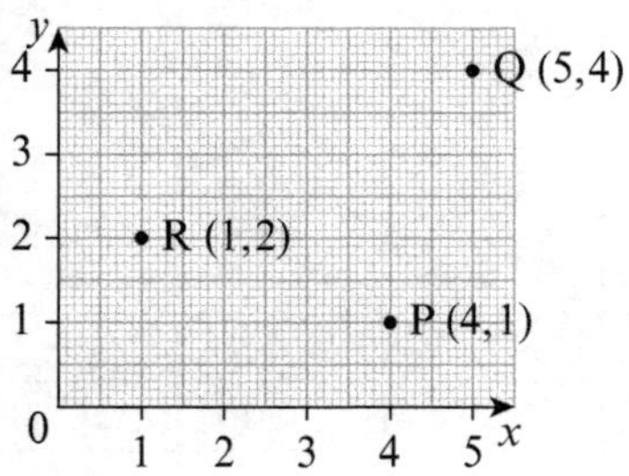

b 3, $\frac{-1}{3}$ c $\sqrt{10}, \sqrt{10}, \sqrt{20}$ d yes

6 The gradients of PQ, QR and PR are $\frac{1}{4}$, $^-4$ and $\frac{5}{3}$.
PQ and QR are perpendicular

Exercise 12H

1 $V = 8.3I + 20$

2 a

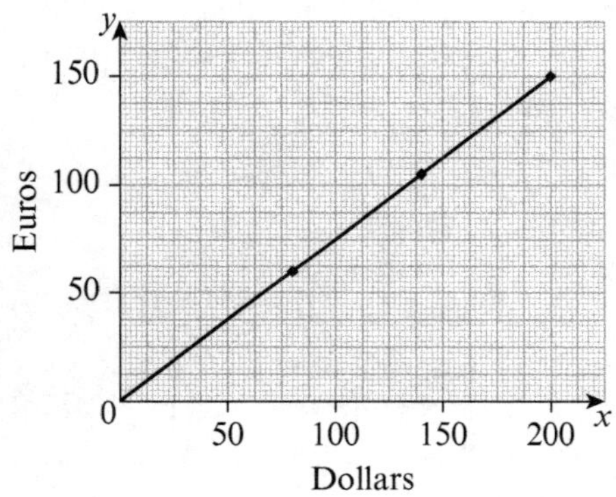

b 0.75 c the exchange rate \$1 = €0.75

3 a

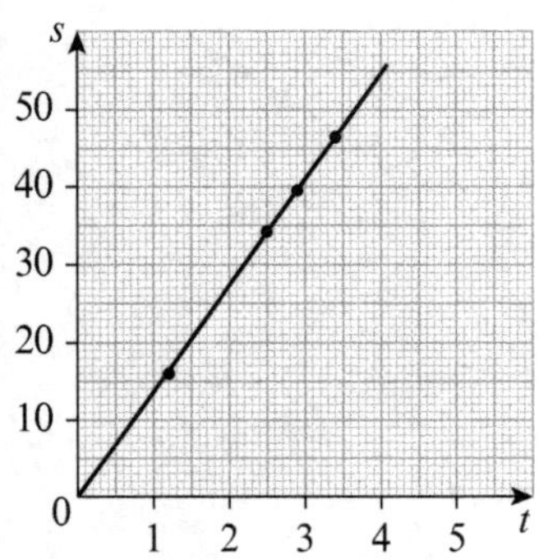

$s = 13.6t$

b

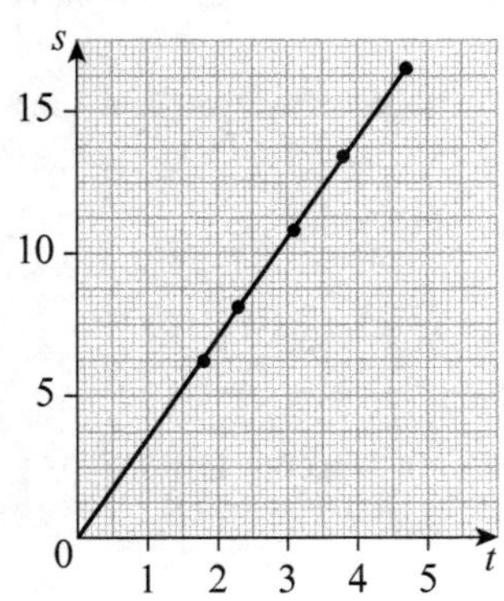

$s = 3.5t$

4 a, d

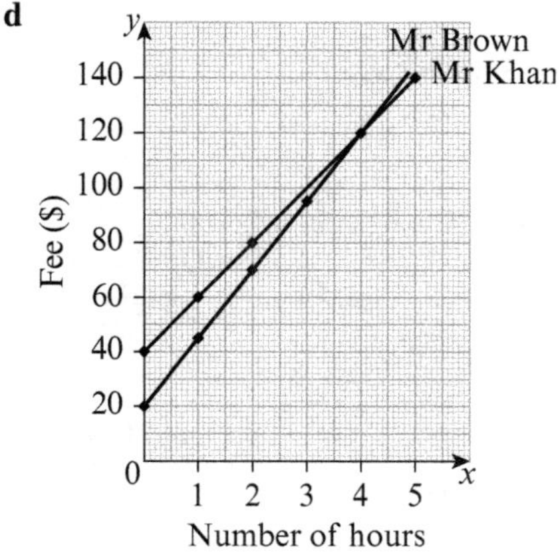

b \$20, the hourly fee

c \$40, the call out fee before any work has taken place

e (4, 120) f Mr Brown

5

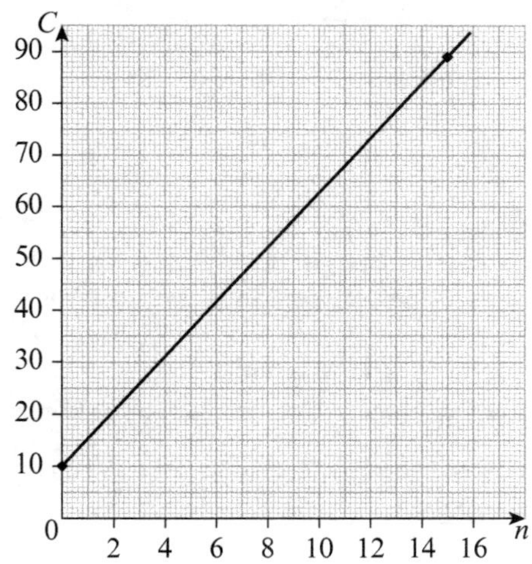

$I = 10, P = 5.3$

6 a (19, 11.1) b 12.7

7 a

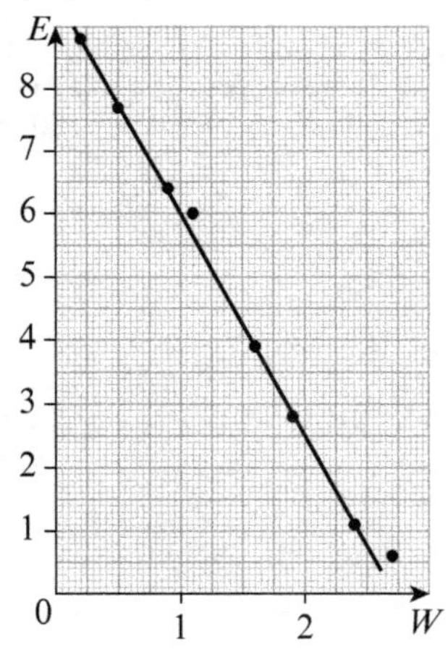

(1.1, 6.0), (2.7, 0.6)

b $a = {}^-3.5, b = 9.5$, (1.1, 5.65), (2.7, 0.05)

8 a \$10 b \$5

c, d

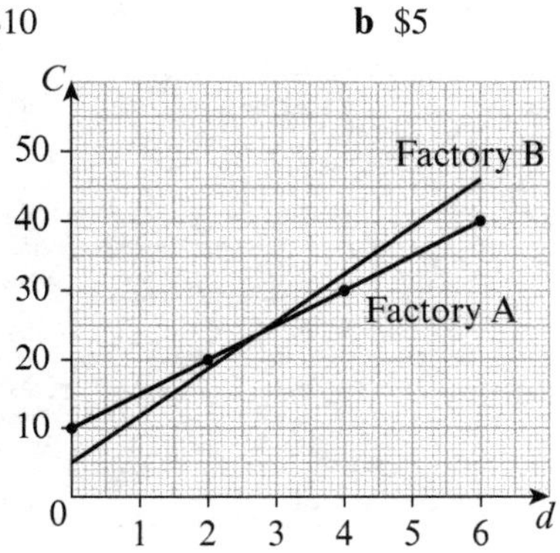

e 6; \$6 cost per metre
f 5, \$5 delivery charge
g $C = 6d + 5$

Exercise 12I

1 a 20 km/h, 40 km/h, 40 km/h, 20 km/h, 0 km/h
b 20 km/h, 40 km/h, 40 km/h, 20 km/h, 0 km/h
c During the third hour, it is the steepest section
d The graph is horizontal
e 40 km/h

2 a 10 m/s^2 b 20 m/s^2 c It stopped accelerating, 0 m/s^2

3 a i 2.5 m/s ii 9 m/s iii 14 m/s iv 12 m/s
b 8 m/s

4 a i 25 m/s^2 ii $6\frac{2}{3}$ m/s^2 iii 0 m/s^2
b 17.5 m/s^2

5 a

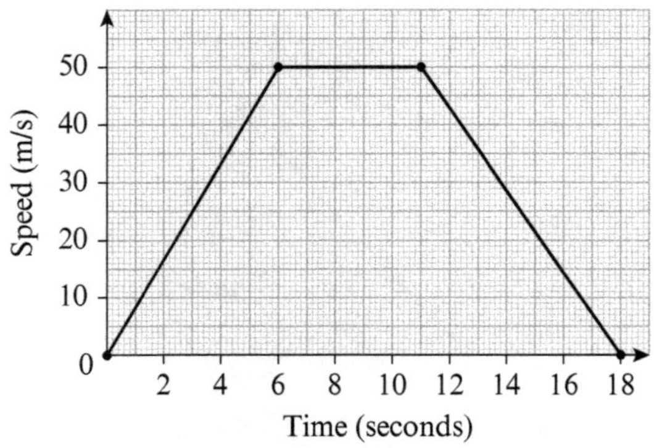

b 8.3 m/s^2 c 7.1 m/s^2

6 a 250 km b Area under graph

7 a between 2011 and 2012
b Mr Patel's
c \$6000
d Mr Patel's lost \$1812.50 per year, Mrs Templeman's lost \$750 per year

8 a Red has a fixed rate of \$5 if you don't go over 200 text messages, then text messages are charged at a rate of \$5 per 100 text messages. Yellow charges \$4 per 100 text messages.
b Yellow
c If you send fewer than 125 text messages

Investigation

84 km/h

Consolidation

Exercise 12

1 a −9, −6, −1, 6, 15, …
b $\frac{1}{5}, \frac{2}{5}, \frac{3}{5}, \frac{4}{5}$, 1, …
c 11, 14, 19, 26, 35

2 a 3, 8, 63, 3968 b 2, 5, 26, 677

3 a $n^2 + 1$ b $n^2 + 16$ c $n^2 - 6$ d $\frac{n}{9}$

4

Position-to-term rule	Position	Term
Square then add 2	10	102
Square then subtract 1	5	24
Cube	7	343
Cube	5	125
Square then add 5	12	149
Square then subtract 6	10	94
Divide by 3	60	20

5

x	−3	−2	−1	0	1	2	3
y	11	6	3	2	3	6	11

Parabola which crosses the y-axis at (0, 2) and does not cross the x-axis.

6

x	−3	−2	−1	0	1	2	3
y	5	0	−3	−4	−3	0	5

Parabola which crosses the y-axis at (0, −4) and does not cross the x-axis.

7 a i

x	−2	−1	0	1	2	3
y	−8	−5	−2	1	4	7

ii

x	−2	−1	0	1	2	3
y	3	3.5	4	4.5	5	5.5

iii

x	−2	−1	0	1	2	3
y	10	7	4	1	−2	−5

b

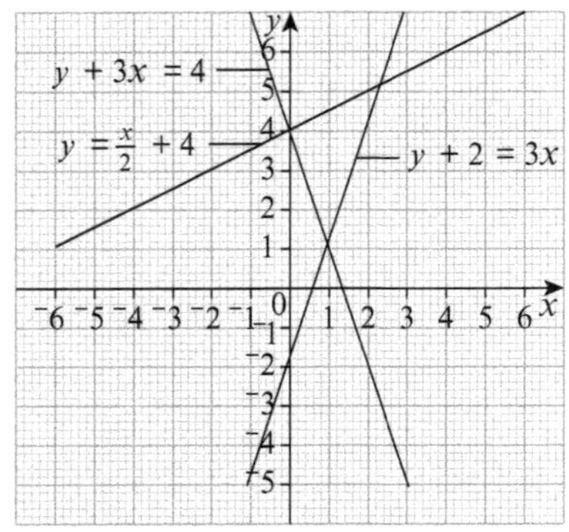

c i gradient = 3, intercept = −2
ii gradient = 0.5, intercept = 4
iii gradient = −3, intercept = 4

8 a $x = 1, y = 5$ b $x = 1, y = 1$ c $x = 2, y = 1$
d $x = 1, y = -1$ e $x = -2, y = -1$ f $x = 2, y = 2$
g $x = 4, y = 1$ h $x = 2, y = 1$

9 a

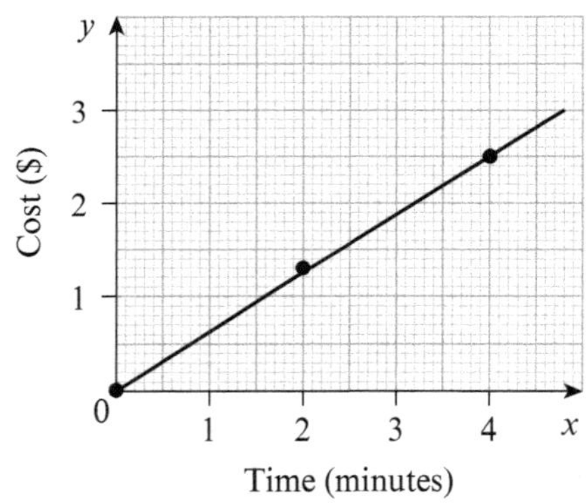

b \$0.10 c \$0.60 d $C = 0.6t + 0.1$
e intercept is the connection fee and the gradient is the cost per minute

10 a $3n - 2$ b $18 - 3n$ c $5n + 9$ d $2.5n + 0.5$
e $8n - 7$ f $0.75n + 14.5$

11 a i 20 km/h ii 10 km/h iii 5 km/h
b 10 km/h

12 a B b 11 km/h c $2\frac{1}{4}$ hours
d first hour e half an hour

13 a Constant speed of 25 m/s for 8 seconds followed by deceleration to rest over the next 12 seconds
b 2.1 m/s^2

Check out

1 **a** 1, 2, 5, 26 **b** 1, 8, 27 **c** $^{-}3, 0, 5, 12, 21$

2 **a** $5n + 7$ **b** $4n + 5$ **c** $62 - 10n$ **d** $17 - 2n$

3 **a** $n^2 - 5$ **b** $n^2 + 10$

4 U-shaped parabola, with minimum point at (0, –3), x-axis intercepts at (1.73, 0) and (–1.73, 0)

5

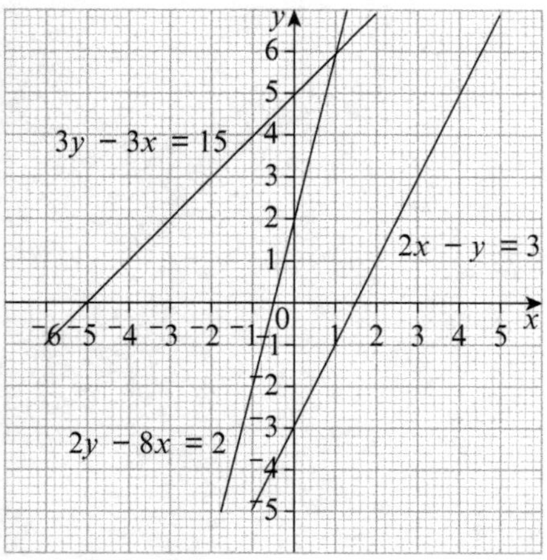

6 **a** $x = 5, y = 1$ **b** $x = 4, y = 5$

7 **a** 4 hours **b** 500 km/h **c** 800 km/h **d** 650 km/h

Chapter 13

Check in

1

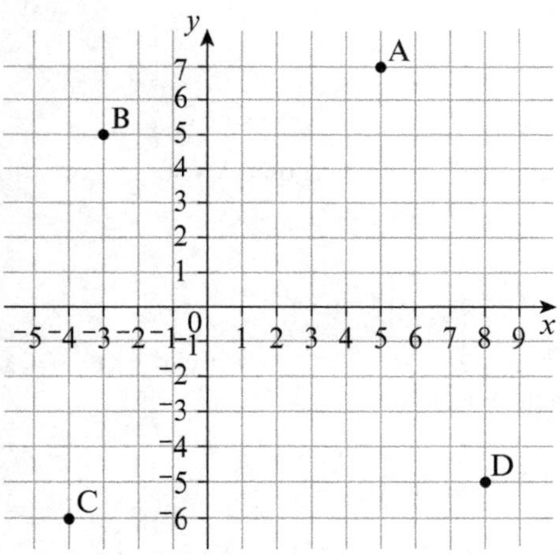

2

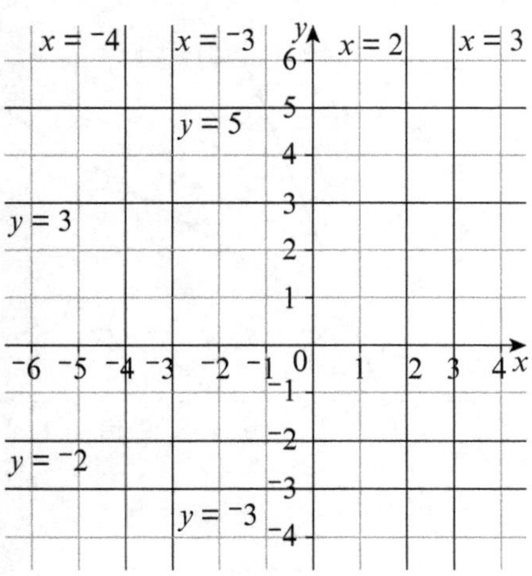

3

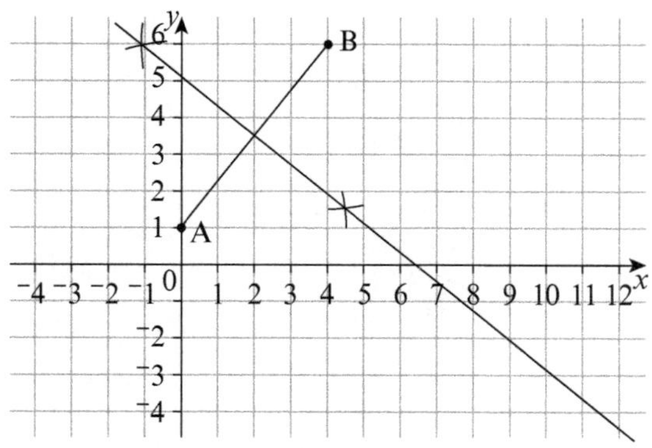

4 **a** $y = 2x + 3$ **b** $y = {}^{-}3x + 1$

Exercise 13A

1 **a**

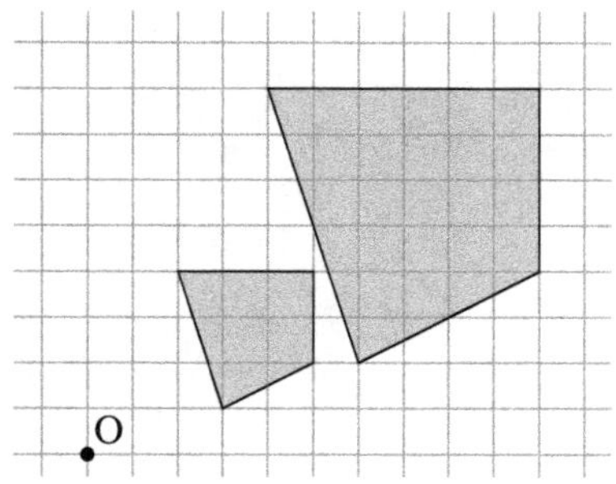

b

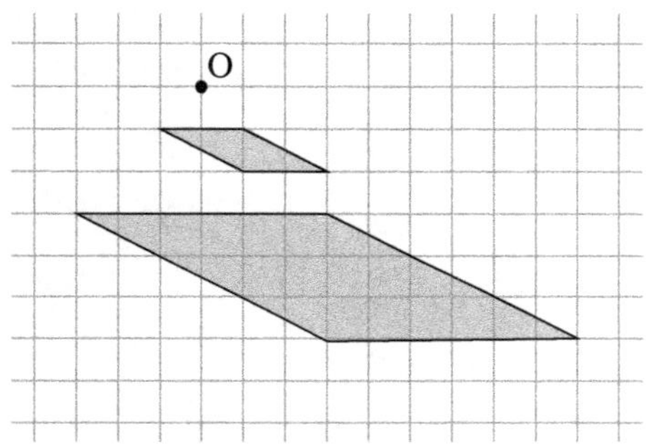

2 **a** P = 24 cm A = 24 cm^2

b P = 36 cm A = 54 cm^2

3 Students' own work

4 No, the area is multiplied by the scale factor squared not just the scale factor (so it will be 576 cm^2)

5 3

6 6

7 **a**

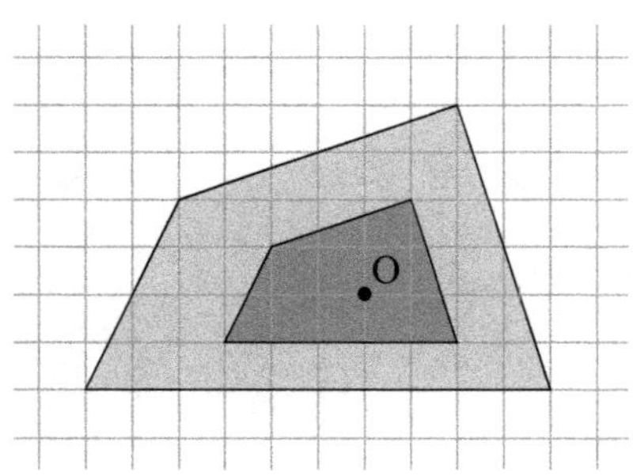

b

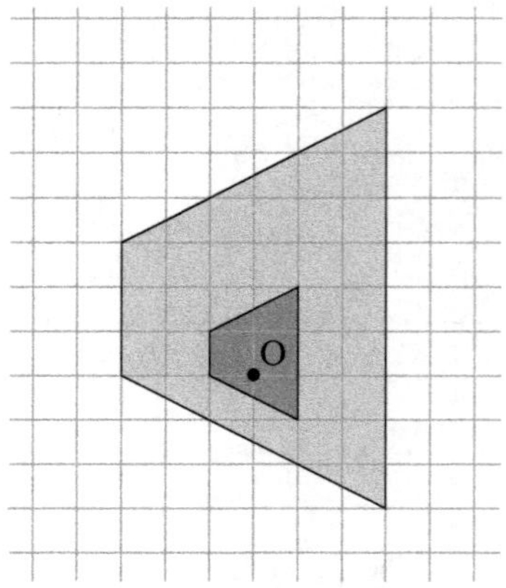

c

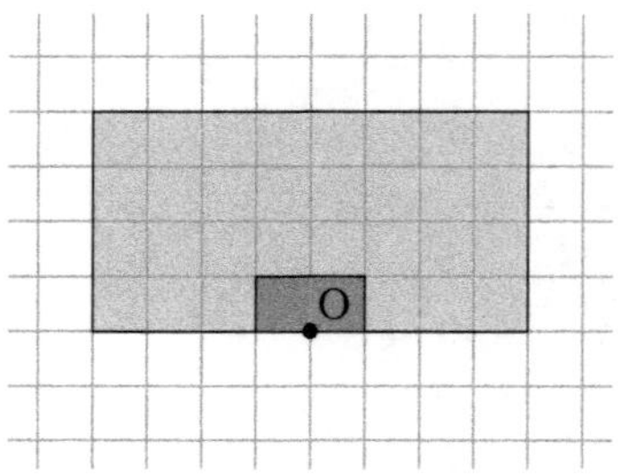

d

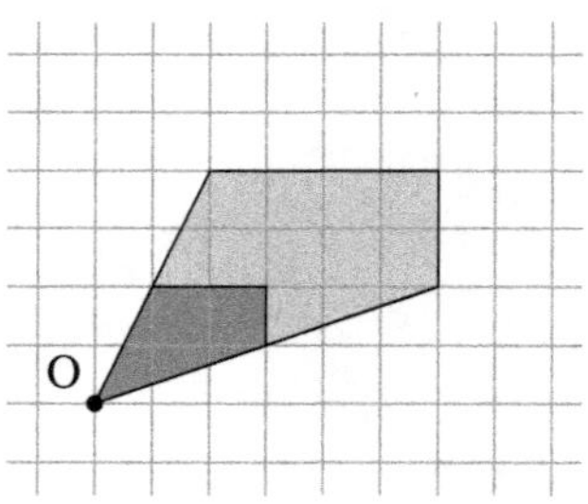

8 5

9 a 8 cm **b** 63 cm²

10 5

11 a 1 : 3 **b** 9 : 1

Exercise 13B

1 a

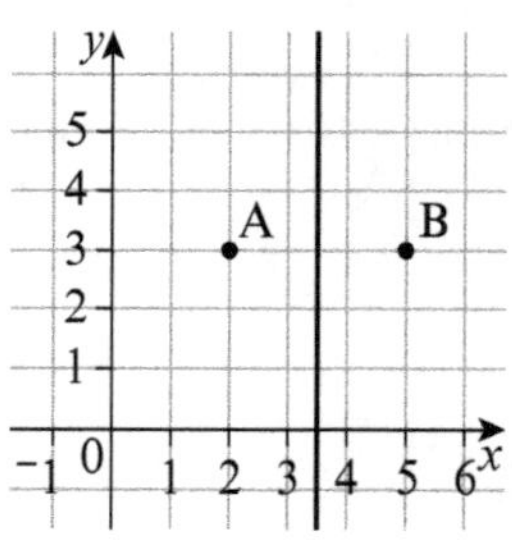

b

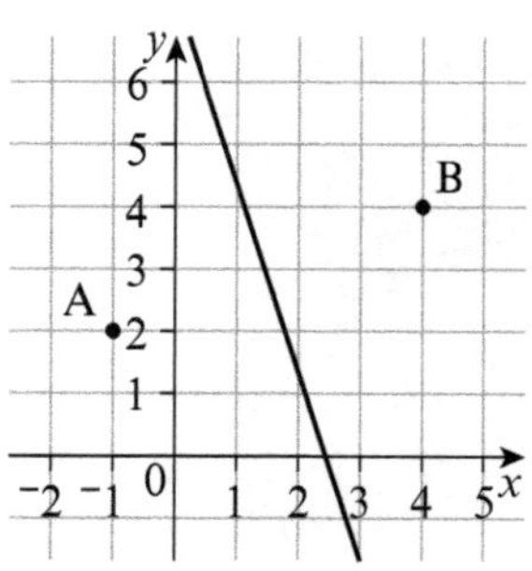

c

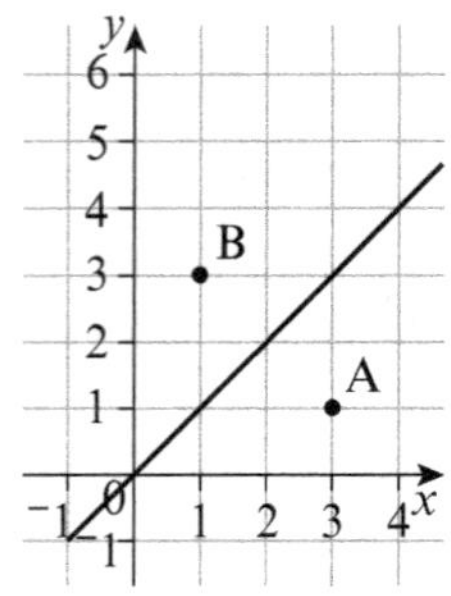

d

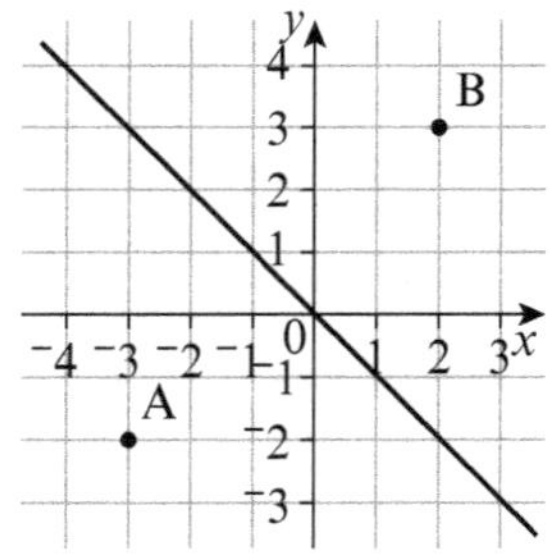

e

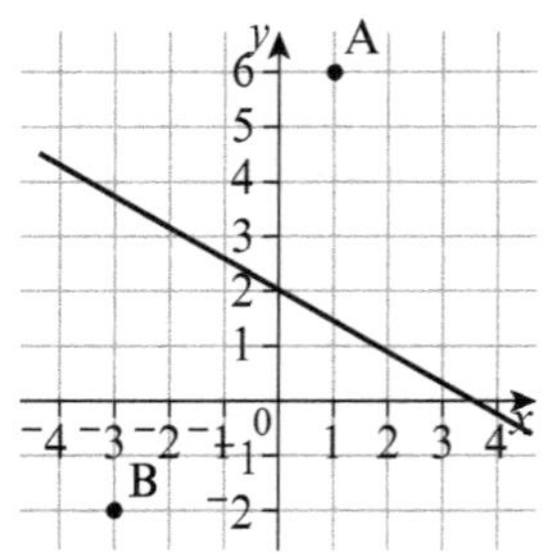

2 c $y = {}^{-}x$

4 a Yes **b i** 90° **ii** 90°, yes

5 centre (7, 5) angle 90° direction anticlockwise

6 Enlargement, centre (6, 4) scale factor 3

7 Translation, vector $\begin{pmatrix} 3 \\ {}^{-}4 \end{pmatrix}$

8 a $y = 4$ **b** $x = 2$ **c** $y = 1\frac{1}{2}$ **d** $y = x$ **e** $y = {}^{-}x$

9 Reflection in the line $y = x$

10 Translation, vector $\begin{pmatrix} {}^{-}1 \\ 4 \end{pmatrix}$

11 Enlargement, centre (⁻2, 3) scale factor 2

12 a, b

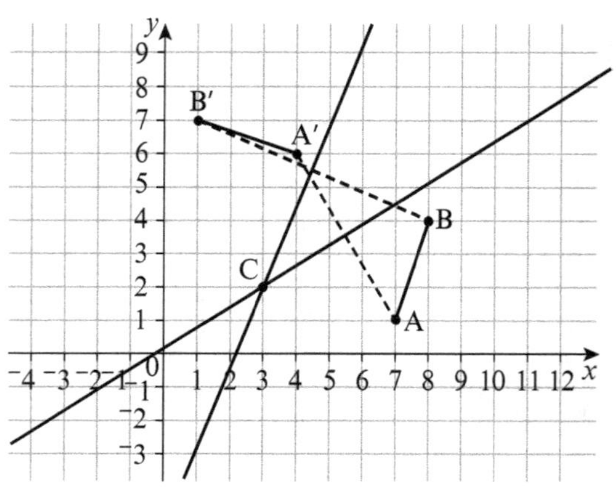

c (3, 2) **d** 90° **e** 90° anticlockwise

13 Translation, vector $\begin{pmatrix} 5 \\ 0 \end{pmatrix}$

Exercise 13C

1 a

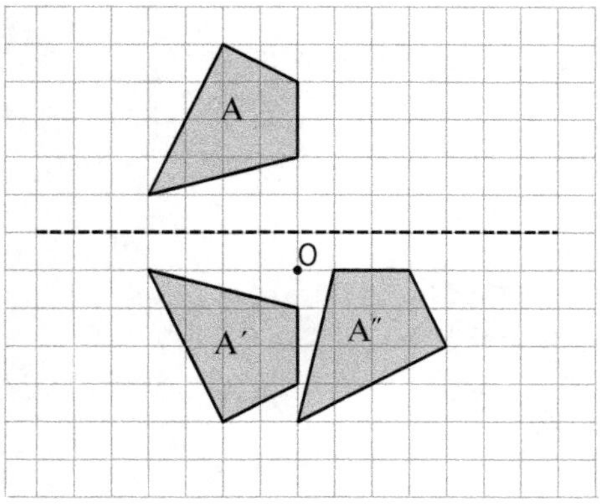

b

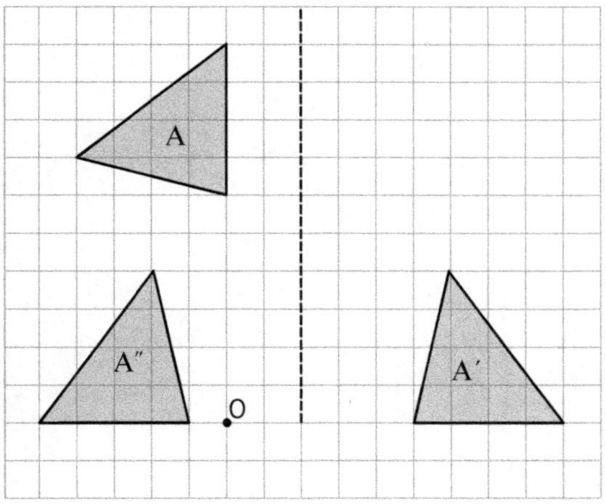

c

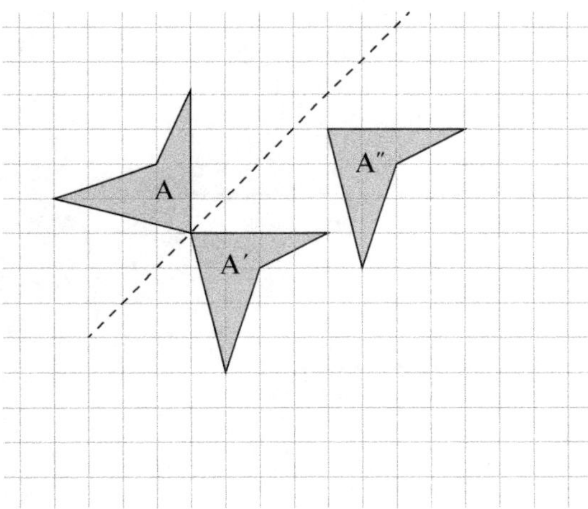

2 Yes, any combination of transformations involving rotations, reflections or translations will always produce a congruent image.

3

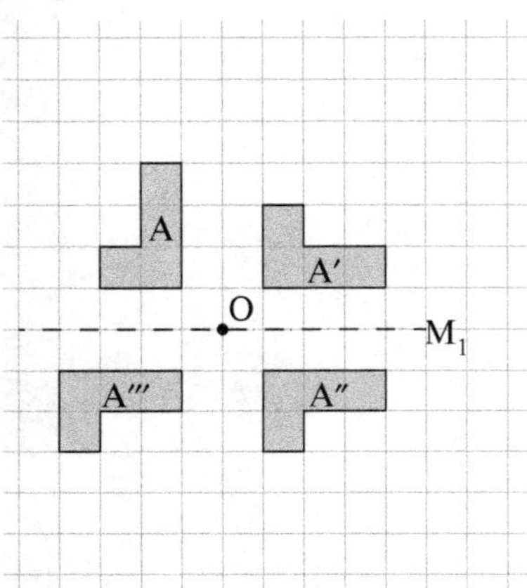

4

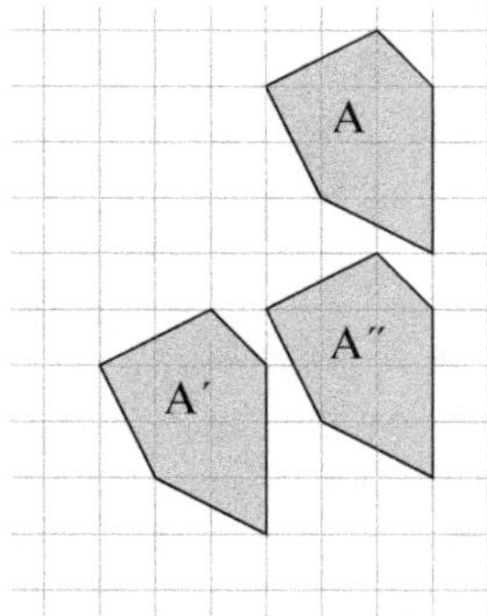

5

First transformation, A to A′	Second transformation, A′ to A″	Is A″ congruent or not congruent to A?
Reflection in $y = x$	Rotation 90° clockwise about the origin	congruent
Translation through $\begin{pmatrix}2\\4\end{pmatrix}$	Reflection in the line $y = 3$	congruent
Enlargement scale factor 2, centre (3, 4)	Translation through $\begin{pmatrix}-1\\0\end{pmatrix}$	Not congruent
Rotation 90° anticlockwise about (7, 2)	Reflection in the line $x = 3$	congruent

6 a Translation **b** $\begin{pmatrix}4\\0\end{pmatrix}$

7 a Yes **b** $\begin{pmatrix}2\\0\end{pmatrix}$

c The magnitude of the vector is twice the distance between the lines of reflection

8 a $\begin{pmatrix}8\\0\end{pmatrix}$

9

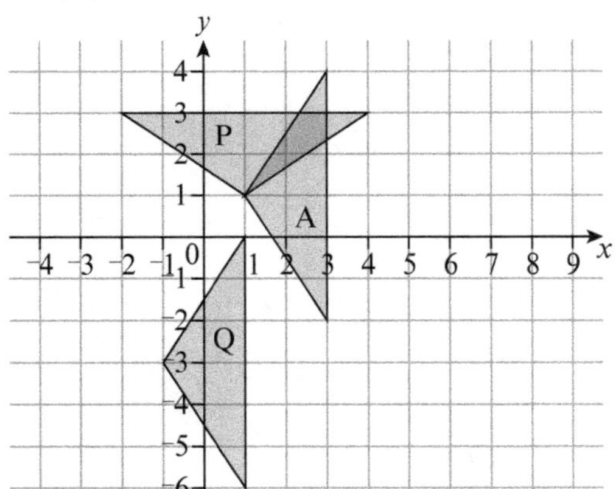

10 Reflection in $x = 4$

11 b (3, 1) **c** (8, 6)

12

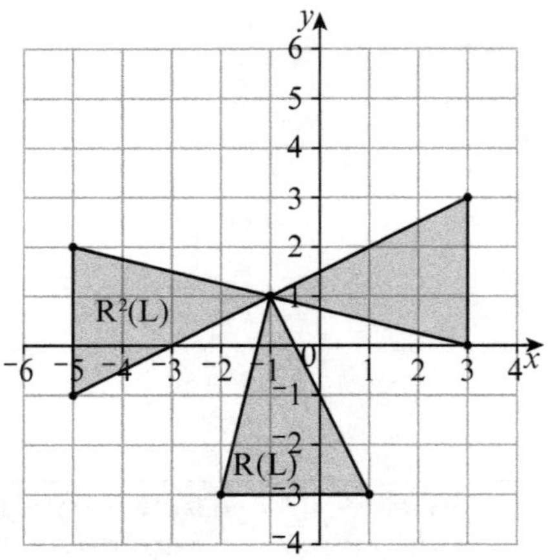

13 a, b

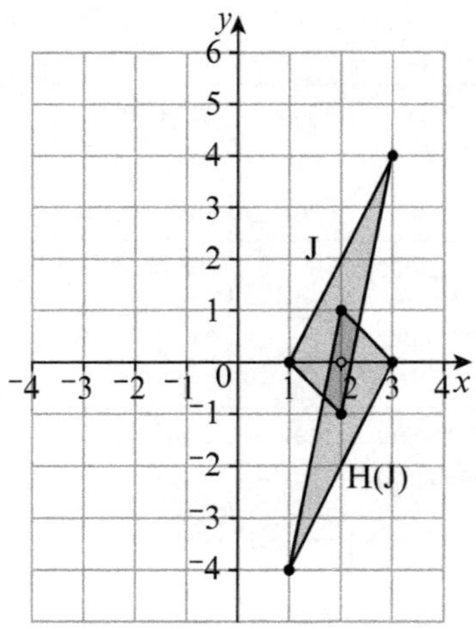

c

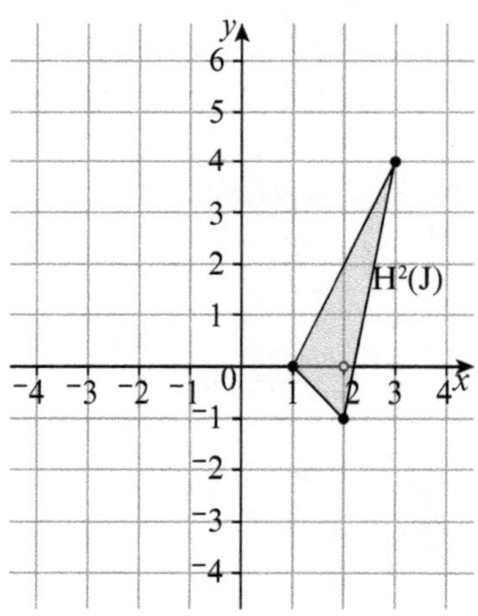

14 a, b, c

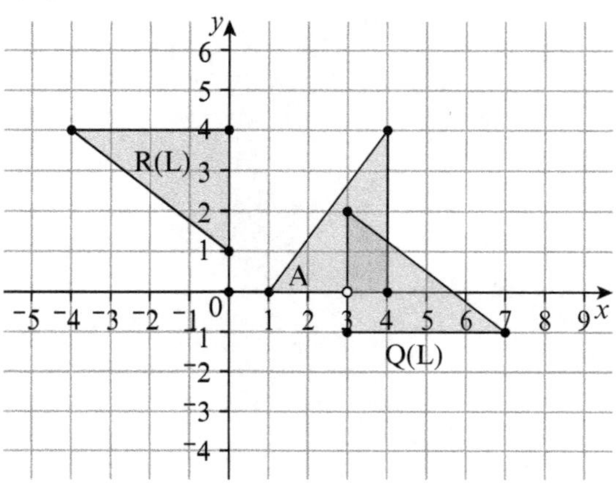

Consolidation

Exercise 13

1 enlargement, scale factor 3, centre (7, 4)

2 a

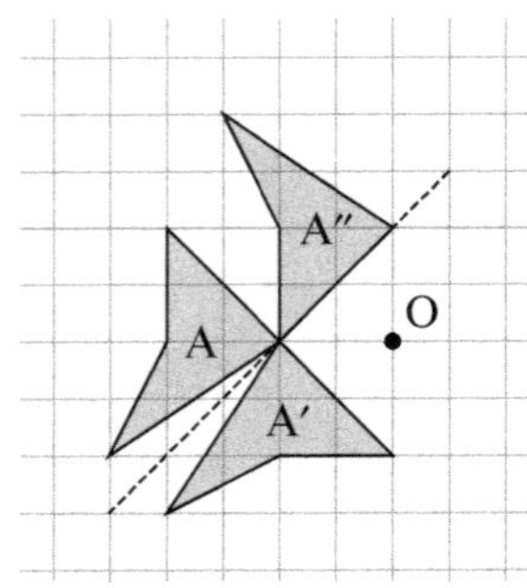

b

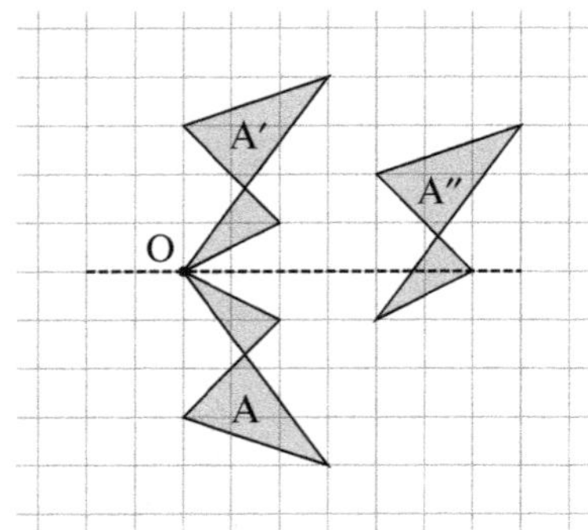

3 Perimeter = 72 cm, area = 320 cm²

4 a

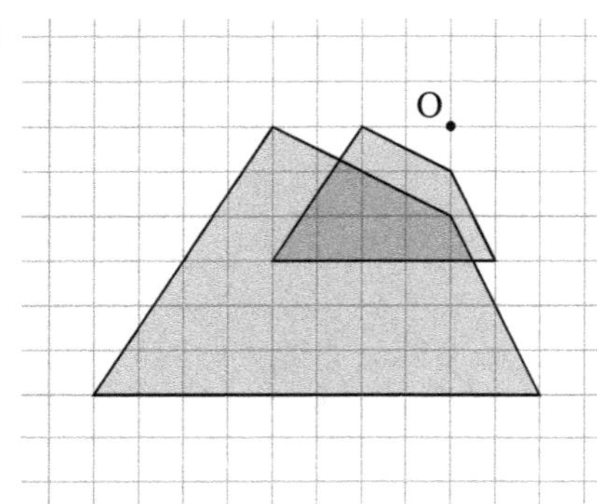

b

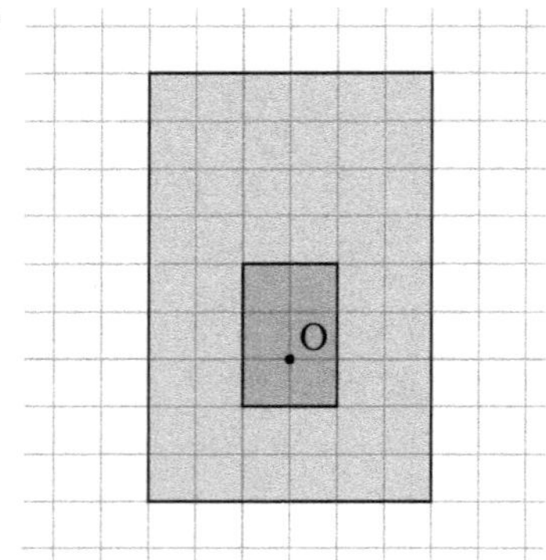

5

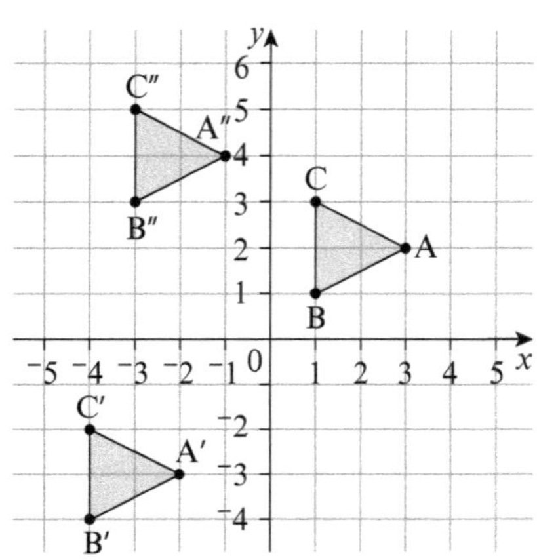

6 a P′(−4, −2), Q′(−3, 0), R′(−2, −1)
  b P′(3, 6), Q′(2, 4), R′(1, 5)
  c P′(7, 8), Q′(6, 6), R′(5, 7)
  d P′(3, −4), Q′(1, −3), R′(2, −2)

7 $n = 9$

8 a
  b iii

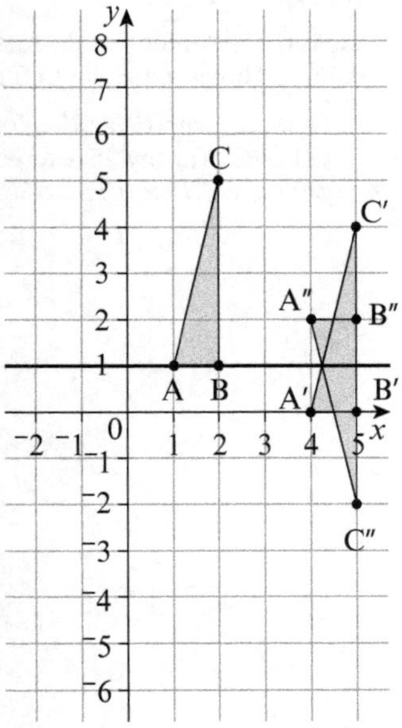

  c No, the translation isn't parallel to the line of reflection

9 rotation 90° anticlockwise about (−1, 1)

10 a i Corners will be A′(1, 8), B′(7, 5), C′(4, 2)
  ii Corners will be A′(9, −1), B′(8, 1), C′(7, 0)
  iii Corners will be A′(−1, −3), B′(−2, −1), C′(−3, −2)
  b For **i**, enlargement scale factor 3, centre (−0.5, −1). For **ii** and **iii**, none.
  c i not congruent – it involves enlargement which changes the size
  ii congruent, rotation and reflection maintain shape and size
  iii congruent, reflection and translation maintain shape and size

Check out

1 a

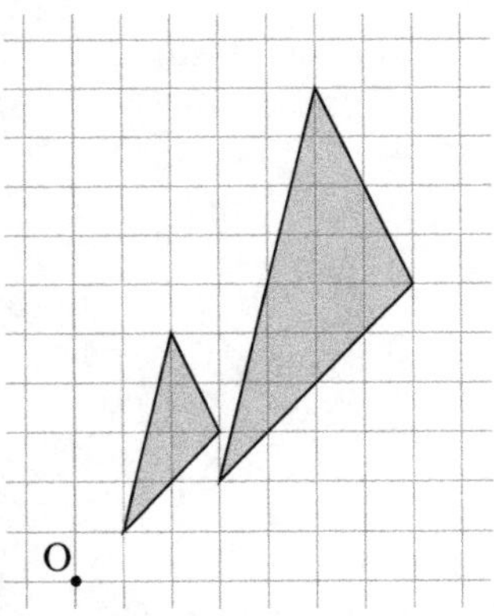

  b P = 50 m, A = 100 m²

2

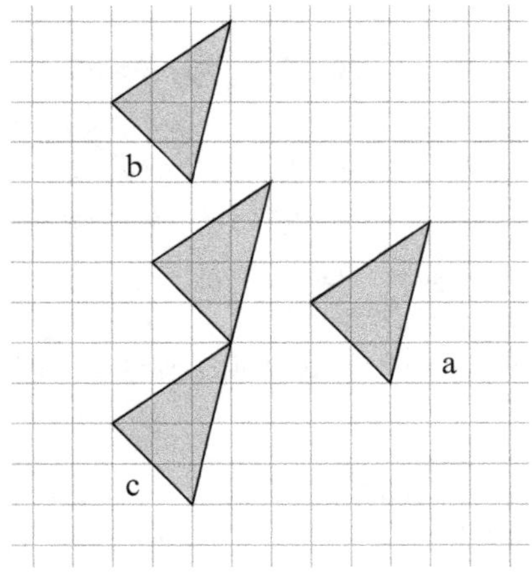

3 a (6, 3)   **b** $y = {}^{-}x$

4 90° clockwise about (−2, 3)

5 a b c

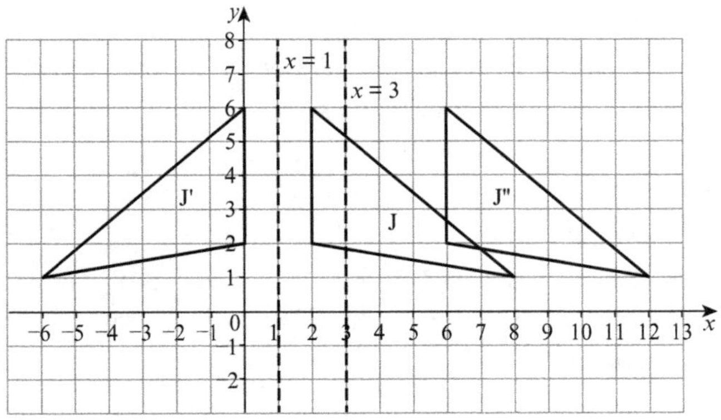

  d Translation through vector $\begin{pmatrix} 4 \\ 0 \end{pmatrix}$

Chapter 14

Check in

1 a 12 **b** 20 **c** 800 **d** 4

2 a −12 **b** −15 **c** −77 **d** 6
  e 16 **f** −6 **g** −3 **h** −7

3 a 84 m **b** \$83.20 **c** 2736 g **d** 32.8 cm²

Exercise 14A

1 a 6 **b** 50 **c** 20 **d** 50
  e −42 **f** 16

2 a 64 **b** −54 **c** 80 **d** 21
  e 0.01 **f** 0.009 **g** 0.003 **h** 0.0056

3 a 51 **b** 4.8 **c** −10 **d** 23

Exercise 14B

For Questions **1** and **2**, other answers are possible.

1 a 4 **b** 2 **c** 4 **d** −3
  e −8 **f** 8

2 a 1.8 **b** $\frac{2}{3}$ **c** $\frac{1}{2}$ **d** 1
  e −62 **f** 0.02

4 a 51.28 **b** 0.86 **c** 0.08 **d** 1316.54

5 a 20 ÷ 20 = 1, so incorrect **b** 0.5 × 18 = 9, could be correct
  c $\frac{5+4}{2} = \frac{9}{2} = 4.5$, so incorrect

Exercise 14C

1 $30 \div 20$ 30×0.04 $30 \div 2\frac{2}{5}$ 30×0.5
$30 \div 30$ $30 \times \frac{1}{3}$ $30 \div 2.7$ $30 \div \frac{5}{2}$

2 $50 \div 0.5$ 50×1.7 $50 \div \frac{1}{8}$ $50 \div 0.29$
$50 \times 4\frac{3}{8}$ $50 \times \frac{5}{3}$ $50 \div 0.23$ $50 \div 0.8$

3 **c** the answer should be greater than 3.8
d the answer should be greater than 17

4 Multiplying 43.5 by a number smaller than 1 would result in a number less than 43.5

5 **a** any number greater than 1
b any number between 0 and 1
c any number greater than 1

Exercise 14D

1 **a** 128.8 **b** $^-128.8$ **c** 1.288 **d** 12.88
e 0.012 88 **f** 1.288 **g** 0.012 88 **h** 1.288

2 **a** $^-1.8$ **b** $^-0.78$ **c** 7.05 **d** 2.66
e 55.62 **f** 87.6 **g** 1.69 **h** 23.606

3 **a** 15.64 m^2 **b** 160.19 cm^2 **c** 87.9248 m^2

4 **a** \$517.80 **b** \$263.22 **c** \$19.42

5 **a** 21 km **b** 5.46 km **c** 2.52 km

6 **a** $^-12$ **b** 1.8 **c** 1.69 **d** 2.468

7 54.25 cm^3

8 **a** 0.75 **b** 4.96 **c** 0.24 **d** 0.504
e $^-3.12$ **f** 2.914 **g** 13.803 **h** 11.696

Exercise 14E

1 **a** 40 **b** 20 **c** $^-2.1$ **d** 3.48
e $^-30$ **f** 300 **g** 30 **h** 200

2 **a** 17 **b** 16 **c** 160 **d** $^-1.6$
e 26.8 **f** $^-59.2$ **g** 4.56 **h** 56

3 135

4 153

5 **a** 7 m **b** 9 m **c** 16 m **d** 288 m

6 \$5.50

7 **a** 40 **b** 17 **c** 5 **d** 30
e 0.4 **f** 0.48 **g** 190 **h** 60

8 **a** 4.3 **b** 8.4 **c** 2.3 **d** 100

9 **a** 2.4 **b** 3.2 **c** 7.3 **d** 0.315
e 20 **f** 0.14 **g** 60.25 **h** 0.3
i 200 **j** 350

Exercise 14F

1 **a** 1.12 **b** 0.3 **c** 0.928 **d** 1.954
e 2.45 **f** 4

2 **a** 4838 km **b** 35.1 kg **c** 222 litres **d** 46 seconds
e 0 hectares **f** \$115

3 **a** \$52 **b** \$208 **c** \$260 **d** \$520
e \$10 400 **f** \$52 000

4 \$555.66

5 \$887.60

6 **a** The multiplier should be 1.057 **b** 5.2 kg

7 **a** \$800 **b** \$1600 **c** \$4000

8 **a** \$120 **b** \$300 **c** \$300
d \$720 **e** \$1215

9 **a** 149.80 **b** \$374.50 **c** \$322.50
d \$763.20 **e** \$1327.63

10 Compound interest is always more than simple interest.

11 **a** \$10 067 **b** \$8859
c During 2027 (it is \$6861 at the end of 2026 and \$6037 at the end of 2027)

12 **a** Any description of a percentage increase of 8%, 5 times on an initial value of 4000 e.g. \$4000 investment with 8% compound interest over 5 years.
b Any description of a percentage decrease of 19%, 3 times on an initial value of 250 e.g. an item originally costs \$250, it is then reduced by 19% each month for 3 months.

13 The multiplier for 25% decrease is 0.75. So, for two decreases of 25%, this is $0.75 \times 0.75 = 0.5625$. This is not equal to 0.5.
Alternative explanations are possible, e.g. choose a price for something, say \$100, reduce this chosen price by 25% to get \$75, then reduce \$75 by 25% again using $\$75 \times 0.75 = \$56.25$ not \$50.

14 $\$400 \times 1.2 = \$480$
$\$480 \times 0.8 = \$384$

15 10 years

16 **a** \$25 **b** 5%

17 1.37 m

18 **a, c** True **b, d** False

19 14 minutes

20 5.06%

Consolidation

Exercise 14

1 $15 \div 12$, 15×0.32, 15×17, $15 \div 1\frac{1}{9}$, 15×30

2 **a** 128.8 **b** $^-128.8$ **c** 12.88 **d** 1.288
e 12.88 **f** 0.01288 **g** 0.01288 **h** 0.001 288

3 **a** $^-0.2$ **b** 0.8 **c** 0.4 **d** 0.16
e $^-0.02$ **f** 0.312 **g** 70 **h** 800
i 23 000 **j** 40 **k** 200 **l** 600

4 $34 \div 0.5$, 34×2.7, $34 \div \frac{7}{8}$, $34 \div 0.2$, $34 \times 1\frac{3}{4}$

5 **a** 28 **b** 3 **c** 6 **d** 25

6 **a** 4 **b** 8 **c** $^-12$ **d** 40
e 80 **f** 270 **g** 4 **h** 9
i 0.4 **j** 1.52 **k** 0.8 **l** 32.5

7 **a** \$2060 **b** From \$2060 to \$2100

8 **a** 540 g **b** \$0.58

9 **a** \$16 **b** \$24 **c** \$40

10 **a** gives the better return (\$1800 versus \$1440)

11 \$1343.46

12 4.6 kg

13 39.8 million

14 \$69.24

15 4 months

Check out

1 **a** 48 **b** 5 **c** 5 **d** 0

2 **b, c, d** are incorrect

3 **a** 0.888 **b** 24.514 **c** 48 **d** 5.8

4 **a** \$1187.50 **b** \$1223.13

5 **a** \$120 **b** 3 years

6 **a** \$367.55 **b** \$145.99

Chapter 15

Check in

1 a $\frac{2}{5}$ b 0.4 c 47% d only **i** and **iii**

2 a 0.61 b 0.65 c 0.25 d 0.92

3 a $\frac{13}{15}$ b $\frac{5}{8}$ c $\frac{1}{5}$ d $\frac{7}{10}$

Exercise 15A

1 a

		Dice 1			
		1	2	3	4
Dice 2	1	1, 1	2, 1	3, 1	4, 1
	2	1, 2	2, 2	3, 2	4, 2
	3	1, 3	2, 3	3, 3	4, 3
	4	1, 4	2, 4	3, 4	4, 4

b $\frac{1}{4}$ c $\frac{3}{4}$

2 a

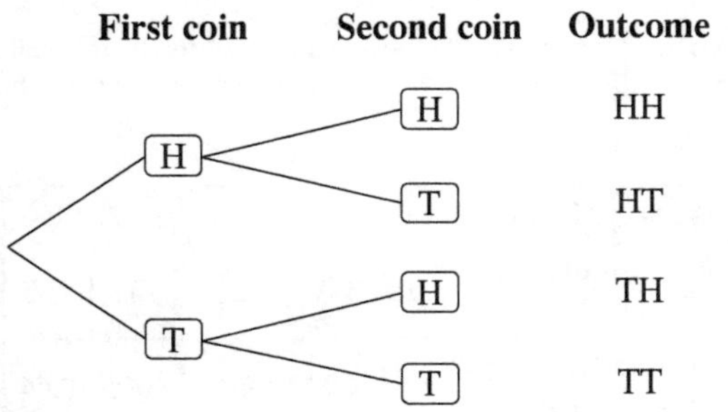

b $\frac{1}{4}$ c $\frac{1}{4}$

3 $\frac{1}{4}$

4 a

		Dice 1					
		1	2	3	4	5	6
Dice 2	1	2	3	4	5	6	7
	2	3	4	5	6	7	8
	3	4	5	6	7	8	9
	4	5	6	7	8	9	10
	5	6	7	8	9	10	11
	6	7	8	9	10	11	12

b i $\frac{5}{36}$ ii $\frac{1}{36}$ iii $\frac{1}{18}$

c 7 d $\frac{5}{18}$ e $\frac{7}{12}$ f yes

5 0.38

6 a 0.19 b 0.58 c 0.42

7 a i 0.28 ii 0.58

b You know the probability of 3 and 4 combined, but you don't know the probability of them individually.

8 Probabilities add to more than 1. A shirt can be, for example, short sleeve and white, so they can both happen at the same time.

9 a

		Mrs Boyd's blouses		
		Blue	White	Red
Mr Wood's shirts	Blue	BB	BW	BR
	Purple	PB	PW	PR
	White	WB	WW	WR
	Green	GB	GW	GR

b i $\frac{1}{12}$ ii $\frac{1}{12}$ iii $\frac{5}{6}$

10

		Spinner 1				
		1	2	3	4	5
Spinner 2	1	2	3	4	5	6
	2	3	4	5	6	7
	3	4	5	6	7	8
	4	5	6	7	8	9
	5	6	7	8	9	10
	6	7	8	9	10	11
	7	8	9	10	11	12
	8	9	10	11	12	13

a 13 b 40 c $\frac{1}{10}$ d $\frac{3}{20}$ e $\frac{3}{40}$

11 a

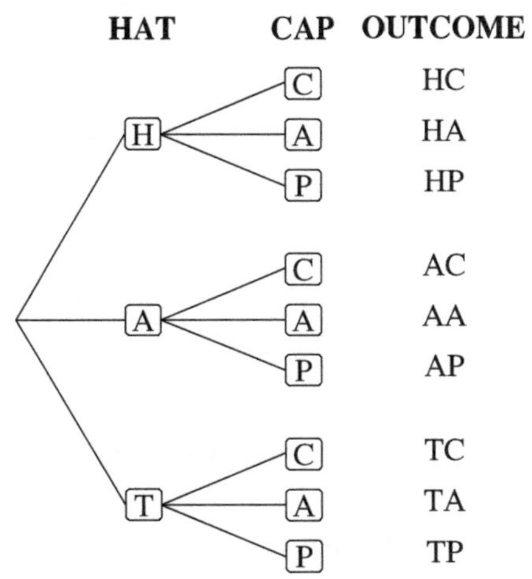

b i $\frac{1}{9}$ ii $\frac{4}{9}$ iii $\frac{5}{9}$

12 a $\frac{1}{6}$

b Diagram with red = 90°, green = 60°, blue = 150°, white = 60°

c because the probabilities add up to more than 1

13 $\frac{3}{5}$

Exercise 15B

1 a false b true c true d true

2 a dependent b independent c independent

d dependent e dependent

3 a AC, AD, BC, BD b AB, CD

4 He is correct that three sixes are very rare but he is not correct about the probability changing. The previous rolls do not affect the probability of the third roll, as all rolls are independent.

5 a less than b greater than

6 No, any set of four numbers is equally likely because picks are independent of each other.

Exercise 15C

1 a

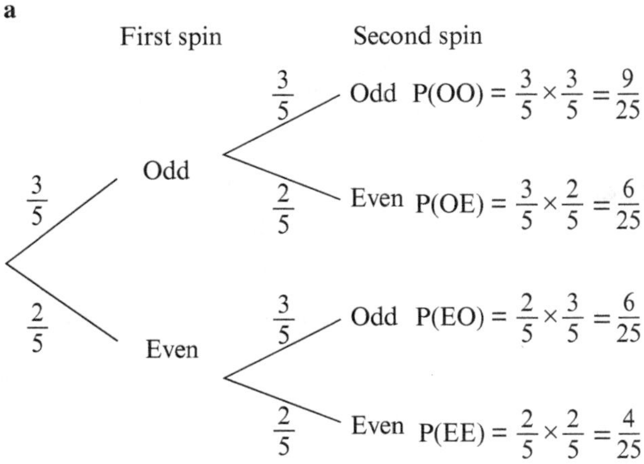

b $\frac{9}{25} + \frac{6}{25} + \frac{6}{25} + \frac{4}{25} = 1$

c $P(OE) + P(EO) = \frac{6}{25} + \frac{6}{25} = \frac{12}{25}$

d $\frac{2}{7} + \frac{4}{7}$ does not add up to 1

2 a

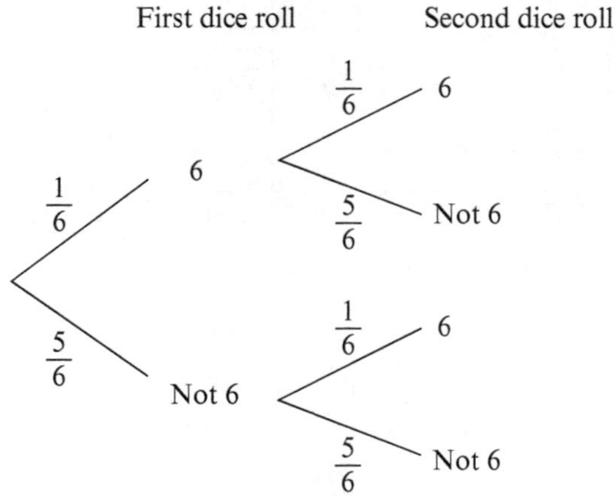

b $\frac{5}{18}$

3 $\frac{5}{12}$

4 $\frac{17}{32}$

5 a $\frac{2}{7}$ **b** Subtract the answer from 1 to get $\frac{5}{7}$

6 0.001

7 Any spinner where the probability of even is $\frac{3}{5}$ and the probability of odd is $\frac{2}{5}$, e.g.

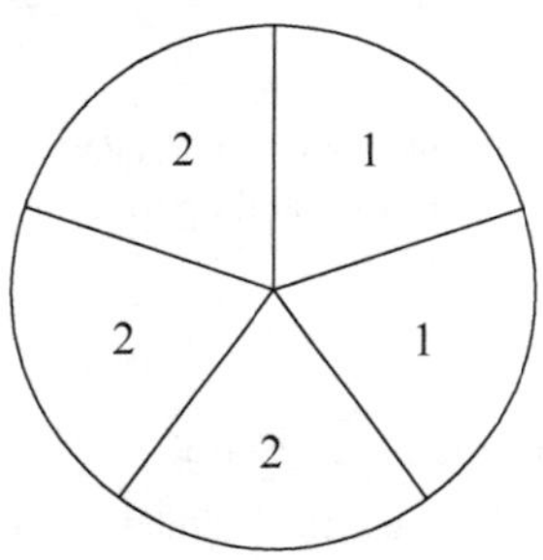

8 a $\frac{3}{16}$ **b** $\frac{1}{4}$

9 Any spinner where the probability of 1 is 0.3, the probability of 2 is $\frac{1}{2}$ and the probability of 3 × the probability of 4 = 0.01, e.g.

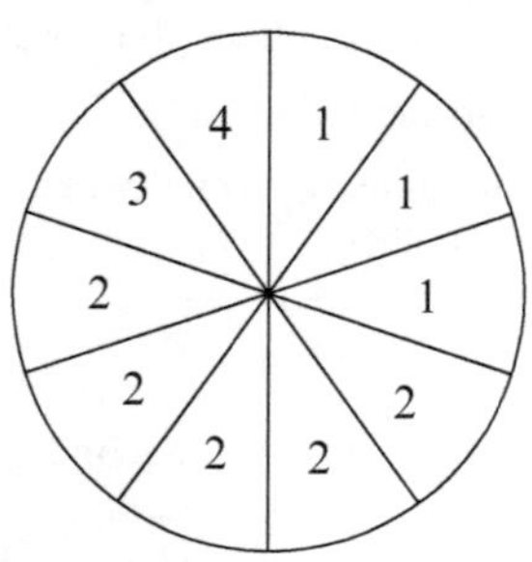

10 a $\frac{5}{11}$ **b** $\frac{19}{33}$

Exercise 15D

1 a i 0.23 **ii** 0.1125 **iii** 0.8875 **iv** 0.29

b i 161 **ii** 147 **iii** 161

2 The dice might be biased, but there are too few trials to be sure.

3 a i 0.256 **ii** 0.388 **iii** 0.848

b vanilla

c Cannot tell because the 'Other' column does not give details of the flavours.

d 63

4 a 18 **b i** 17 **ii** 34 **c** 87

5 a 50

b No. Experimental data is often close to, but not exactly the same as, theoretical data.

c The results for 1 and 5 are a lot higher and lower, respectively, than 50, and 300 rolls is a reasonable sample size, so it looks to be that the dice is biased.

6 a i 4 **ii** 12

b He needs to pick more counters until the graph starts to level off.

c 450

7 a 105 **b** 0.23 **c** 1150

d No, he has not used the best estimate of the theoretical probability. If he uses 0.23 × 2000 he would expect 460 reds.

8 a 0.48

b

Number of throws	200	240	280	320	360	400
Number of times point up	**92**	104	126	141	159	**172**
Relative Frequency (to 2 d.p.)	0.46	**0.43**	**0.45**	**0.44**	**0.44**	0.43

c

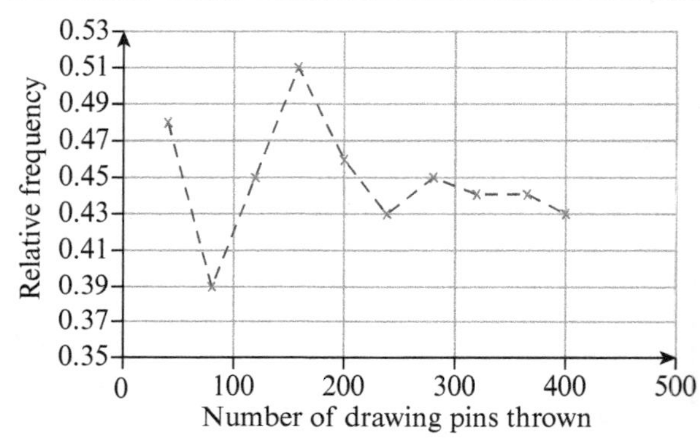

d Close to 0.43

e i 0.42 **ii** part **e i** is likely to be the closest, as it involves the most trials.

9 20 blue, 13 red and 17 green

Consolidation

Exercise 15

1 a 0.36 **b** 0.54 **c** 0.9

2 BC and AD

3 a i 0.11 **ii** 0.54 **iii** 0.46

b

Colour	White	Black	Grey	Blue
Number of counters	40	44	140	176

4 a

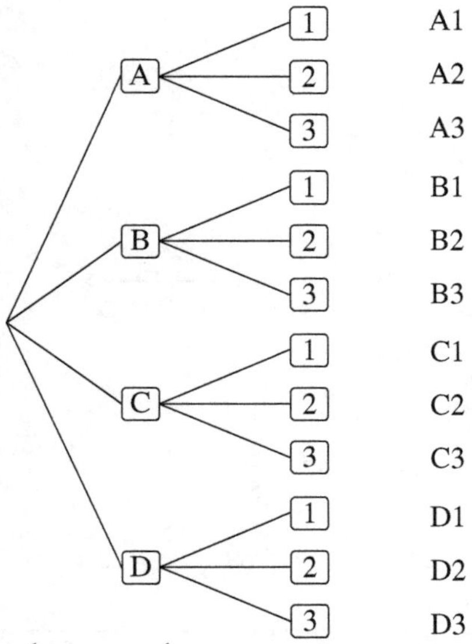

b i $\frac{1}{12}$ **ii** $\frac{1}{6}$

c i 15 **ii** 30

5

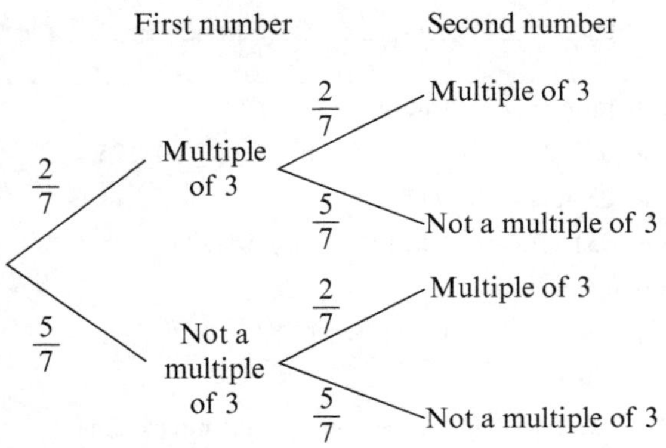

b i $\frac{4}{49}$ **ii** $\frac{25}{49}$ **iii** $\frac{20}{49}$

c i false **ii** true

d All outcomes are equally likely and, as there are 49 possible outcomes, it will be difficult to draw this on a tree diagram but easier to use a two-way table.

6 a The probability of the colour of the second ball is not affected by which ball is selected first because the ball is replaced.

b 0.52

c 0.48

7 a 156 **b** 0.45 **c** 700

8 a

		Blue card				
		1	2	3	4	5
Red card	2	2, 1	2, 2	2, 3	2, 4	2, 5
	4	4, 1	4, 2	4, 3	4, 4	4, 5
	6	6, 1	6, 2	6, 3	6, 4	6, 5
	8	8, 1	8, 2	8, 3	8, 4	8, 5

b i $\frac{1}{20}$ **ii** $\frac{1}{20}$ **iii** 0

Check out

1 $\frac{7}{20}$

2 a dependent **b** independent

3 a

		Letter		
		H	E	N
Spinner	1	H1	E1	N1
	2	H2	E2	N2
	3	H3	E3	N3

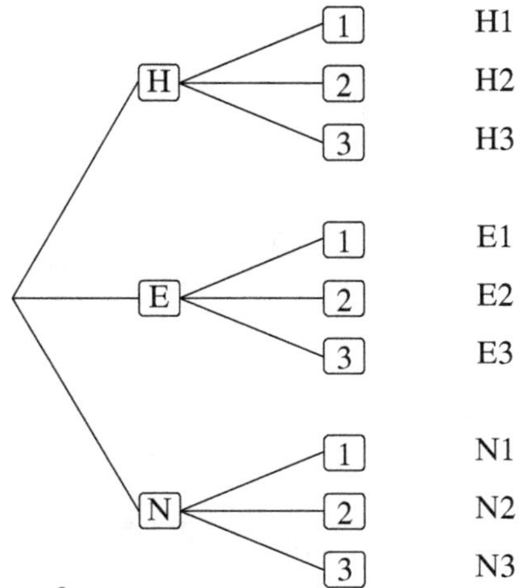

b $\frac{2}{9}$

4 a 20 **b** 40 **c** 60

Review C

1 a 9.72 **b** 0.0972 **c** ⁻9720 **d** 0.972

e 0.0972 **f** ⁻0.000 972

2 52

3

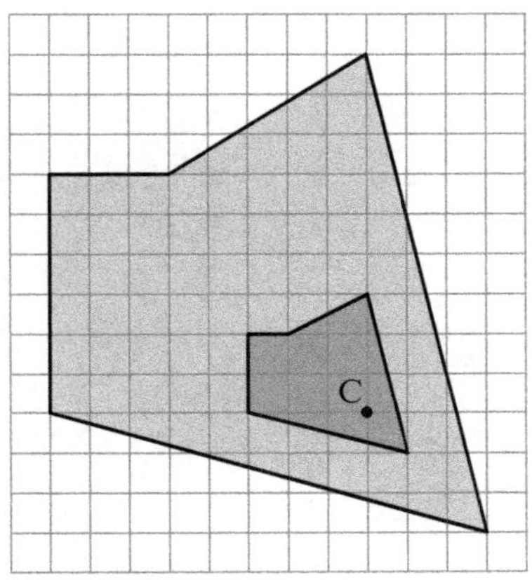

4 a \$28 and \$98

b 60 g and 100 g

c 20 cm, 10 cm and 50 cm

5 a

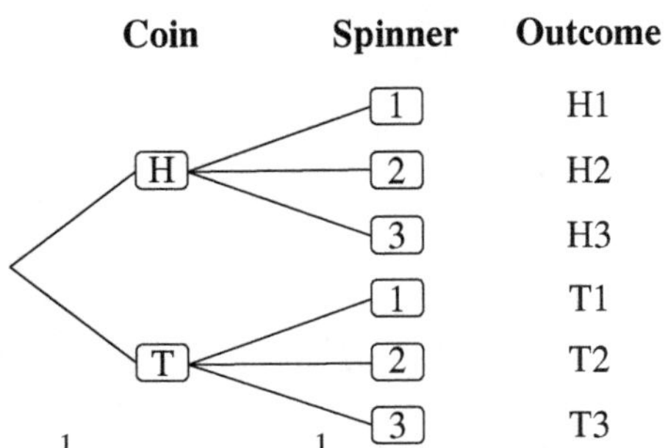

b i $\frac{1}{6}$ **ii** $\frac{1}{3}$

6 a

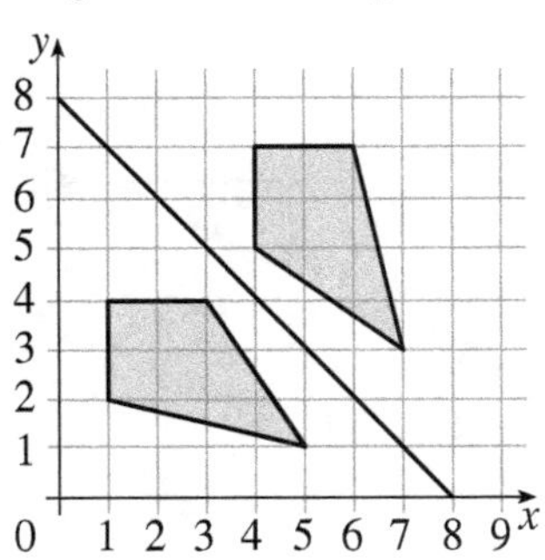

b Any coordinates on the line $y = 8 - x$, e.g. (1, 7), (6, 2), etc

c $y = 8 - x$

7

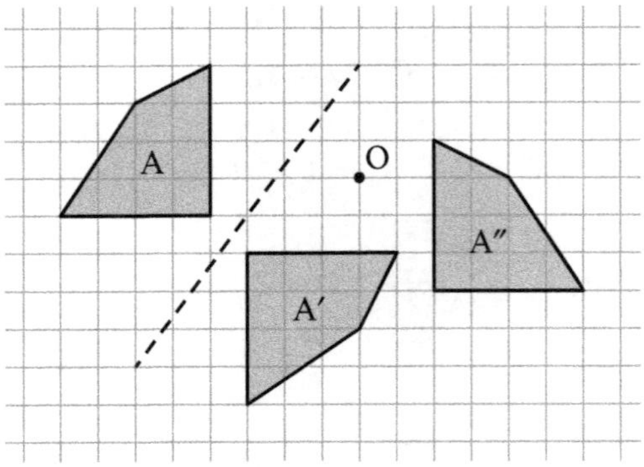

8 48 cm³

9 a

		Letter 1				
		B	**O**	**O**	**K**	**S**
Letter 2	**W**	BW	OW	OW	KW	SW
	O	BO	OO	OO	KO	SO
	R	BR	OR	OR	KR	SR
	K	BK	OK	OK	KK	SK

b i $\frac{1}{20}$ **ii** $\frac{1}{10}$

10 a First $\frac{1}{2}$ hour, 24 km/h

b 1.5 hours

c 32 km

11 a 0.24 **b** 0.156 **c** ⁻17.02 **d** 21.58

e 6 **f** ⁻0.6 **g** 41 **h** 40

12

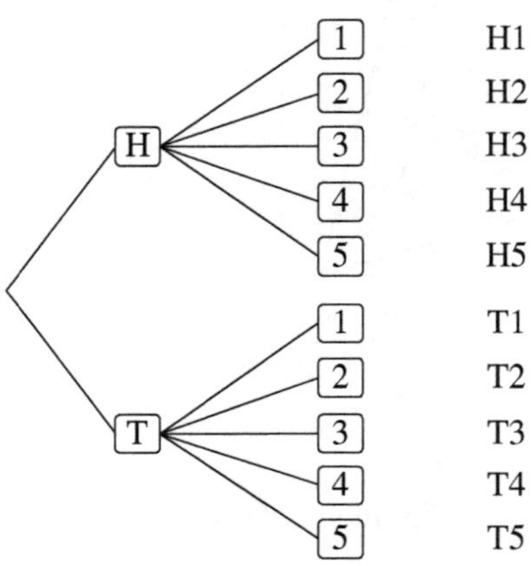

a $\frac{1}{10}$ **b** $\frac{1}{5}$ **c** $\frac{4}{5}$

13 a

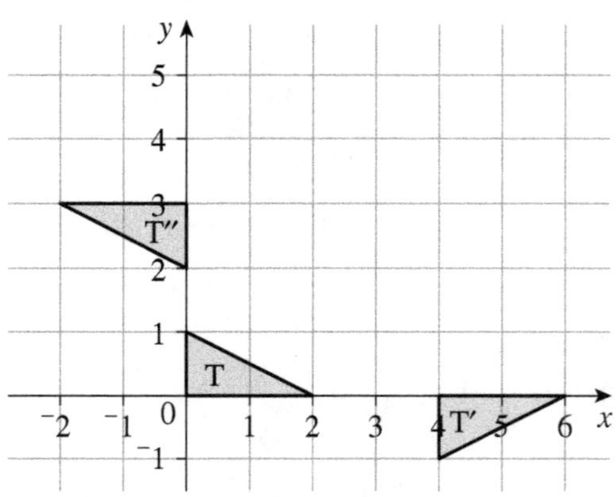

b rotation 180° centre (0, 1.5)

14 a \$11 025 **b** \$4593.75 **c** \$14 700

15 a 0.34 **b** 0.48

16

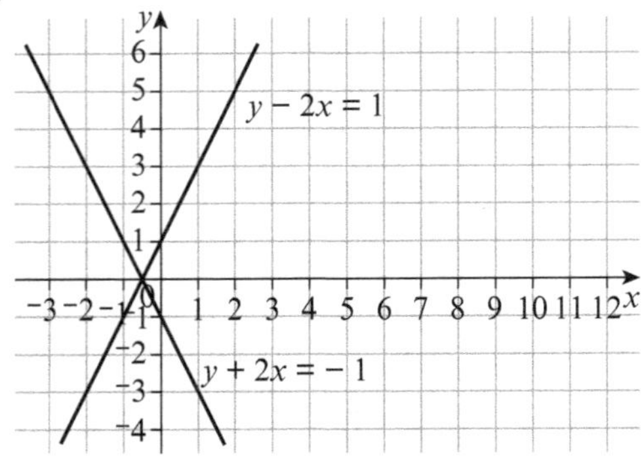

17 $57 \div 0.9$, $57 \div \frac{3}{8}$, $57 \div 0.8$, $57 \div 1\frac{2}{5}$

18 276

19 a ⁻9 **b** 0.6 **c** 1.69 **d** 2.442

e 2.9 **f** 74.1

20 a

x	5	2	90
y	35	14	630

b

x	2	15	7.5
y	32	240	120

21 a false **b** true **c** true

22 a 0, 3, 8, 15, 24 **b** 1, 8, 27, 64, 125

23 a 63 hours **b** 9 hours

24 a \$51,750 **b** \$48,127.50 **c** \$41,400

25 6 minutes

26 a $y = 2.5x$ **c i** 17.5 **ii** 17

27 \$8539.70

28
- reflection – give the equation of the **mirror line**
- rotation – give the **angle** and **direction** of the rotation and the coordinates of the **centre of rotation.**
- translation – give the **vector**
- enlargement – give the coordinates of the **centre of enlargement** and **the scale factor** of the enlargement

29 Amanda \$72, Donna \$126

30 a

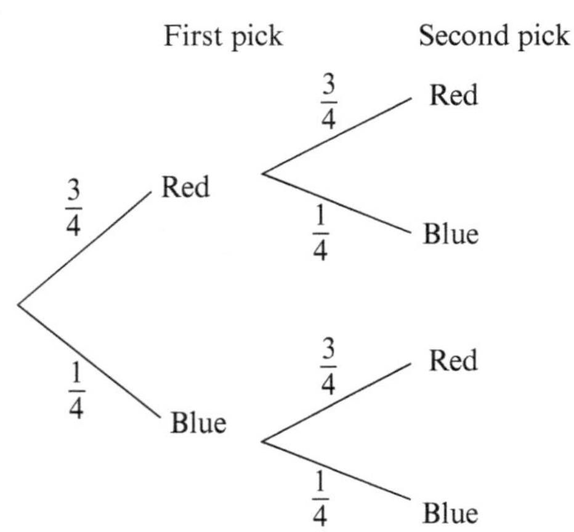

b $\frac{3}{16}$ **c** $\frac{5}{8}$

31 a (⁻1, 7) **b** (11, 9) **c** (0, 4)

32 Enlargement scale factor 2, centre (⁻4, ⁻1)

33 9 : 5

34 \$24,920

35 a 74 km/h **b** 126 km **c** 48 minutes

36

x	12	**8**	**5**
y	40	**60**	96

37 50 cm^2

38

Position-to-term rule	Position	Term
Square then subtract 2	11	**119**
Cube	**6**	216

39 4%

40 The letter made is the letter H, with 2 lines of symmetry.

41 $5348

42 a 50 **b** 250

43 a $^-3$, 0, 5, 12 **b** $\frac{1}{2}$, 1, $1\frac{1}{2}$, 2 **c** 4, 7, 12, 19

44 0.6 and 3.75

45 a, b

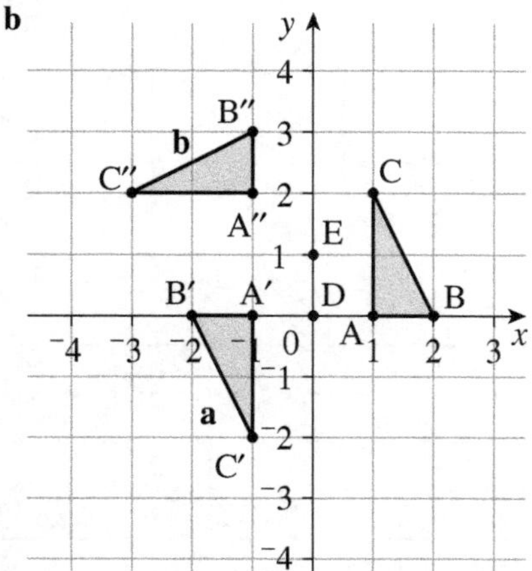

c Rotate 90° anticlockwise about ($^-1$, 0)

46 a 36

b

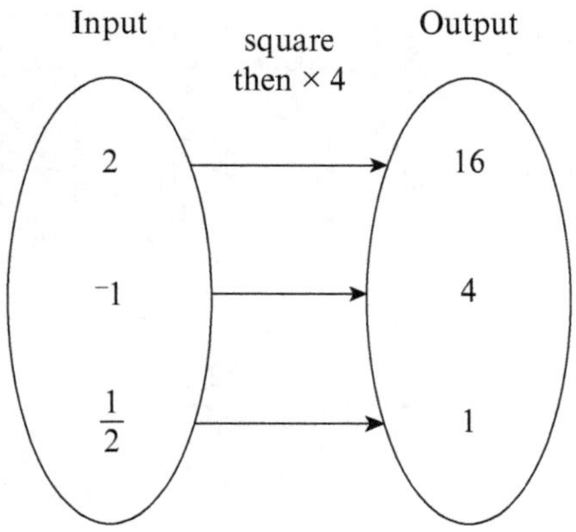

c

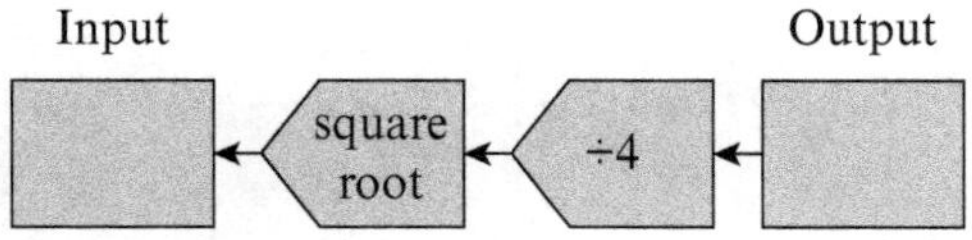

d 5 and $^-5$

47 a

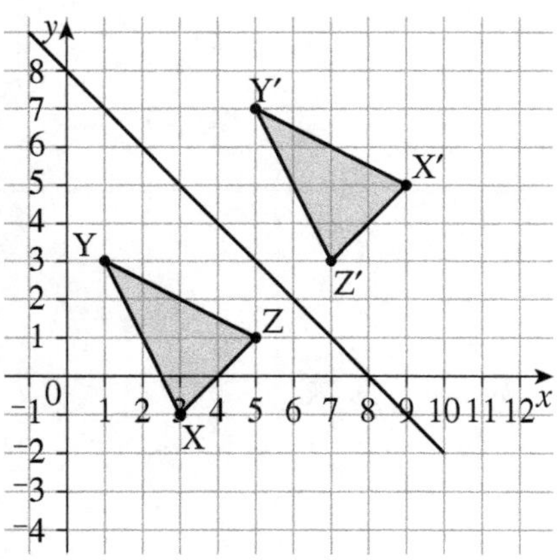

b reflection in line $y = 8 - x$

48 $y = \frac{x}{2} + 7$

49 9 cm

50 a

x	$^-3$	$^-2$	$^-1$	0	1	2	3
y	11	6	3	2	3	6	11

b

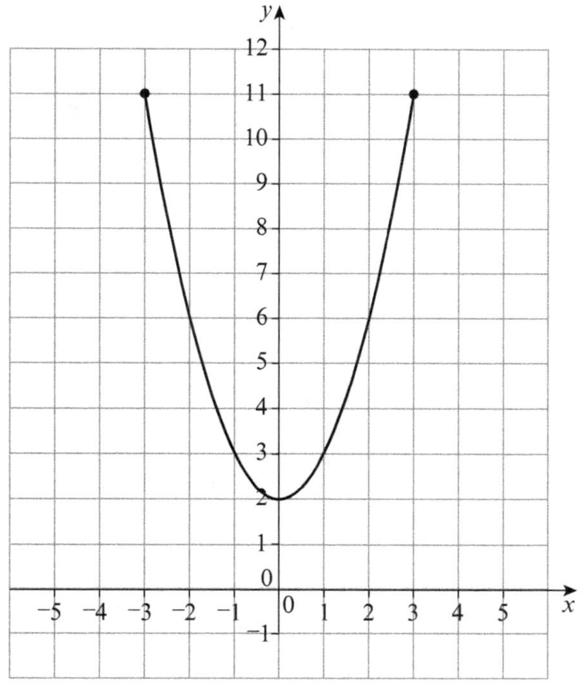

51 a $\frac{n}{9}$ **b** $n^2 + 5$

52 a i 14 **ii** 16 **b** 800

53 a

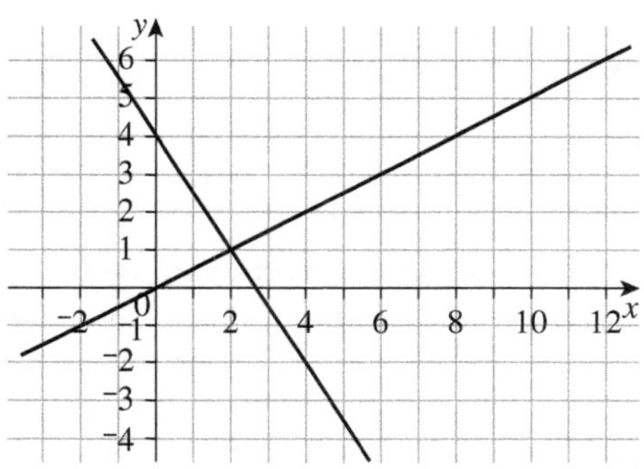

$x = 2$, $y = 1$

b

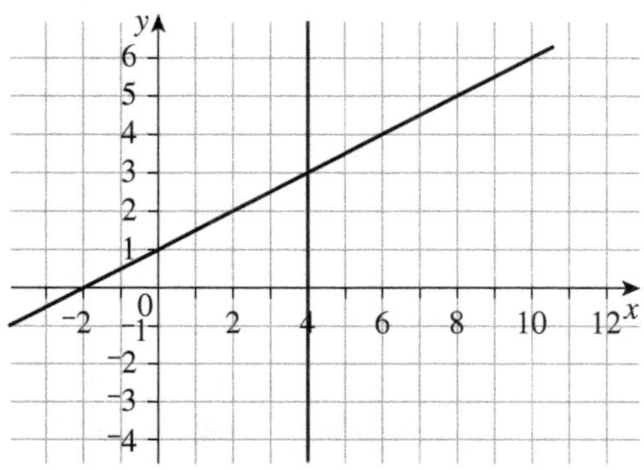

$x = 4$, $y = 3$

c

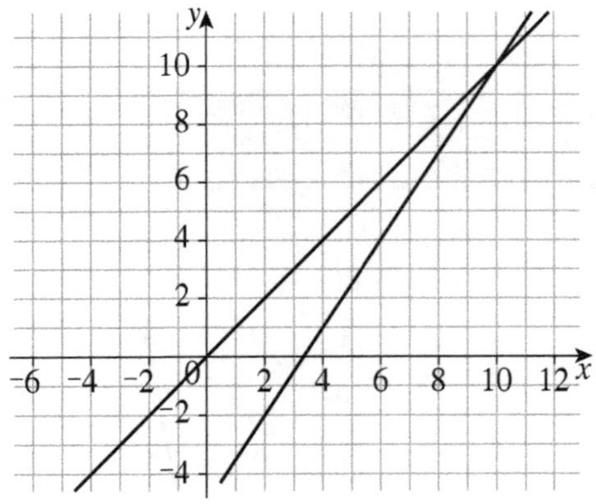

$x = 10$, $y = 10$

d

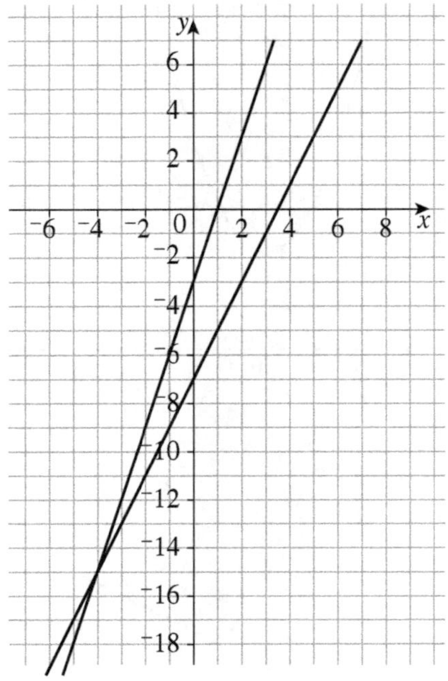

$x = -4$, $y = -15$

54 a congruent **b** not congruent **c** congruent

55 a $a + (n - 1)d$ **b** $7n + 10$ **c** 150

d 32nd

Chapter 16

Check in

1 a $4x$

b $3(x^2 - y^2)$

c $b(7a - 4)$

2 a $x = 5$

b $x = -1$

c $x = 1$

Exercise 16A

1

x	−4	−3	−2	−1	0	1	2
x^2	16	9	4	1	0	1	4
2x	−8	−6	−4	−2	0	2	4
$y = x^2 + 2x$	8	3	0	−1	0	3	8

b

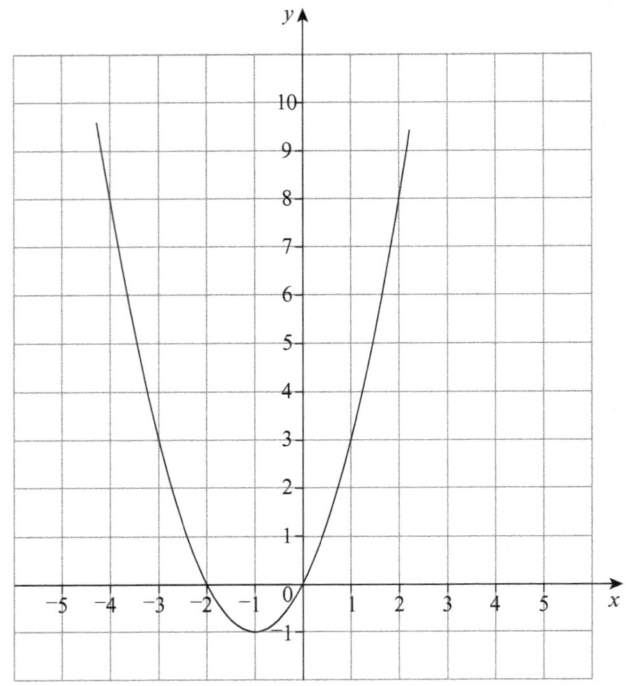

2 a

x	−3	−2	−1	0	1	2	3
x^2	9	4	1	0	1	4	9
−3x	9	6	3	0	−3	−6	−9
y	18	10	4	0	−2	−2	0

b Parabola which crosses the y-axis at (0, 0) and crosses the x-axis at (0, 0) and (3, 0).

c (0, 0) and (3, 0)

3 a

x	−5	−4	−3	−2	−1	0	1	2
x^2	25	16	9	4	1	0	1	4
3x	−15	−12	−9	−6	−3	0	3	6
−2	−2	−2	−2	−2	−2	−2	−2	−2
$y = x^2 + 3x - 2$	8	2	−2	−4	−4	−2	2	8

b

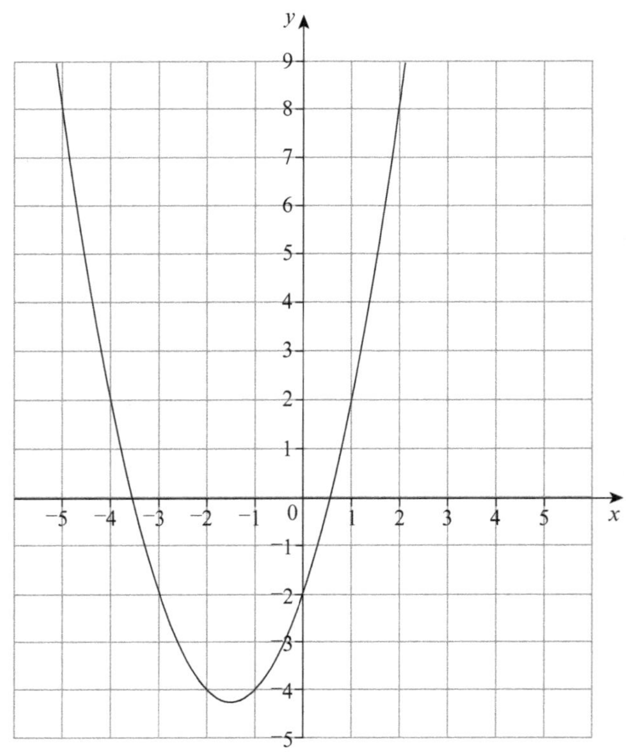

4 Points won't form a symmetric curve because point F is in the wrong place.

5

x	$^{-}2$	$^{-}1$	0	1	2	3	4
x^2	4	1	0	1	4	9	16
$y = 6 - x^2$	2	5	6	5	2	$^{-}3$	$^{-}10$

Upside-down parabola which crosses the y-axis at (0, 6) and crosses the x-axis at ($^{-}2.4$, 0) and (2.4, 0).
The graph is inverted (a dome rather than a trough)

Exercise 16B

1 **a** $2x^2 + 13x + 20$ **b** $6x^2 + 19x + 10$
c $a^2 + 2ab + b^2$

2 **a** $2x^2 + 21x + 27$
b $12x^2 + 26x + 10$
c $15x^2 - 37x - 8$
d $24x^2 - 26x + 6$
e $70x^2 + 73x - 8$
f $48x^2 - 130x + 77$

5 $ac + bc + ad + bd$

6 **a** Yes **b** Yes

7 **a** $2pr + 6ps + qr + 3qs$ **b** $2ac + ad + 6bc + 3bd$
c $18x + 15mx + 6y + 5my$ **d** $21p + 2mp + 51q + 5mq$
e $20ac + 5ad + 4bc + bd$ **f** $xp + 7qx + 3py + 21qy$

Exercise 16C

2 **a** 1681 **b** 2601 **c** 529 **d** 5184

4 **a** 1521 **b** 2401 **c** 729 **d** 4624

6 **a** 55 **b** 640

7 **a** 121 **b** 1000

8 **a** 441 **b** 2704 **c** 841 **d** 5929

9 **a** 40 **b** 160 **c** 7800 **d** 460 **e** 1720 **f** 3420

10 **a** 14 **b** 62 **c** 4.8 **d** 18.8 **e** 28.6 **f** 54
g 3.2 **h** 65.8

Exercise 16D

1 **a** $(p - q)(p + q)$ **b** $(l - m)(l + m)$
c $(p - 2q)(p + 2q)$ **d** $(3l - m)(3l + m)$
e $(3p - 2q)(3p + 2q)$ **f** $(4l - 3m)(4l + 3m)$

2 **a** $(x - y)(x + y)$ **b** $(s - t)(s + t)$
c $(2a - b)(2a + b)$ **d** $(l - 3m)(l + 3m)$
e $(4a - b)(4a + b)$ **f** $(p - 5q)(p + 5q)$
g $(4a - 5b)(4a + 5b)$ **h** $(7p - 4q)(7p + 4q)$

3 **a** $(b - c)(b + c)$ **b** $(9x - y)(9x + y)$
c $(p - 10q)(p + 10q)$ **d** $(10a - 9b)(10a + 9b)$
e $(xy - z)(xy + z)$ **f** $(a - bc)(a + bc)$
g $(2p - qr)(2p + qr)$ **h** $(3lm - n^2)(3lm + n^2)$

4 **a** 5 cm **b** 24 cm **c** 40 cm

5 **a** 60 cm **b** 84 cm **c** 3.6 cm **d** 2.97 cm

6 **a** 14 cm² **b** 128 cm² **c** 40.32 cm²

Exercise 16E

1 **a** $(x + 6)(x + 1)$ **b** $(x + 1)(x + 15)$ **c** $(x + 5)(x + 3)$
d $(x + 5)(x + 4)$ **e** $(x + 3)(x + 2)$

2 **a** $(x - 2)(x - 4)$ **b** $(x - 3)(x - 5)$ **c** $(x - 1)(x - 4)$
d $(x - 8)(x - 1)$ **e** $(x - 3)(x - 3)$ **f** $(x - 2)(x - 3)$

3 **a** $(x + 11)(x - 1)$ **b** $(x + 7)(x - 3)$ **c** $(x - 7)(x + 5)$
d $(x + 3)(x - 7)$ **e** $(x + 2)(x + 30)$ **f** $(x + 1)(x - 2)$

4 **a** $(x + 11)(x - 2)$ **b** $(x + 6)(x - 2)$ **c** $(x + 17)(x - 1)$
d $(x + 3)(x - 11)$ **e** $(x + 3)(x - 8)$ **f** $(x + 7)(x - 7)$

Exercise 16F

1 **a** $(3x + 1)(x + 4)$ **b** $(5x + 2)(x + 1)$ **c** $(3x + 1)(2x + 3)$
d $(3x - 1)(2x + 4)$ **e** $(5x + 3)(3x + 4)$ **f** $(2x - 4)(3x + 1)$

2 **a** $(3x + 1)(5x + 1)$ **b** $(11x + 1)(x + 1)$ **c** $(12x - 1)(2x + 1)$
d $(2x + 1)(8x - 1)$ **e** $(6x + 1)(5x - 1)$ **f** $(7x - 1)(7x - 1)$

Exercise 16G

1 **a** 3 and 4 **b** 5 and 6 **c** 5 and $^{-}2$ **d** 2 and $^{-}5$
e $^{-}4$ and 7 **f** $^{-}6$ and 3

2 **a** 3 and 4 **b** $^{-}5$ and 2 **c** $^{-}2$ and $^{-}9$ **d** $\frac{1}{2}$ and 3
e $\frac{1}{4}$ and $\frac{1}{5}$ **f** 0 and 3 **g** $\frac{1}{7}$ and $\frac{1}{4}$ **h** $\frac{3}{2}$ and $\frac{2}{3}$

3 **a** 1 and 5 **b** 4 and 6 **c** 7 and 8 **d** 2 and 4
e 4 and 8 **f** 4 and 5

4 **a** 2 and 5 **b** 3 and 4 **c** 3 and 7
d 6 and 7 **e** 4 and 9 **f** 2

5 **a** $^{-}3$ and 7 **b** $\frac{1}{3}$ and $^{-}4$ **c** 0 and $\frac{^{-}1}{3}$ **d** $\frac{2}{7}$ and $\frac{^{-}4}{3}$
e 5 and 2 **f** $^{-}4$ and 3

6 **a** $^{-}1$ and 8 **b** $^{-}5$ and 3 **c** $\frac{1}{3}$ and $\frac{1}{2}$ **d** $\frac{1}{5}$ and 1
e 1 and $\frac{5}{3}$ **f** 1 and $\frac{7}{5}$ **g** $^{-}3$ and $\frac{3}{2}$ **h** $\frac{^{-}4}{7}$ and 2
i $\frac{^{-}1}{4}$ and 13 **j** $\frac{^{-}18}{5}$ and 1 **k** $\frac{1}{3}$ **l** $\frac{^{-}2}{5}$
m $\frac{^{-}1}{9}$ and 1 **n** $\frac{5}{2}$

7 **a** 2 **b** 1 **c** 3 **d** 5
e $^{-}4$ **f** $^{-}6$ **g** $^{-}2$ **h** $^{-}3$
i $\frac{1}{2}$ **j** $\frac{^{-}1}{3}$

8 **a** $^{-}4$ and $^{-}3$ **b** $^{-}7$ and 2 **c** $^{-}3$ and $^{-}2$ **d** $^{-}1$ and 9
e $^{-}6$ and $^{-}1$ **f** $^{-}4$ and 6

9 **a** $^{-}6$ and 3 **b** $^{-}7$ and 7 **c** $^{-}5$ and 4
d 11 and $^{-}8$ **e** 4 and $^{-}15$ **f** $^{-}11$ and $^{-}10$

10 **b** $^{-}2$ and $^{-}1$

Exercise 16H

1 $x(x + 10) = 56$, 4 cm and 14 cm

2 5 cm, 12 cm, 13 cm

3 3 cm, 4 cm, 5 cm

4 5 cm, 12 cm, 13 cm

5 5 and 7

6 4, 5, 6

7 5, 6, 7

8 50 km/h

Exercise 16I

1 **a** $x(p + q) + y(p + q) = (x + y)(p + q)$
b $x(a - b) + y(a - b) = (x + y)(a - b)$
c $u(t^2 + s) + v(t^2 + s) = (u + v)(t^2 + s)$
d $x(2l + m) - y(2l + m) = (x - y)(2l + m)$

2 **a** $(x + y)(r + s)$ **b** $(x + y)(2r + s)$ **c** $(x + y)(3a + b)$
d $2(a + b)(p + q)$ **e** $2(4a + b)(p + q)$

3 **a** $(x + 2y)(z + 3t^2)$ **b** $(x - y)(2a + 3b)$ **c** $(m + n)(p^2 - 2q)$
d $(x + 2y)(g - 2h)$ **e** $(a^2 + b^2)(l - m)$

Exercise 16J

1 $x = 4.2$

2 $m = 3.66$

3 $x = {}^{-}1.8$

4 $x = 4.7$ or $^-2.7$

5 $x = 2.12$ or $^-4.52$

6 $x^2 + 2x = 13.7$; $x = 2.83$ cm

7 $x = 4.7$ or $^-2.6$ or 0.4

8 **a** $x = \pm 1.4$ **b** $x = 7.5$ or $^-3.5$ **c** 1.7 or $^-9.7$

9 11.7

Consolidation

Exercise 16

1 **a** Parabola which crosses the y-axis at (0, 5) and does not cross the x-axis.
b Upside-down parabola which crosses the y-axis at (0, 8) and crosses the x-axis at ($^-2.8$, 0) and (2.8, 0).
c Parabola which crosses the y-axis at (0, 4) and does not cross the x-axis.

2 **a** $2x^2 - 12x - 54$
b $8x^2 + 26x - 24$

3 **a** $(x - y)(x + y)$ **b** $(6x - y)(6x + y)$
c $(p - 3q)(p + 3q)$ **d** $(2p - q^2)(2p + q^2)$

4 **a** $(x + 1)(x + 3)$ **b** $(x - 1)(x - 3)$
c $(x - 3)(x + 1)$ **d** $(x - 1)(x + 3)$

5 **a** 2, 3 **b** 1, 4 **c** −2, 5 **d** $\frac{^-1}{2}$, 3

6 **a** $2x^2 + 3x - 2$ **b** $4x^2 - 1$
c $6x^2 - 5x - 6$ **d** $12x^2 + 13x - 4$
e $9x^2 - 15x + 4$ **f** $-6x^2 + 35x - 49$

7 **a** $x = {^-1}$ or $^-2$ **b** $x = 1$ or 6
c $x = 2$ or 5 **d** $x = {^-2}$ or $^-5$

8 **a** $(x - a)(x + a)$
b **i** 9800 **ii** 996004 **iii** 99940009

9 **a** **i** $(9 - 5r)(9 + 5r)$ **ii** $(4x - 5)(3x - 2)$
b $^-2.5$ or 1

10 8, 10

11 **a** $(x + y)(d + e)$ **b** $(d + f)(2x + m)$
c $(h - j)(3y + t)$ **d** $(a + b)(1 - c)$

12 **a** $x = 4.3$ **b** $t = 3.3$ **c** $x = {^-1.8}$ **d** $x = 5.8$

Check out

1 Parabola which crosses the y-axis at (0, 0) and crosses the x-axis at ($^-2$, 0) and (0, 0).

2 **a** $2x^2 + 9x - 5$ **b** $9x^2 - 30x + 25$

3 **a** $(x + 3)(x + 1)$ **b** $(x + 13)(x - 1)$ **c** $(x - 5)^2$
d $(2x - 3)(x + 4)$ **e** $(10m - 7)(10m + 7)$

4 **a** $x = {^-1}$ or $^-2$ **b** $x = 4$ **c** $x = \frac{^-1}{3}$ or $\frac{1}{2}$

5 **a** $x = 2.7$ **b** $y = 4.3$ **c** $x = 21.6$ **d** $x = 2.2$

6 **a** $(x + m)(y + k)$ **b** $(3d + f)(x + y)$ **c** $(w - 2r)(y + 5s)$

Answers for Homework Book 9

1A

1 i $^-7, \sqrt{9}, \frac{7}{11}, 8, \sqrt{2.25}, \sqrt[3]{^-27}, 1\frac{4}{5}$

ii $\sqrt{8}$, 2π

iii $\sqrt{^-7}$

2 a $^-6, \sqrt{16}$ **b** $\frac{3}{4}, 0.24, \frac{6}{7}$ **c** $\sqrt{10}$, 4π

3 a Irrational **b** Either rational or irrational **c** Rational

4 a True **b** False **c** False **d** True

5 A and C

6 a $2^2 \times 5^2$ **b** $\sqrt{2} \times \sqrt{5} \times \sqrt{10} = \sqrt{100} = 10$

1B

1 a 9 **b** 10 **c** 9.6

2 a 11 **b** 12 **c** 11.4

3 a ±4.6 **b** ±8.5 **c** ±3.5

4 a 3 **b** 4 **c** 3.4

5 a 4.1 **b** 3.7 **c** 4.5

6 a 10, 11, 12, 13, 14, 15

b 1, 2, 3, 4, 5, 6, 7, 8, 9

7 a ±0.89

b ±0.59

8 a 6.3 cm **b** 4.3 cm

9 a 10.7 **b** 3.1 **c** 6.7

1C

1 a $64 = 4^3 = 4 \times 4 \times 4$ **b** $125 = 5^3 = 5 \times 5 \times 5$

c $64 = 2^6 = 2 \times 2 \times 2 \times 2 \times 2 \times 2$

d $3^3 \times 3^2 = 3 \times 3 \times 3 \times 3 \times 3 = 3^5$

e $10^5 \times 10^2 = 10 \times 10 \times 10 \times 10 \times 10 \times 10 \times 10 = 10^7$

f $2^4 \times 2^3 = 2 \times 2 \times 2 \times 2 \times 2 \times 2 \times 2 = 2^7$

2 a 2^{12} **b** $2^3 \times 3^2 \times 5^5$

c $2^{7-4} = 2^3$ **d** $10^{8-5} = 10^3$

e $6y^2$ **f** $\frac{5t^3}{x}$

3 a $10^7 \div 10^{-2} = 10^9$ **b** $x^5 \div x^5 = x^0 = 1$

c $\frac{1}{4} = \frac{1}{2^2} = 2^{-2}$ **d** $3^2 \div 3^4 = 3^{-2}$

e $5^0 = 1$ **f** $\frac{1}{8} = 2^{-3}$

4 a False **b** True **c** False **d** True

5 a 7^{-1} **b** 5^{-2} **c** 6^2

6 a $10^7 \times 10^2 = 10^9$ **b** $5^4 \times 5^3 = 5^7$

c $2^{-6} \div 2^{-3} = 2^{-3}$ **d** $3^1 \div 3^5 = 3^{-4}$

e $7^2 \div \frac{1}{7^2} = 7^4$ **f** $2^2 \div 2^3 \times 2^4 = 2^3$

7 a 5^{-3} **b** 2^3 **c** 3^2 **d** 3

e $\frac{125}{8}$ **f** $\frac{(4^5)^2}{4^8} = \frac{4^{10}}{4^8} = 4^2 = 16$

2A–2B

1 a x^7 **b** d^{12} **c** y^{a+b+c} **d** x

e $5k^5$ **f** $2m^5$ **g** $30x^8$ **h** $5s^{-1}$

i $12t^{12}$

2 a x^{12} **b** $4x^{10}$ **c** x^6 **d** x^3

e $x^{-2} = \frac{1}{x^2}$ **f** x^7

3 a $7x + 9y$ **b** $^-2a - 3b$

c $^-xy + 8pq$ **d** $2x + x^2 - 2x^3$

e $7b^2 + 5b$ **f** $7y + y^2 + 3y^3$

g $4abc - xy^2 - 7xy$ **h** $-pqr - 3abc$

i $x^2y^2 - 2xy + 6xy^2 + x^2y$

4 a $p^0 = 1$ but then she needs to multiply by 4. Correct answer: 4

b She has forgotten to square the 7. Correct answer: $49x^6$

c She has multiplied the powers rather than adding them. Correct answer: $6y^5$

d She has subtracted the two numbers rather than dividing them. Correct answer: $5a^3$

5 a ^-48x **b** $21y$ **c** $40x^2$ **d** $70a^3$

6 a $4x$ **b** 3 **c** $2a$ **d** x^2

7 a $2x + 10y$ **b** $10b - 15c$

c $3xy - 6xz$ **d** $4x - 12y + 8z$

e $p^2 + 2q^2p - 3rp$ **f** $^-2x^3 - 2x^2 + 6x$

8 a $^-7a^3b^2$ **b** $15x^3y^4$ **c** $3a^2b$ **d** $\frac{3r}{4pq^4}$

9 a $x^2y^2z^2$ **b** $\frac{x^3}{yz}$ **c** $3x^4y^6$ **d** $6x^2y^2z^7$

e x^2 **f** $4x^6 + 4x^9$

2C–2D

1 a $1\frac{3}{14}$ **b** $\frac{27}{40}$ **c** $\frac{35}{36}$

2 a $\frac{9a}{20}$ **b** $\frac{17a}{20}$ **c** $\frac{29a}{30}$ **d** $\frac{6x + 7y}{14}$

e $\frac{6x + 4y + 3z}{12}$ **f** $\frac{10y + 12z + 15x}{30}$

3 a $3x + 4y$ **b** $8x + 10y$ **c** $\frac{1}{3x+4y}$ **d** $x + 3y$

4 Equivalent: $\frac{12x+16}{4}, \frac{1.5x+2}{0.5}, \frac{8x+15+7x+5}{5}, \frac{1}{6}(18x+24)$

Not equivalent: $\frac{3x+12}{3}, \frac{1}{3x+4}, \frac{10x-(-20x+40)}{10}$

5 Jamil has not fully simplified the expression. The fully simplified answer is $2x - 7$

6 a $\frac{10y}{7}$ **b** $\frac{2y + 3}{y}$ **c** $\frac{7x - 2}{x}$ **d** $\frac{8 + 6x}{12}$

e $\frac{2xy + z}{xyz}$ **f** $\frac{3x^2y + 2yz^2}{6xz}$

7 a $\$\frac{2.5}{x} = \$\frac{5}{2x}$ **b** $\$\frac{3.75}{y} = \$\frac{15}{4y}$

c $\$\frac{15x + 10y}{4xy}$

8 a $\frac{5x - 1}{6}$ **b** $\frac{^-x - 4}{(x + 1)(x + 2)}$

2E–2F

1 a $x^2 + 5x + 6$ **b** $x^2 + 11x + 30$

2 a $x^2 + 5x + 4$ **b** $x^2 + 8x + 12$

c $x^2 + 9x + 20$ **d** $x^2 - 4x - 5$

e $x^2 + 3x - 18$ **f** $x^2 - 3x - 28$

3 a $x^2 - 7x + 10$ **b** $x^2 - 13x + 22$

c $x^2 - 10x + 24$ **d** $x^2 - 5x - 36$

e $x^2 - 15x + 26$ **f** $x^2 - x - 42$

4 a $x^2 + 6x + 9$ **b** $x^2 - 4x + 4$

c $16 - 8x + x^2$

5 a $x^2 - 16x + 55$ **b** $x^2 + 5x - 14$

6 a $x^2 - 9$ **b** $x^2 - 144$

7 a 6 **b** 15 **c** 8, 8 **d** a, b, a, b

8 Students' answers, e.g.

$$(x+1)(x+2)(x+3) = (x+1)\left(x^2+5x+6\right)$$
$$= x^3 + 6x^2 + 11x + 6$$

2G

1 a $^{-}150$ **b** $^{-}112$ **c** 900 **d** 1.75

e $^{-}3\frac{2}{3}$ **f** $^{-}1\frac{1}{2}$

2 a 0.12 **b** $^{-}0.6$ **c** 0.000576

d 22.5 **e** $^{-}1\frac{2}{3}$ **f** $^{-}0.65$

3 a $V = 15$ volts **b** $V = 216$ volts

4 a 133.5 minutes **b** 169.5 minutes

c 106.5 minutes

5 a 14 cm **b** 31.4 cm **c** 26.8 cm

6 a $\frac{25}{9}$ **b** 36 **c** 0.62

2H

1 a $x^2 + x - 12$ **b** $x^2 - 49$ **c** $5x + 21$

2 $(x^2 + 14x + 49)\text{cm}^2$

3 $(x^2 + x - 2)\text{cm}^2$

4 a $3n^2 + 24n + 48$ **b** 8

2I–2J

1 a $t = \frac{2 + ag}{T}$ **b** $g = \frac{Tt - 2}{a}$ **c** $a = \frac{Tt - 2}{g}$

2 a $x = \frac{Rc}{a + b}$ **b** $x = \pm\sqrt{z + 8pq}$

c $x = \pm\sqrt{t^2 + w^2} + s$ **d** $x = \frac{36a}{G}$

3 Both students are correct. Beverley divided by 3 first whereas Jez expanded the brackets.

The two answers are equivalent.

4 $Q = \frac{1}{2}\left(\frac{P}{y - x} + R\right)$ or equivalent

3A

1 Students' constructions

2 Students' constructions

3 a – d Students' constructions

e BC = 9.2 cm

4 See original questions

3B

1–4 Students' constructions

3C

1 West ≈ 51 km; South ≈ 61 km

2 West ≈ 139 km; North ≈ 80 km

3 a 58 km **b** 259°

3D

1

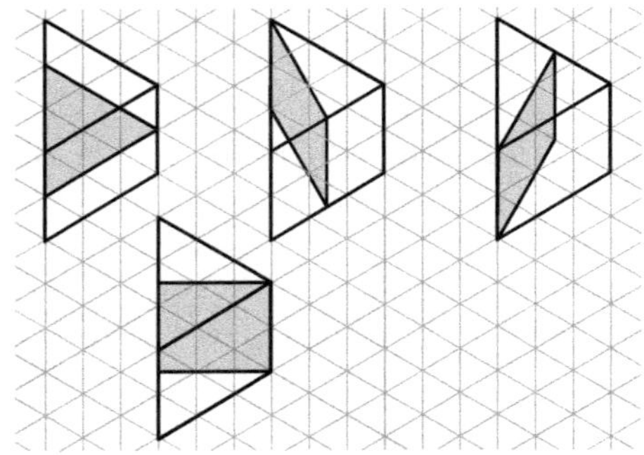

2

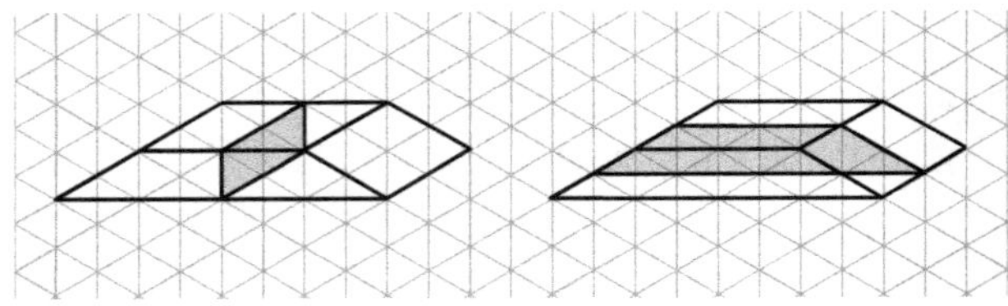

3

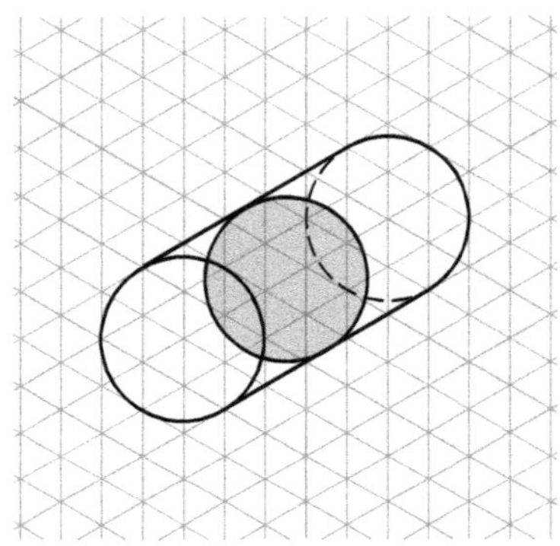

Horizontally: infinite planes of symmetry

4 a 9 **b** 11 **c** 8 **d** Infinite

4A

1 a 400 000 **b** 5400 **c** 6400 **d** 76

e 41 **f** 26 478

2 a 87 **b** 0.1483 **c** 0.00764 **d** 65

e 1450 **f** 768 000

3 a $6.2 \times 10^3 = 6200$ **b** $4\,480\,000 \div 10^5 = 44.8$

c $53\,000 \times 10^{-3} = 53$ **d** $0.043 \times 10^4 = 430$

e $0.17 \div 10^{-5} = 17\,000$ **f** $0.0089 \div 10^{-6} = 8900$

4 a 8 030 000 **b** 52 440

c 9001.9

4B–4C

1 a 5×10^3 **b** 2×10^2 **c** 7×10^1 **d** 6×10^5

e 8×10^6 **f** 4×10^8

2 a 7000 **b** 800

c 50 000 **d** 6 000 000

e 900 000 000 **f** 2 000 000 000

3 a 0.04 **b** 0.7 **c** 0.003 **d** 0.00001

e 0.000005 **f** 0.0000009

4 a 2.7×10^{-1} **b** 9.03×10^{-1}

c 1.6×10^{-2} **d** 5.71×10^{-3}

e 5.04×10^{-2} **f** 1.04×10^{-2}

5 **a** 8×10^2 **b** 7.32×10^5 **c** 9.5×10^4 **d** 4×10^3

e 3×10^{-2} **f** 8.1×10^{-2} **g** 8.5×10^1 **h** 2×10^{-7}

6 **a** The number is greater than 10. Correct answer: 3.71×10^6

b The number is less than 1. Correct answer: 5×10^5

c The powers of 10 need simplifying. Correct answer: 8.1×10^2

d The number is less than 1 and the powers of 10 need simplifying. Correct answer: 3×10^{-10}

7 2.7×10^{-26} kg

8 1.08×10^9 km/h

4D

1 345, 348, 351, 354.9

2 **a** $855\le n<865$ **b** $1350\le n<1450$

c $16.5\le n<17.5$ **d** $23.75\le n<23.85$

e $19.405\le n<19.415$

3 $74.5\le h<75.5$

4 **a** He has given the bounds for rounding to 3 significant figures.

b $715\le x<725$

5 Option **b**

5A

1 **a** 3 km **b** 4.72 km **c** $0\le d<2$

d 10 km **e** $4\le d<6$

2 **a** 3.5 kg **b** 4.6 kg **c** $4\le m<5$

d 4 kg **e** $4\le m<5$

3 **a** 4.5 seconds **b** 9.84 seconds

c $12\le t<15$ **d** 12 seconds

e 9 seconds

5B–5C

1 *Children have too much homework, e.g.*

a Year group, name, amount of homework per week, collect for 4 weeks in a row or year group, name, amount of homework per day, collect for 5 days in a row

b 10 students per year group would cover different ages and provide a sample, the bigger the sample the more accurate the data

c Rounding to the nearest hour would be too inaccurate, measuring in 5 minute timeslots might be too time consuming, measuring in quarter-hours would be probably the most efficient and informative

d

Year Group	Name	Average amount of homework per week			
		Week 1	Week 2	Week 3	Week 4

2

	Advantages	Disadvantages
Primary data	Reliable	Can be time consuming and expensive to collect
Secondary data	Saves time and money	Can be unreliable

3 **a** It is not pertinent to the survey

b Leading question

c The option for 4 visits is missing and it is unclear which box to tick for exactly 10 visits

How many times did you visit the doctor last year?

☐ None ☐ Between 1 and 3 ☐ 4–8 ☐ 9 or more

4 **a** Qualitative **b** Quantitative **c** Quantitative
d Qualitative **e** Quantitative

6A

1 **a** **i** 40.72 cm² **ii** 22.62 cm

b **i** 63.62 cm² **ii** 28.27 cm

c **i** 1.13 cm² **ii** 3.77 cm

2 **a** **i** Arc length is 8.64 cm **ii** Sector area is 23.76 cm²

b **i** Arc length is 16.49 cm **ii** Sector area is 28.86 cm²

3 4.07 cm

4 **a** **i** area is 84.57 cm² **ii** perimeter is 48.57 cm

b **i** area is 64.27 cm² **ii** perimeter is 29.85 cm

5 **a** 30.9 cm² **b** 69.12 cm²

6 **a** **i** Arc length is 3.3 cm **ii** Sector area is 11.55 cm²

b **i** Arc length is 7.2 cm **ii** Sector area is 9 cm²

c **i** Arc length is 43.98 cm **ii** Sector area is 184.73 cm²

6B–6C

1 **a** 7.6 cm **b** 17.0 cm

2 **a** 8.49 cm (3 s.f.) **b** 9.33 cm (3 s.f.)

3 **a** 11.3 cm (3 s.f.) **b** 103 m (3 s.f.)

4 **a** Yes **b** No

5 0.56 cm²

6D

1 **a** 216 cm² **b** 100 cm² **c** 870 cm²

2 342 m²

3 **a** **i** 433.5 cm² **ii** 356.6 cm²

b **i** 659.7 cm² **ii** 472.8 cm²

4 **a** 158.3 cm² **b** 235.3 cm²

5 **a** 16964.60 cm² **b** 22619.47 cm²

c 40 684 cm² (nearest cm²) or 4.1 m² (1 d.p.)

6E

1 **a** 154 cm³

b 220.5 m³

2 **a** 6 cm **b** 4.15 m (3 s.f.)

3 **a** 339.29 cm³ **b** 1563.18 cm³

4 **a** **i** 361.9 cm³ **ii** 485.5 cm³

b **i** 7.5 cm **ii** 15.0 cm

5 **a** 942477.8 cm³ **b** 942.5 ℓ

c 1108

6 1.12 cm

7 **a** 2448.9 cm³ **b** 21.87 kg

6F

1 **a** False; 30 g **b** True **c** True

d False; 12 000 000 B **e** True **f** True

2 **a** **i** 3 B **ii** 3 TB

b **i** 0.65 nm **ii** 0.65 Gm

c **i** 3.4 nl **ii** 3.4 kl

3 200 000 000

4 **a** 9.6 g **b** 2000 μg

5 **a** 20 000 **b** 20 000 000 **c** 0.000 02 **d** 20 000 000 000

6 **a** 0.5 **b** 9500 000

7A

1 **a** 1, 2, 3, 6, 9, 18 **b** 1, 3, 5, 9, 15, 45
c 1, 2, 4, 8, 16, 32 **d** 1, 2, 4, 5, 8, 10, 20, 40

2 **a** 6 **b** 2 **c** 3 **d** 3

3 **a** HCF = 54, $\frac{2}{3}$ **b** HCF = 6, $\frac{12}{25}$
c HCF = 42, $\frac{1}{6}$

7B

1 **a** $\frac{5}{9}$ **b** $9\frac{3}{4}$ **c** $8\frac{19}{35}$ **d** $2\frac{1}{12}$
e $2\frac{2}{3}$ **f** $\frac{27}{49}$

2 **a** $1\frac{4}{7}$ m **b** $2\frac{1}{32}$ kg **c** $\frac{1}{8}$ **d** $\frac{16}{25}$

3 $8\frac{6}{35}$ cm^2

4 53

7C

1 **a** $5\frac{1}{16}$ **b** $20\frac{1}{3}$ **c** $2\frac{1}{10}$ **d** $3\frac{9}{20}$
e $5\frac{11}{25}$ **f** $3\frac{8}{15}$

2 **a** $^{-}\frac{3}{10}$ **b** $1\frac{11}{24}$ **c** $1\frac{5}{58}$ **d** $\frac{20}{39}$

3 \$25.50

7D

1 **a** $0.1\dot{6}$ **b** $0.41\dot{6}$ **c** 3.56
d $0.\dot{2}\dot{3}$ **e** 5.375 **f** $0.\dot{8}\dot{1}$

2 **a** $0.\dot{7}$ **b** $0.\dot{5}\dot{3}$ **c** $0.\dot{1}2\dot{7}$

3 Group A: $\frac{15}{18}, \frac{19}{12}, \frac{49}{120}, \frac{83}{90}$
Group B: $\frac{7}{10}, \frac{17}{16}, \frac{81}{125}, \frac{79}{500}, \frac{21}{14}$

8A

1 **a** $x = 4$ **b** $s = 7$ **c** $x = 14$ **d** $x = 3$
e $t = 4$ **f** $t = 4$

2 **a** $x = 10$ **b** $t = {}^{-}30$ **c** $x = 14$ **d** $x = {}^{-}40$
e $x = {}^{-}20$ **f** $x = 1\frac{1}{5}$

3 **a** $x = 6$ **b** $x = 4$ **c** $d = 7$ **d** $r = 2$
e $s = 3$ **f** $x = {}^{-}2$

4 **a** $x = 4$ **b** $x = 25$ **c** $x = 98$ **d** $x = 21$
e $x = {}^{-}16$ **f** $x = {}^{-}14$

5 **a** $x = 2$ **b** $a = 8$ **c** $t = {}^{-}3$ **d** $d = 4$
e $p = 3$ **f** $x = {}^{-}3\frac{1}{2}$

6 **a** $x = {}^{-}3$ **b** $z = 3$ **c** $p = {}^{-}\frac{2}{3}$ **d** $x = 4$
e $z = {}^{-}11$ **f** $e = {}^{-}3$

7 **a** $d = {}^{-}8$ **b** $y = 1\frac{1}{2}$ **c** $f = {}^{-}3$ **d** $x = {}^{-}1\frac{1}{2}$
e $a = {}^{-}18$ **f** $k = {}^{-}7$

8 **a** $x + 7$ **b** $2x + 7 = 43$
c $x = 18$, Gracie is 18 years old and Sheena is 25 years old

9 **a** $(2x - 5) + (3x + 2) + (2x - 1) = 59$, $7x - 4 = 59$
b $x = 9$, sides 13 cm, 29 cm, 17 cm

10 **a** $2(2x + 5) + 2(x + 3) = 40$; $6x + 16 = 40$
b $6x = 24$; $x = 4$, sides 13 cm and 7 cm

11 **a** $6x = 42$ **b** $x = 7$, length is 7 cm

12 **a** $12x = 24$ **b** $x = 2$, length is 6 cm

13 **a** $x + 2x + 1 + 3x + 2 = 39$; $6x + 3 = 39$
b $x = 6$, so 6 yellow, 13 green and 20 black

14 $2p + p + 7 = 97$, $p = 30$, so a pencil costs \$0.30, an eraser costs \$0.37

15 $2c + 3(c + 5) = 5c + 15 = 95$; $c = 16$

16 $2x + 2x + 2 + 2x + 4 + 2x + 6 = 92$
$8x + 12 = 92$, $x = 10$, so numbers are 20, 22, 24 and 26

17 27, 28, 29, 30 , 31

8B

1 **a** 2 **b** 5 **c** 6 **d** 5

2 **a** 1 **b** 3 **c** 12 **d** 9

3 $\frac{195}{x + 8} = 13$; $x = 7$

4 **a** Equation **i** **b** 20

8C–8D

1 **a** $x > 2\frac{1}{2}$ **b** $x < 4$ **c** $x \leq 2$ **d** $x < 4$
e $x > 8$ **f** $x \geq {}^{-}6$

2 **a** $x > 2$ **b** $x \leq 2$ **c** $x \leq 3$ **d** $x > {}^{-}1\frac{1}{3}$
e $x > 9$ **f** $x > 2$

3 Question 1:

a
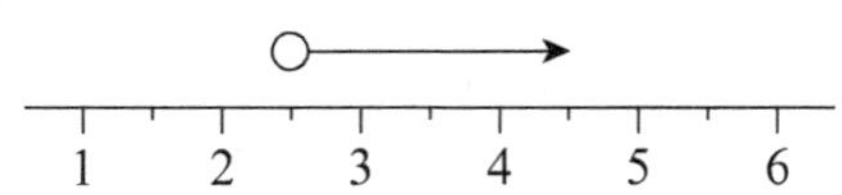

b
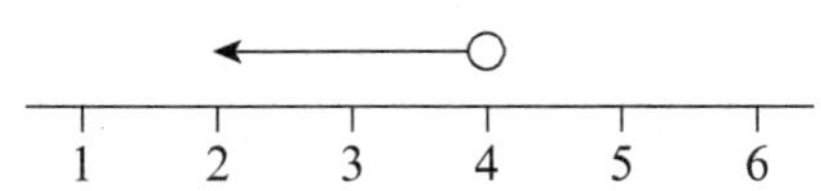

c
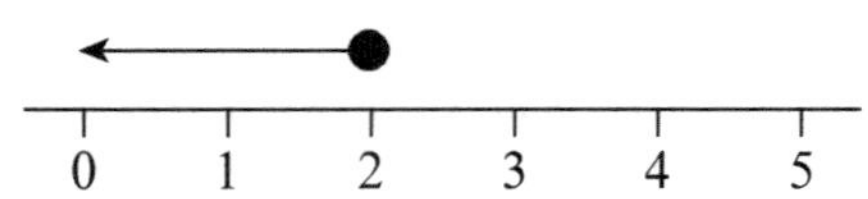

d
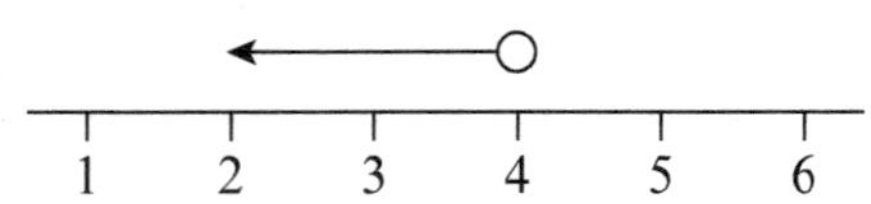

e
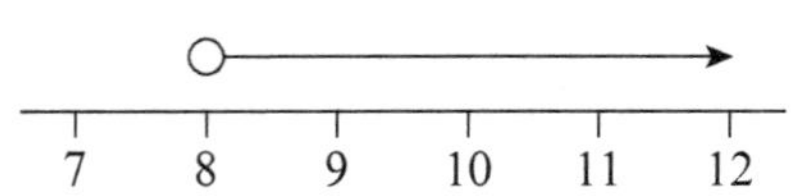

f
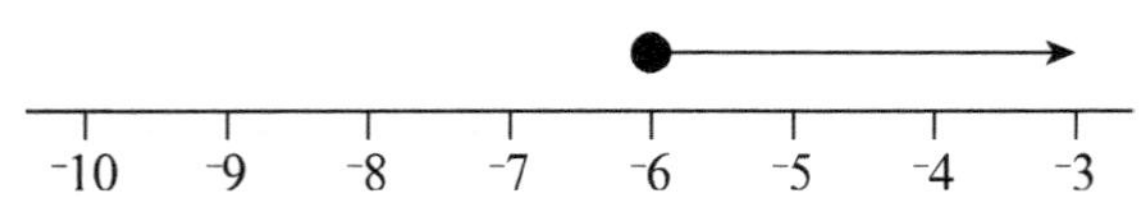

Question 2:

a
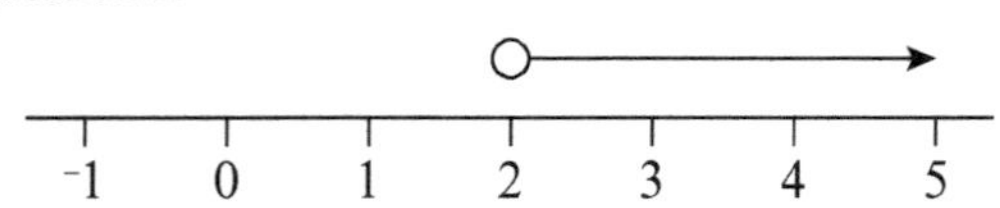

b
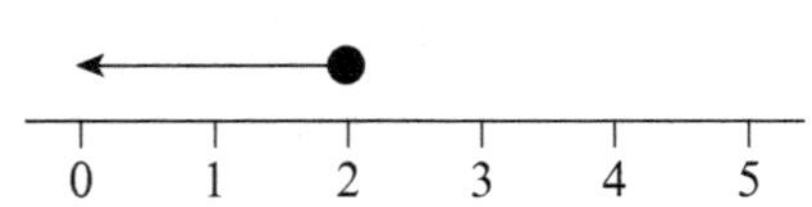

c

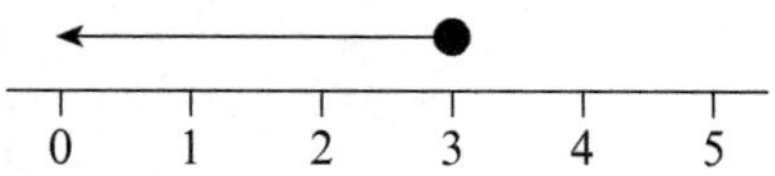

d

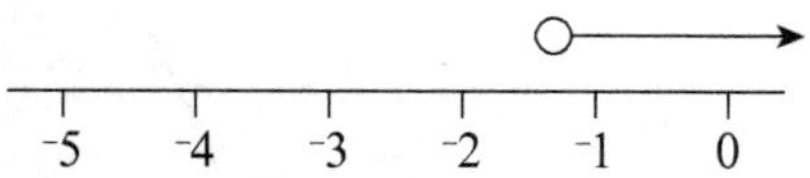

e

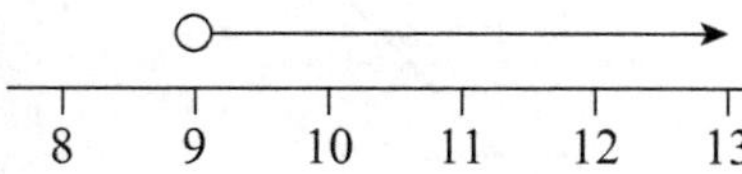

f

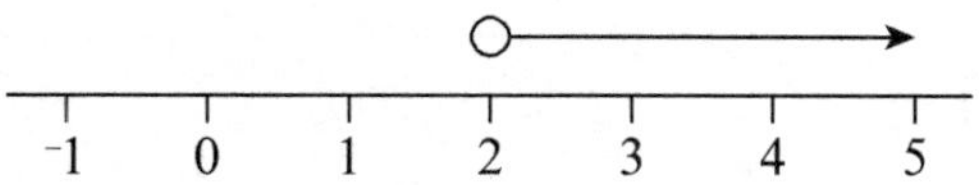

4 **a** $x \geq {}^{-}1$ **b** ${}^{-}4 \leq x < 2$ **c** ${}^{-}7 < x < 6$

5 **a** $3 < x < 5$ **b** $1 \leq x < 6$ **c** $2 < x < 6$

6 **i** **a**

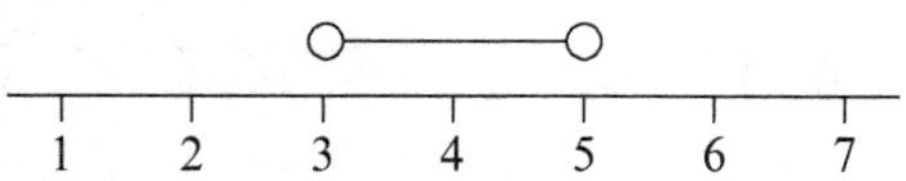

b

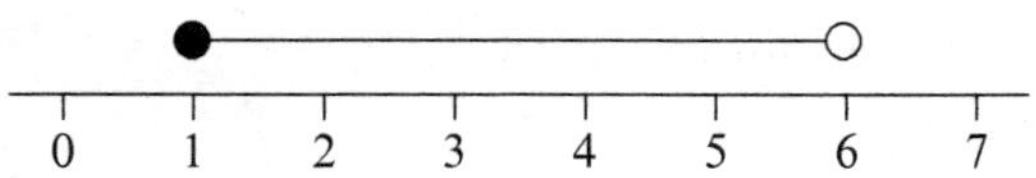

c

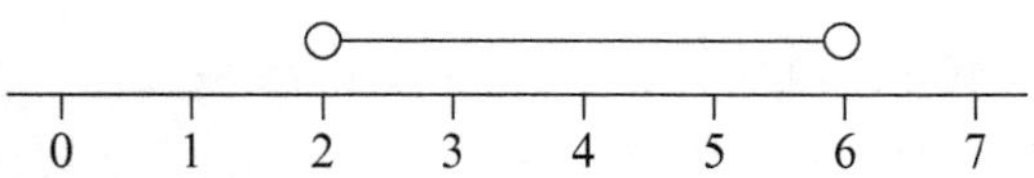

ii **a** 4 **b** 1, 2, 3, 4, 5

c 3, 4, 5

7 **a** $x < 2$ **b** $x \geq {}^{-}3$ **c** $x \geq 7$ **d** $x < {}^{-}6$

e $x \leq {}^{-}3$ **f** $x < {}^{-}2$

8 **a** $x < \frac{1}{2}$ **b** $x \leq 10$ **c** $x < 6$ **d** $x > 11$

e $x < 7\frac{1}{2}$ **f** $x > {}^{-}2$

9 **a** 4, 5 **b** 3, 4, 5, 6, 7, 8

10 $w > 5$ cm

11 Rani is < 9 years of age

12 Different answers possible, e.g.

a $x + 2 \geq 9$ **b** $4x + 1 < 9$ **c** $2x - 1 < 3x + 1 < 2x + 4$

13 $x > 3$

8E

1 **a** $y = 2x + 1$ **b** $y = 6 - x$ **c** $y = 3x - 5$

2 **a** $y = 2x + 1$ **b** $y = 3x + 1$

c $y = 1 - x$ **d** $y = 1 - \frac{1}{2}x$

3 **a** $y = 2x + 2$ **b** $y = 2x - 1$

c $y = 2x - \frac{1}{2}$ **d** $y = 2x + 0.1$

4 **a** 3 **b** 5 **c** ${}^{-}2$ **d** $\frac{{}^{-}1}{3}$

5 **a** b

b a and d

8F–8H

1 **a** $x = 3, y = 2$ **b** $x = 4, y = 1$

c $x = 3\frac{1}{2}, y = {}^{-}1$

2 **a** $x = 3, y = 4$ **b** $x = 3, y = 1$

c $x = {}^{-}2, y = {}^{-}5$

3 **a** $x = 4, y = 1$ **b** $x = 3, y = 4$

c $x = 4, y = {}^{-}1$

4 4 adults and 32 children

5 \$10.80

6 No: mass of 7 piles of bricks and 8 buckets of cement is 428 kg

7 **a** $x = 6, y = 23$ **b** $x = 7, y = 1$

9A–9B

1 **a** $a = 120°$ **b** $b = 115°$ **c** $c = 75°$

2 $d = 72°, e = 54°, f = 72°$

3 36°; 144°

4 $x = 56°$

5 **a** 15° **b** 24 sides

6 No; the exterior angle would be 25° and 25 does not divide 360 exactly.

7 $x = 24.5°$

9C–9D

1 **a** Students' diagrams

b Students' diagrams

2 Students' diagrams

3 $a = 68°, b = 112°, c = 112°, d = 68°, e = 68°$

4 $a = 60°, b = 120°, c = 60°$

5 Different shapes possible, e.g.

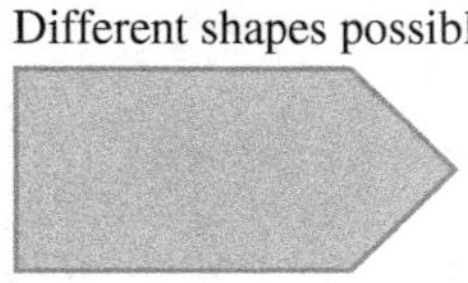

6 **a** Squares and hexagons

b

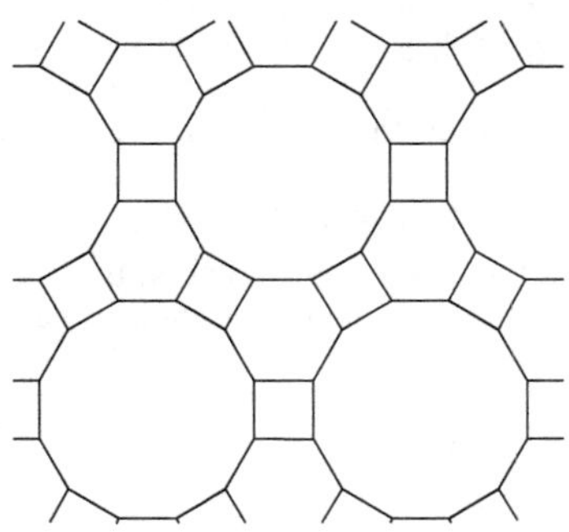

9E

1 **a** $e = 59°$ **b** $f = 62°, g = 118°$

2 **a** $e = 95°, f = 85°, g = 95°$ **b** $h = 109°, i = 98°$

c $j = 31°$

3 $x = 28°$, so $5x = 140°, 4x = 112°, 3x = 84°$

4 $w = 75°, x = 32°, y = 73°, z = 79°$

5 $a = 115°, b = 65°, c = 127°$

6 $y = 30°$

9F

1 **a** (5.4, 12) **b** (12, 9.25)

2 (8, 17)

10A

1 **a** 41 **b** 2.5 kg

c

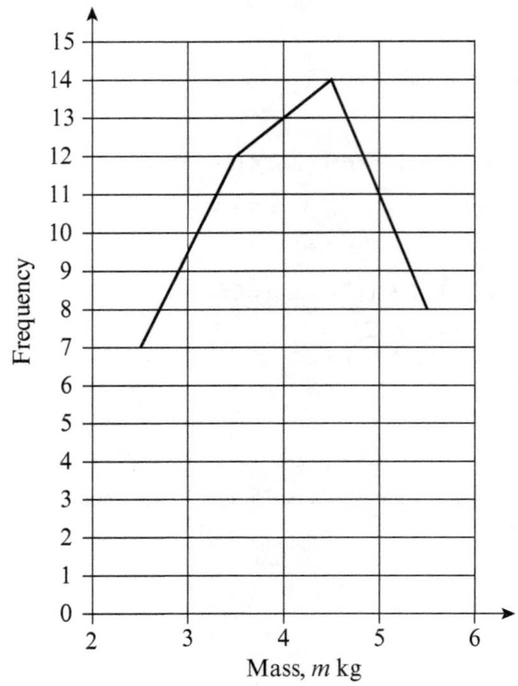

2 a 45% (or 45.5%)

b Students' graphs, with points at (5.5, 0), (15.5, 2), (25.5, 15), (35.5, 24), (45.5, 45), (55.5, 50), (65.5, 46), (75.5, 19), (85.5, 7), (95.5, 2)

10B

1 a

Police Station A		Police Station B
5	1	3 4 8 9 9
9 8 8 7 6 5 4 4	2	0 1 2 3 4 6 6 9 9
7 5 5 4 3 2 1 1 0	3	0 1 1 3 4 7
3 2 2	4	1 2 3 9
8 6 4	5	

Key: | 4 | 1 means 4.1 seconds and 5 | 1 | means 1.5 seconds

b Police station B: more entries in the first row

c Police Station A: 3.15 seconds, Police Station B: 2.75 seconds. So yes Police Station B answers calls more quickly

2 a

Geometry		Algebra
9 9 8 6 3	1	4 8
6 4 2	2	1 5 7 7
3 2 1	3	0 6 7
2	4	0 1 3

Key: 3 | 1 | 8 means 13 in geometry and 18 in algebra

b Geometry: more low scores

c Geometry: 23; Algebra: 28.5 – this supports the conclusion in part **b**

10C

1 a Positive correlation: the older a person is normally the longer they would take

b Students' graphs

c Yes in general, but the runners aged 35 and 42 years have faster running times than would be expected

d Students' graphs

e i Using graph, about 2 hours 50 mins

ii Using graph, 34 or 35 years old

2 a, c

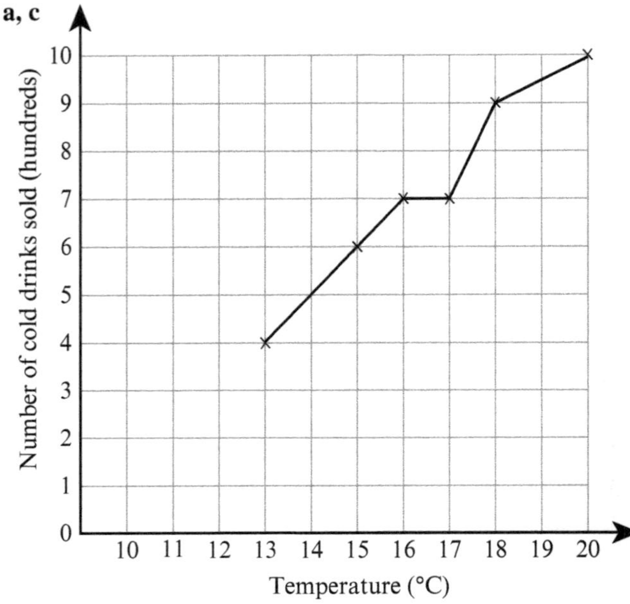

b (Strong) positive correlation

d 940

10D

1 a Two variables so a scatter graph would be best

b Positive correlation

c Students' graphs

d Yes – would expect people who are good at one language to be good at a different one

e The mean – uses all the results and although it would possibly give answers that were not whole numbers it would be representative of the data

2 a Pie chart

b Largest section for manufacturing and smallest section for marketing and management

c Students' graphs

d Yes, because factories usually have more staff in the manufacturing departments

3 a A bar chart

b Highest bar around day 1 or 2

c Students' graphs

d Yes, the bar chart shows the highest bars around day 1 and 2, students are likely to be off sick at least once or twice.

e The mode or mean. The mode would show a whole number but the mean would include all the data and probably produce a non-whole number.

11A

1 a \$240 : \$300 **b** 0.85 m : 2.55 m : 5.1 m

c 0.18 kg : 0.27 kg : 0.36 kg **d** 48 min : 80 min : 112 min

2 240 kg

3 12 cm, 18 cm, 27 cm

4 30, 75, 120

5 \$7

6 42%

11B–11C

1 a

x	5	35	0.8
y	20	140	3.2

b

x	0.3	1.2	8.4
y	1.5	6	42

2 $2.85

3 a 625 g b 30 g

4

x	3	5	8
y	12	20	32

a 4 b Students' graphs c Yes

d 4 e $y = 4x$ f 36 g 18

5 $a = 70.40$, $b = 102.27$

6 a $283.69 b R$4117.20

11D

1 a Direct b Inverse c Inverse

2 a 35 hours b 4.375 hours

3 a 18 hours b 2.4 hours

4 128

5 4

6 a

x	30	20	10
y	7	10.5	21

b $xy = 210$ (or equivalent)

12A–12B

1 a geometric b linear

c geometric d linear

e linear f other

2 a 32, 64 b 26, 30 c $\frac{1}{3}, \frac{1}{9}$ d 21, 25

e 5.2, 5.9 f 16, 22

3 a double b add 4 c divide by 3

d add 4 e add 0.7

f add on the previous difference plus one

4 a 512 b 42 c $\frac{1}{243}$ d 37

e 8 f 46

5 a $7n - 1$ b $3n + 7$ c $25 - 3n$

6 a 4, 7, 12, 18 b 2, 8, 18, 32

7 a $n^2 + 1$ b $n^2 - 2$

8 $\frac{n}{13}$

9 a 4, 11, 18, 25 b $7n - 3$

10 a 4, 7, 10, 13, 16 b $3n + 1$

12C

1 a

x	-3	-2	-1	0	1	2	3
y	6	1	-2	-3	-2	1	6

b

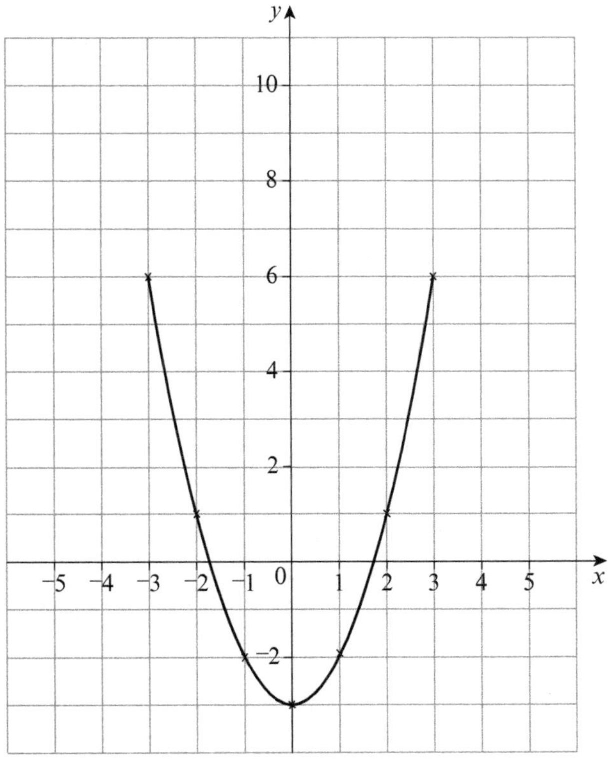

2 a

x	-3	-2	-1	0	1	2	3
y	13	8	5	4	5	8	13

b

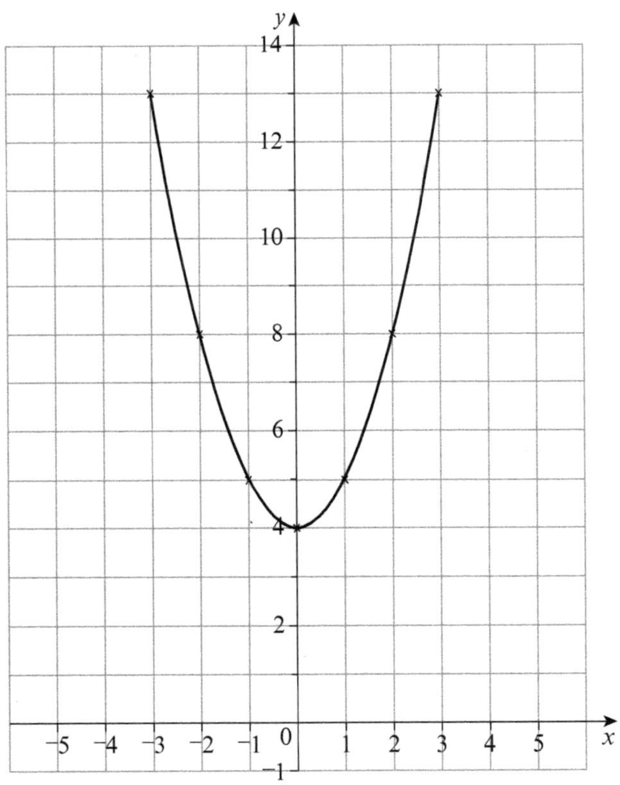

12D–12E

1 a

x	-2	-1	0	1	2	3	4
$3x$	-6	-3	0	3	6	9	12
-2	-2	-2	-2	-2	-2	-2	-2
y	-8	-5	-2	1	4	7	10

b Students' graphs

2 a

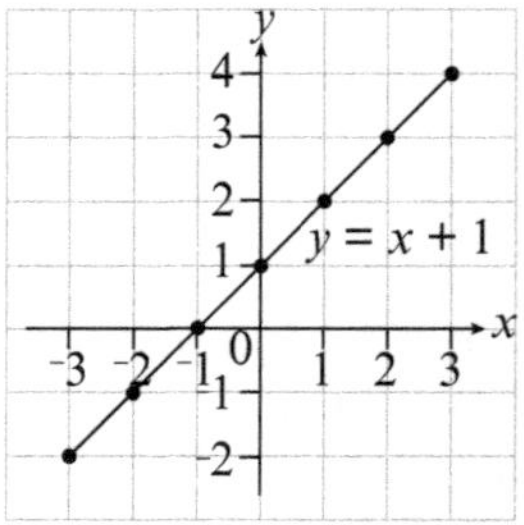

b

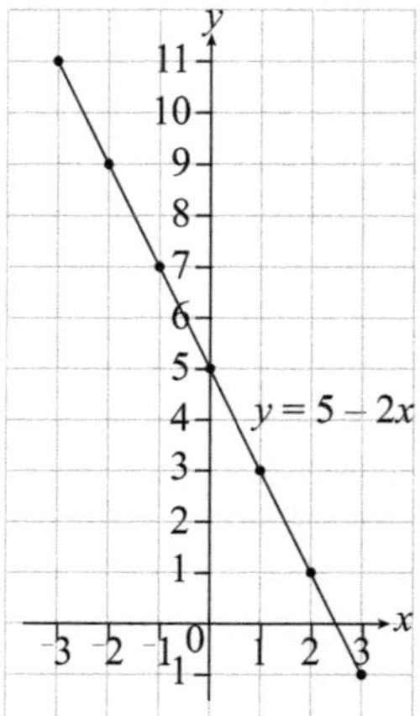

c

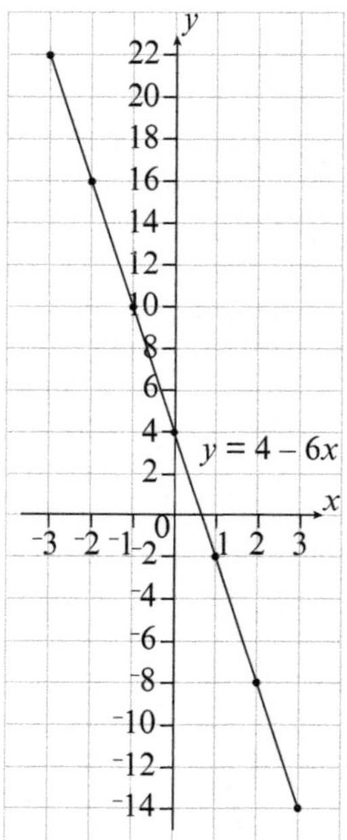

12F–12G

1 a

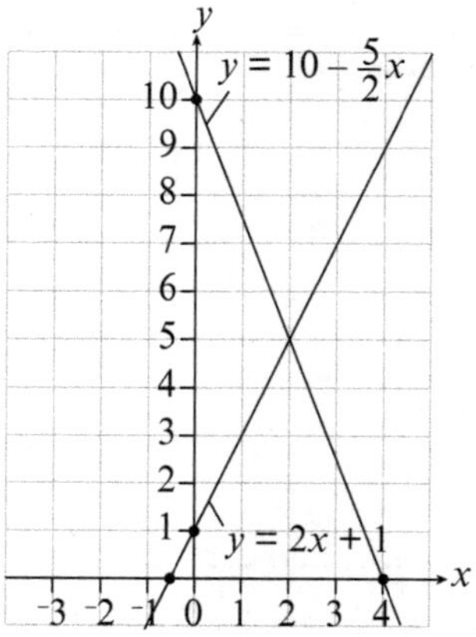

Solution is (2, 5)

b

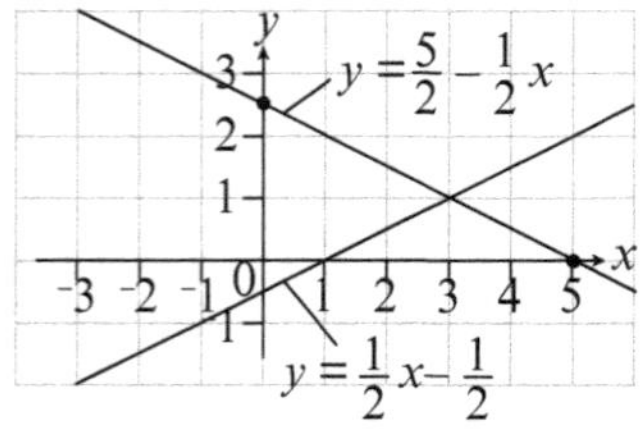

Solution is (3, 1)

c

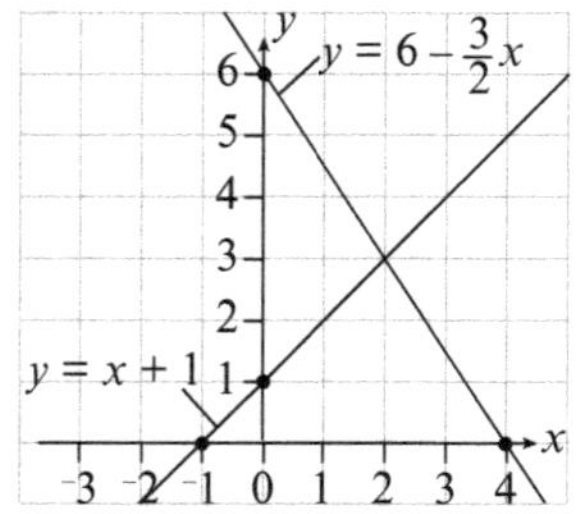

Solution is (2, 3)

2 a

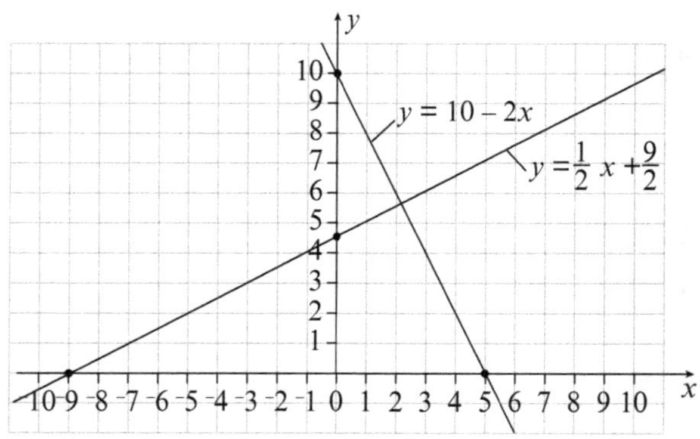

Solution is (2.2, 5.6)

b

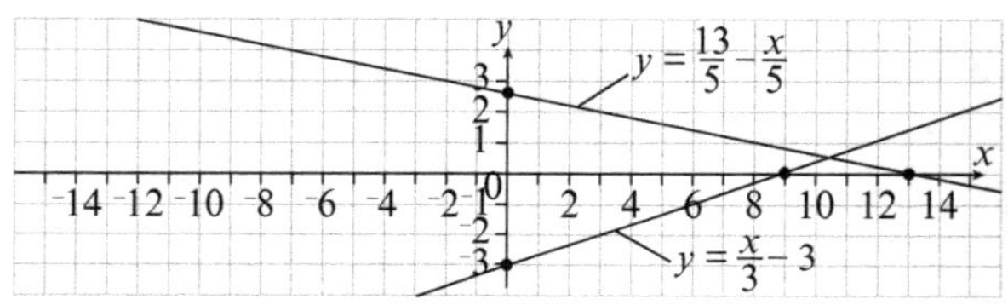

Solution is (10.5, 0.5)

3

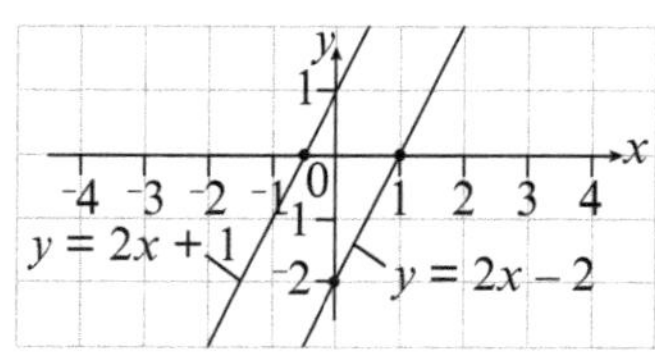

The equations have the same coefficient of x and hence they have the same gradient and are parallel.

4 $y = {}^-2x - 1,\ y = \dfrac{-x}{2} + 2$

5 $y = 6 - x$, $y = x - 2$: point of intersection is (4, 2)

12H–12I

1 a

Number of days, d	1	2	3
Cost in \$, C	15	20	25

b \$47.50

c Students' graphs

2 a Students' graphs

Yes, the graph shows a linear relationship

b 1.8 **c** 104 °F **d** 32 **e** $y = 1.8x + 32$

3 a 15 km/h **b** 60 km/h

c The motorcyclist stopped **d** 40 km/h

4 a $2\,\text{m/s}^2$ **b** 0 **c** $1\,\text{m/s}^2$

13A

1

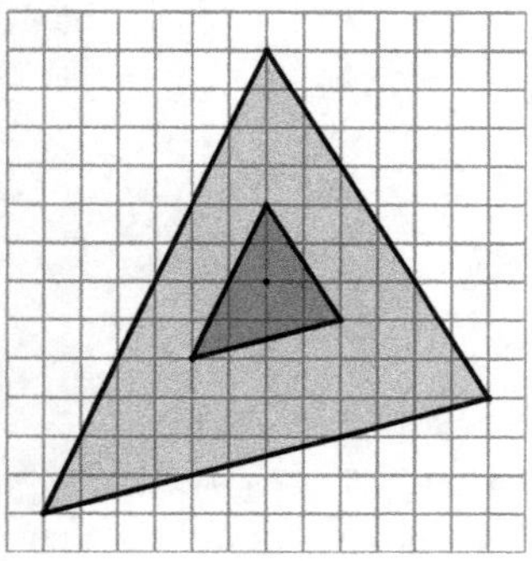

2 a Yes

b

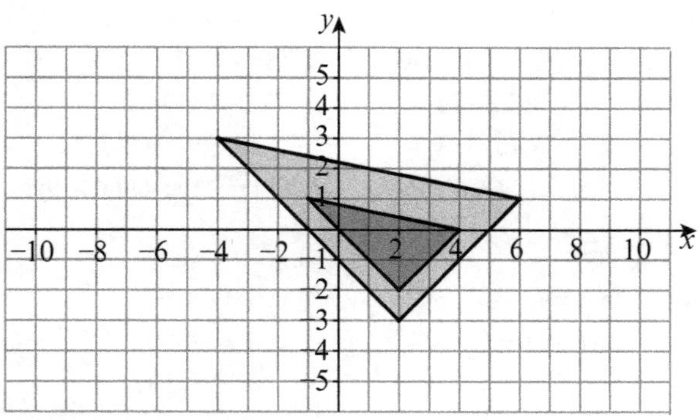

Centre of enlargement (2, $^{-}1$)

c 2

3 a 26 cm b 48 cm^2

13B

1 Students' drawings; translation by $\begin{pmatrix} 5 \\ 1 \end{pmatrix}$

2 Students' drawings; reflection in *x*-axis

3 Students' drawings; reflection in line $x = {}^{-}1$

4 Students' drawings; rotation by 90° anticlockwise about (1, 0)

5 Students' drawings; enlargement with centre (5, 0) by scale factor 3

6 Students' drawings; enlargement with centre (1, 7) and scale factor 3

13C

1 Translation by $\begin{pmatrix} 8 \\ 0 \end{pmatrix}$

2 a, b Students' drawings c ($^{-}2$, $^{-}5$)

3 Students' drawings

Translation by $\begin{pmatrix} ^{-}4 \\ 3 \end{pmatrix}$

14A–14B

1 a 20 b 100 (or 120)
  c 16 d 28
  e 64 f 36

2 a 8 b 42
  c 21 d 1
  e 0 (or 0.028) f 1

3 a 15 b 6 c 4

4 a 4 b 7 c 3 d 30
  e 0.25 f 100

5 a 9 b 3 c 500

6 a $^{-}15$ b $^{-}64$ c $^{-}105$ d $^{-}80$

7 a Incorrect: $20 \div 20 = 1$
  b Likely to be correct: $0.4 \times 20 = 8$
  c Incorrect: $\frac{^{-}8}{0.2} = {}^{-}40$

8 36 cm^3

14C–14E

1 50×0.9, $50 \div 50$, $50 \div 2.5$, $50 \times \frac{1}{2}$

2 20×1.9, $20 \div 0.97$, 20×1.05, $20 \div \frac{1}{2}$, $20 \div 0.2$, 20×2

3 b $78.1 \div 6.2 = 21.2$ c $24.1 \div 0.59 = 0.48$

4 $40 \times 0.03 = 40 \times 3 \times 0.01 = 120 \times 0.01 = 1.2$

5 a 296.1 b 29.61 c $^{-}2.961$ d 2961
  e $^{-}29610$ f 296.1

6 a 4.2 b $^{-}51.39$ c 59.36 d $^{-}33.11$
  e 1.21 f $^{-}3.726$

7 a 0.15 b $^{-}0.84$ c 6.48 d 3.096
  e 6.171 f $^{-}2.4462$

8 a \$521.22 b \$201.04 c \$24.20

9 a 22.4 km b 14 km c 3.73 km

10 a 20 b 1.3 c $^{-}3.68$ d 50
  e $^{-}500$ f 200

11 a 13 b $^{-}27$ c 270 d $^{-}2.7$
  e 5.47 f 68.5

12 47.5 (47 complete lengths)

13 a 18.5 m b 11.11 m c 230 m

14 a 580 b $^{-}0.58$ c 0.58 d 5800
  e $^{-}0.00058$ f $^{-}58$

14F

1 a 1.13 b 2.55 c 0.92 d 0.826

2 a 470.4 km b \$157.50 c 64 kg d 72 litres

3 a \$15 b \$45 c \$9

4 a \$315 b \$345 c \$309

5 a \$28 b \$42 c \$49

6 a \$156 b \$259.20 c \$2250 d \$144

7 5 years

8 Simple interest :12 years

9 \$1605.87

10 \$9966.63

15A–15B

1 a

		Spinner 1				
		1	2	3	4	5
Spinner 2	1	2	3	4	5	6
	2	3	4	5	6	7
	3	4	5	6	7	8
	4	5	6	7	8	9
	5	6	7	8	9	10

b 10 c $\frac{1}{5}$ d $\frac{3}{25}$

2 Tree diagram with the following outcomes: A1, A2, A3, B1, B2, B3, C1, C2, C3.

3 a $\frac{1}{11}$ b $\frac{4}{11}$ c $\frac{4}{11}$

4 a

Colour	Red	Green	Yellow	Blue	Brown
Probability of spinning colour	$\frac{1}{3}$	$\frac{1}{6}$	$\frac{3}{16}$	$\frac{1}{8}$	$\frac{3}{16}$

b Spinner with sectors labelled as follows: Brown = 67.5%, Red = 120°, Green = 60°, Yellow = 67.5%, Blue = 45°

5 $\frac{2}{3}$

6 a True; there are more even numbers than multiples of 3

b False; there are 5 even numbers so the probability is $\frac{5}{10} = \frac{1}{2}$

c True; there are 3 multiples of 3

d True; since the first card is replaced, the probability of B is not dependent on the outcome of A

15C

1 a Spin 1 Spin 2

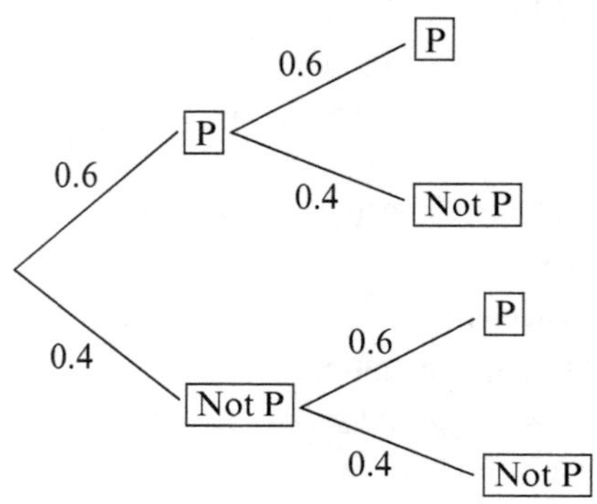

b i 0.36 ii 0.48

2 a 0.343 b $\frac{1}{64}$

15D

1 12

2

	Oranges	Bananas	Pears	Apples	Other
Number of people	36	28	44	20	16

a i $\frac{7}{36}$ ii $\frac{5}{36}$ iii $\frac{3}{4}$ b Pears

3 a 150 b 45 c 75

4 12 black, 22 white, 6 red

16A

1 a

x	$^-3$	$^-2$	$^-1$	0	1	2	3
x^2	9	4	1	0	1	4	9
+2	2	2	2	2	2	2	2
y	11	6	3	2	3	6	11

b, c Students' graphs d (0, 2)

16B

1 a $2x^2 + 10x + 12$ b $10x^2 + 19x + 6$

2 a $3x^2 + 17x + 10$ b $8x^2 + 18x + 10$

c $6x^2 + 7x - 20$ d $33x^2 - 4x - 12$

e $6x^2 - 11x + 3$ f $30x^2 - 52x + 14$

3 a $ab + 2ad + 3bc + 6cd$

b $12qs + 9qt + 8rs + 6rt$

c $10ln + 2lp - 15mn - 3mp$

d $15wy - 3wz + 10xy - 2xz$

e $3ac - 2ad - 9bc + 6bd$

f $24dfh - 28dgh - 30ef + 35eg$

16C–16D

1 a 529 b 2704 c 961 d 3721

2 a $(8 - 3)(8 + 3) = 55$ b $(17 - 7)(17 + 7) = 240$

c $(26 - 6)(26 + 6) = 640$ d $(78 - 22)(78 + 22) = 5600$

e $(8.6 - 1.4)(8.6 + 1.4) = 11.2$

3 a 6 b 660 c 47.6

4 a $(x - y)(x + y)$ b $(2a - b)(2a + b)$

c $(3m - 2n)(3m + 2n)$ d $(3c - b)(3c + b)$

5 a $(3a - b)(3a + b)$ b $(3c - 2d)(3c + 2d)$

c $(5x - 3y)(5x + 3y)$ d $(7s - 10t)(7s + 10t)$

e $(3cd - 4e)(3cd + 4e)$ f $(11x^2 - 9yz)(11x^2 + 9yz)$

16E–16F

1 a $(x + 6)(x + 2)$ b $(x + 1)(x + 7)$

c $(x + 3)(x + 6)$ d $(x + 2)(x + 7)$

e $(x + 4)(x + 7)$ f $(x + 9)(x + 1)$

2 a $(x + 5)(x - 1)$ b $(x + 12)(x - 2)$

c $(x + 6)(x - 2)$ d $(x + 1)(x - 17)$

e $(x - 3)(x + 5)$ f $(x + 3)(x - 9)$

3 a $(x + 1)(x + 5)$ b $(x + 14)(x - 2)$

c $(x + 11)(x - 11)$ d $(x - 11)(x - 2)$

e $(x - 6)(x - 7)$ f $(x - 13)(x + 4)$

4 a $11x^2 - 10x - 1 = (11x + 1)(x - 1)$

b $10x^2 - 3x - 1 = (5x + 1)(2x - 1)$

c $8x^2 + 6x + 1 = (4x + 1)(2x + 1)$

d $14x^2 + 9x + 1 = (7x + 1)(2x + 1)$

e $42x^2 + 13x + 1 = (7x + 1)(6x + 1)$

f $18x^2 - 7x - 11 = (18x + 11)(x - 1)$

16G

1 a $x = 2, x = 3$ b $x = 3, x = {}^-5$

c $x = 2, x = 6$ d $x = 2, x = {}^-4$

e $x = {}^-3, x = {}^-4$ f $x = {}^-5, x = 4$

2 a $(x + 3)(x - 2); x = 2, x = {}^-3$

b $(x + 3)(x + 5); x = {}^-3, x = {}^-5$

c $(x - 5)(x - 2); x = 2, x = 5$

d $(x - 3)(x - 7); x = 3, x = 7$

e $(x + 6)(x - 4); x = {}^-6, x = 4$

3 a $x = 3, x = {}^-7$ b $x = 0, x = \frac{1}{5}$

c $x = {}^-3, x = {}^-6$ d $x = \frac{1}{5}, x = {}^-7$

e $x = \frac{{}^-1}{4}, x = \frac{5}{2}$ f $x = \frac{1}{3}, x = 2$

4 a $(3x + 1)(x - 2) = 0; x = \frac{1}{{}^-3}, x = 2$

b $(5x - 1)(x + 3) = 0; x = \frac{1}{5}, x = {}^-3$

c $(5x - 1)(2x + 3) = 0; x = \frac{1}{5}, x = \frac{3}{{}^-2}$